Advanced Glasses, Composites and Ceramics for High Growth Industries

Advanced Glasses, Composites and Ceramics for High Growth Industries

Special Issue Editors

Milena Salvo
Mike Reece
Aldo R. Boccaccini

MDPI • Basel • Beijing • Wuhan • Barcelona • Belgrade

MDPI

Special Issue Editors

Milena Salvo
Politecnico di Torino
Italy

Mike Reece
Queen Mary University of London
UK

Aldo R. Boccaccini
University of Erlangen-Nuremberg
Germany

Editorial Office
MDPI
St. Alban-Anlage 66
4052 Basel, Switzerland

This is a reprint of articles from the Special Issue published online in the open access journal *Materials* (ISSN 1996-1944) from 2018 to 2019 (available at: https://www.mdpi.com/journal/materials/special_issues/Glasses_Ceramics).

For citation purposes, cite each article independently as indicated on the article page online and as indicated below:

LastName, A.A.; LastName, B.B.; LastName, C.C. Article Title. *Journal Name* **Year**, *Article Number*, Page Range.

ISBN 978-3-03897-960-9 (Pbk)
ISBN 978-3-03897-961-6 (PDF)

Contents

About the Special Issue Editors vii

Preface to "Advanced Glasses, Composites and Ceramics for High Growth Industries" ix

Min Yu, Theo Saunders, Taicao Su, Francesco Gucci and Michael John Reece
Effect of Heat Treatment on the Properties of Wood-Derived Biocarbon Structures
Reprinted from: *Materials* 2018, *11*, 1588, doi:10.3390/ma11091588 1

Bhuvanesh Srinivasan, Alain Gellé, Jean-François Halet, Catherine Boussard-Pledel and
Bruno Bureau
Detrimental Effects of Doping Al and Ba on the Thermoelectric Performance of GeTe
Reprinted from: *Materials* 2018, *11*, 2237, doi:10.3390/ma11112237 10

Gianmarco Taveri, Enrico Bernardo and Ivo Dlouhy
Mechanical Performance of Glass-Based Geopolymer Matrix Composites Reinforced with
Cellulose Fibers
Reprinted from: *Materials* 2018, *11*, 2395, doi:10.3390/ma11122395 19

Katarzyna Placha, Richard S. Tuley, Milena Salvo, Valentina Casalegno and Kevin Simpson
Solid-Liquid Interdiffusion (SLID) Bonding of p-Type Skutterudite Thermoelectric Material
Using Al-Ni Interlayers
Reprinted from: *Materials* 2018, *11*, 2483, doi:10.3390/ma11122483 30

Acacio Rincon Romero, Sergio Tamburini, Gianmarco Taveri, Jaromír Toušek, Ivo Dlouhy
and Enrico Bernardo
Extension of the 'Inorganic Gel Casting' Process to the Manufacturing of Boro-Alumino-Silicate
Glass Foams
Reprinted from: *Materials* 2018, *11*, 2545, doi:10.3390/ma11122545 42

Pablo Lopez-Iscoa, Nirajan Ojha, Ujjwal Aryal, Diego Pugliese, Nadia G. Boetti,
Daniel Milanese and Laeticia Petit
Spectroscopic Properties of Er^{3+}-Doped Particles-Containing Phosphate Glasses Fabricated
Using the Direct Doping Method
Reprinted from: *Materials* 2019, *12*, 129, doi:10.3390/ma12010129 54

Hassan Javed, Antonio Gianfranco Sabato, Ivo Dlouhy, Martina Halasova, Enrico Bernardo,
Milena Salvo, Kai Herbrig, Christian Walter and Federico Smeacetto
Shear Performance at Room and High Temperatures of Glass–Ceramic Sealants for Solid Oxide
Electrolysis Cell Technology
Reprinted from: *Materials* 2019, *12*, 298, doi:10.3390/ma12020298 66

Alessia Masini, Thomas Strohbach, Filip Šiška, Zdeněk Chlup and Ivo Dlouhý
Electrolyte-Supported Fuel Cell: Co-Sintering Effects of Layer Deposition on Biaxial Strength
Reprinted from: *Materials* 2019, *12*, 306, doi:10.3390/ma12020306 78

Cristian Marro Bellot, Marco Sangermano, Massimo Olivero and Milena Salvo
Optical Fiber Sensors for the Detection of Hydrochloric Acid and Sea Water in Epoxy and Glass
Fiber-Reinforced Polymer Composites
Reprinted from: *Materials* 2019, *12*, 379, doi:10.3390/ma12030379 94

Francesca E. Ciraldo, Kristin Schnepf, Wolfgang H. Goldmann and Aldo R. Boccaccini
Development and Characterization of Bioactive Glass Containing Composite Coatings with Ion
Releasing Function for Antibiotic-Free Antibacterial Surgical Sutures
Reprinted from: *Materials* **2019**, *12*, 423, doi:10.3390/ma12030423 . . . . **104**

**Rocío Tejido-Rastrilla, Sara Ferraris, Wolfgang H. Goldmann, Alina Grünewald,
Rainer Detsch, Giovanni Baldi, Silvia Spriano and Aldo R. Boccaccini**
Studies on Cell Compatibility, Antibacterial Behavior, and Zeta Potential of Ag-Containing
Polydopamine-Coated Bioactive Glass-Ceramic
Reprinted from: *Materials* **2019**, *12*, 500, doi:10.3390/ma12030500 . . . . **113**

**Francesco Gucci, Fabiana D'Isanto, Ruizhi Zhang, Michael J. Reece, Federico Smeacetto and
Milena Salvo**
Oxidation Protective Hybrid Coating for Thermoelectric Materials
Reprinted from: *Materials* **2019**, *12*, 573, doi:10.3390/ma12040573 . . . . **126**

Acacio Rincón Romero, Nicoletta Toniolo, Aldo R. Boccaccini and Enrico Bernardo
Glass-Ceramic Foams from 'Weak Alkali Activation' and Gel-Casting of Waste Glass/Fly Ash
Mixtures
Reprinted from: *Materials* **2019**, *12*, 588, doi:10.3390/ma12040588 . . . . **137**

Matteo Cavasin, Marco Sangermano, Barry Thomson and Stefanos Giannis
Exposure of Glass Fiber Reinforced Polymer Composites in Seawater and the Effect on Their
Physical Performance
Reprinted from: *Materials* **2019**, *12*, 807, doi:10.3390/ma12050807 . . . . **151**

About the Special Issue Editors

Milena Salvo (Prof) has a PhD in materials engineering and is an associate professor of materials science and technology at the Politecnico di Torino, Italy. She has considerable experience in international research on advanced materials. Her research activity has been dedicated mostly to advanced materials and composites, including the (i) joining of advanced materials for high-temperature applications; (ii) coating of advanced materials for oxidation and wear protection; (iii) development, production and characterisation of glass-ceramic and composite sealing materials for solid oxide fuel cells; and (iv) vitrification and reuse of waste. She is a member of the American Ceramic Society and a member of J-Tech—Advanced Joining Technology @ POLITO. She is also a coordinator of Advanced Glasses, Composites and Ceramics for High-Growth Industries' (CoACH) MSC European Training Network. She is a co-author of more than 160 papers in international journals, conference proceedings and book chapters and has three patents and two pending patent applications in the field of advanced ceramics.

Michel Reece (Prof) has a BSc and PhD in solid state physics from Essex University. From 1986 to 1989, he was a research assistant at Queen Mary College in the cyclic fatigue of advanced structural ceramics. From 1989 to 1992, he was a senior scientific officer in the National Physical Laboratory working on the development and standardisation of microstructural and mechanical techniques for characterising ceramic and cermet materials. From 1992 to the present, at Queen Mary University of London (QMUL), his group's research has focused on the development of field (electric, magnetic and gravity)-assisted processing of ceramics. A long-term objective of his work is to commercialise materials prepared by field-assisted processing through knowledge transfer and spin-outs. He is a director of Nanoforce Technology Ltd, a spin-out company of QMUL. Nanoforce focuses on developing new structural and functional materials, including dielectrics, ferroelectrics, thermoelectrics and high-entropy ceramics. This includes materials with nanostructure, texture and metastable structures that can be commercialised. He is also a director of the Northwestern Polytechnical University–Queen Mary University of London Joint Research Institute (2017–present). He was awarded the Verulam medal (2010) and a Royal Society Industry Fellowship (2011–2015). He is the Co-Editor-in-Chief of "Advances in Applied Ceramics" (2012–present). He has published over 200 papers and holds two patents.

Aldo R. Boccaccini is professor of biomaterials and head of the Institute of Biomaterials at the University of Erlangen-Nuremberg, Germany. He is also a visiting professor at Imperial College London, UK. He holds a nuclear engineering degree from Instituto Balseiro (Argentina), Dr-Ing. (PhD) from RWTH Aachen University (Germany) and Habilitation from Technical University of Ilmenau (Germany). Prior to his current position, he spent 10 years at Imperial College London in the Department of Materials as a lecturer, reader and professor. He has held post-doctoral positions at University of Birmingham (UK) and University of California, San Diego (USA). His research activities are in the field of ceramics, glasses and composites for biomedical, functional and/or structural applications. He is the author or co-author of more than 800 scientific papers and 25 book chapters. His work has been cited more than 31,000 times, and he was included in the "Highly cited researchers" list in 2018 (Clarivate Analytics). He is the Editor-in-Chief of the journal *Materials Letters*, founding Editor of the journal *Biomedical Glasses* and serves on the editorial board of more than 10 international journals, being the Section Editor-in-Chief (Biomaterials) of the journal *Materials* (MDPI). He is a Fellow of the Institute of Materials, Minerals and Mining (IOM3) (UK) and the American Ceramic Society and the Society of Glass Technology (UK). He is a member of the Council of the European Society for Biomaterials (ESB), the World Academy of Ceramics, the Executive Committee of the Federation of European Materials Societies (FEMS) and the National Academy of Science and Engineering of Germany (acatech).

Preface to "Advanced Glasses, Composites and Ceramics for High Growth Industries"

'Advanced Glasses, Composites and Ceramics for High-Growth Industries' (CoACH) was a European Training Network (ETN) project (http://www.coach-etn.eu/) funded by the Horizon 2020 program. CoACH involved multiple actors in the innovation ecosystem for advanced materials, comprised of five universities and ten enterprises in seven different European countries. The project studied the next generation of materials that could bring innovation in the healthcare, construction, and energy sectors, among others, from new bioactive glasses for bone implants to eco-friendly cements and new environmentally friendly thermoelectrics for energy conversion. The novel materials developed in the CoACH project pave the way for innovative products, improved cost competitiveness, and positive environmental impact.

The present Special Issue contains 14 papers resulting from the CoACH project, showcasing the breadth of materials and processes developed during the project:

(i) Graphitized porous biocarbon monoliths were produced by means of spark plasma sintering (SPS). Their high thermal conductivity makes them candidate materials for thermal energy storage, such as thermal enhancers and containers for phase change materials [1].

(ii) Innovative thermoelectric materials from nontoxic elements and new manufacturing techniques for more efficient thermoelectric devices are discussed in [2,4,12].

(iii) Energy-efficient, low-cost, and eco-friendly materials from industrial wastes with improved strength and fracture resilience were produced and discussed in [3]. Eco-sustainable porous materials with low thermal conductivity that could be exploited for building applications are presented in [5,13].

(iv) The effect of the incorporation of Er2O3-doped particles on the structural and luminescent properties of phosphate glasses was investigated. The obtained results provided evidence that the direct doping method is a promising technique for the development of new active glasses [6].

(v) New glass–ceramic sealants that could increase the reliability of solid oxide electrolysis cells (SOECs) were produced and tested at temperatures up to 850 °C [7]. Furthermore, the effect of the manufacturing process on the final strength of the whole reversible solid oxide cell (SOC) stack was studied in [8].

(vi) Innovative glass fibre sensors and new tests to monitor the degradation of polymer composites in harsh environments are reported in [9,14].

(vii) Novel antibacterial and nanostructured coatings for medical devices and implants for dental, orthopaedic, and tissue engineering applications were developed. They can help to reduce bacterial infections and cut the use of antibiotics by patients, as discussed in [10,11].

References

1. Yu, M.; Saunders, T.; Su, T.; Gucci, F.; Reece, M. Effect of Heat Treatment on the Properties of Wood-Derived Biocarbon Structures. Materials 2018, 11(9), 1588; https://doi.org/10.3390/ma11091588.

2. Srinivasan, B.; Gellé, A.; Halet, J.; Boussard-Pledel, C.; Bureau, B. Detrimental Effects of Doping Al and Ba on the Thermoelectric Performance of GeTe. Materials 2018, 11(11), 2237; https://doi.org/10.3390/ma11112237.

3. Taveri, G.; Bernardo, E.; Dlouhy, I. Mechanical Performance of Glass-Based Geopolymer Matrix Composites Reinforced with Cellulose Fibers. Materials 2018, 11(12), 2395; https://doi.org/10.3390/ma11122395.

4. Placha, K.; Tuley, R.; Salvo, M.; Casalegno, V.; Simpson, K. Solid-Liquid Interdiffusion (SLID) Bonding of p-Type Skutterudite Thermoelectric Material Using Al-Ni Interlayers. Materials 2018, 11(12), 2483; https://doi.org/10.3390/ma11122483.

5. Rincon Romero, A.; Tamburini, S.; Taveri, G.; Toušek, J.; Dlouhy, I.; Bernardo, E. Extension of the 'Inorganic Gel Casting' Process to the Manufacturing of Boro-Alumino-Silicate Glass Foams. Materials 2018, 11(12), 2545; https://doi.org/10.3390/ma11122545.

6. Lopez-Iscoa, P.; Ojha, N.; Aryal, U.; Pugliese, D.; Boetti, N.; Milanese, D.; Petit, L. Spectroscopic Properties of Er3+-Doped Particles-Containing Phosphate Glasses Fabricated Using the Direct Doping Method. Materials 2019, 12(1), 129; https://doi.org/10.3390/ma12010129.

7. Javed, H.; Sabato, A.; Dlouhy, I.; Halasova, M.; Bernardo, E.; Salvo, M.; Herbrig, K.; Walter, C.; Smeacetto, F. Shear Performance at Room and High Temperatures of Glass–Ceramic Sealants for Solid Oxide Electrolysis Cell Technology. Materials 2019, 12(2), 298; https://doi.org/10.3390/ma12020298.

8. Masini, A.; Strohbach, T.; Šiška, F.; Chlup, Z.; Dlouhý, I. Electrolyte-Supported Fuel Cell: Co-Sintering Effects of Layer Deposition on Biaxial Strength. Materials 2019, 12(2), 306; https://doi.org/10.3390/ma12020306.

9. Marro Bellot, C.; Sangermano, M.; Olivero, M.; Salvo, M. Optical Fiber Sensors for the Detection of Hydrochloric Acid and Sea Water in Epoxy and Glass Fiber-Reinforced Polymer Composites. Materials 2019, 12(3), 379; https://doi.org/10.3390/ma12030379.

10. Ciraldo, F.; Schnepf, K.; Goldmann, W.; Boccaccini, A. Development and Characterization of Bioactive Glass Containing Composite Coatings with Ion Releasing Function for Antibiotic-Free Antibacterial Surgical Sutures. Materials 2019, 12(3), 423; https://doi.org/10.3390/ma12030423.

11. Tejido-Rastrilla, R.; Ferraris, S.; Goldmann, W.; Grünewald, A.; Detsch, R.; Baldi, G.; Spriano, S.; Boccaccini, A. Studies on Cell Compatibility, Antibacterial Behavior, and Zeta Potential of Ag-Containing Polydopamine-Coated Bioactive Glass-Ceramic. Materials 2019, 12(3), 500; https://doi.org/10.3390/ma12030500.

12. Gucci, F.; D'Isanto, F.; Zhang, R.; Reece, M.; Smeacetto, F.; Salvo, M. Oxidation Protective Hybrid Coating for Thermoelectric Materials. Materials 2019, 12(4), 573; https://doi.org/10.3390/ma12040573.

13. Rincón Romero, A.; Toniolo, N.; Boccaccini, A.; Bernardo, E. Glass-Ceramic Foams from 'Weak Alkali Activation' and Gel-Casting of Waste Glass/Fly Ash Mixtures. Materials 2019, 12(4), 588; https://doi.org/10.3390/ma12040588.

14. Cavasin, M.; Sangermano, M.; Thomson, B.; Giannis, S. Exposure of Glass Fiber Reinforced Polymer Composites in Seawater and the Effect on Their Physical Performance. Materials 2019, 12(5), 807; https://doi.org/10.3390/ma12050807.

Milena Salvo, Mike Reece, Aldo R. Boccaccini
Special Issue Editors

materials **MDPI**

Article

Effect of Heat Treatment on the Properties of Wood-Derived Biocarbon Structures

Min Yu [1,2], **Theo Saunders** [1,2], **Taicao Su** [1,2], **Francesco Gucci** [1,2] and **Michael John Reece** [1,2,*]

1   School of Engineering and Material Science, Queen Mary University of London, London E1 4NS, UK;
    min.yu@qmul.ac.uk (M.Y.); t.g.saunders@qmul.ac.uk (T.S.); t.su@qmul.ac.uk (T.S.);
    f.f.gucci@qmul.ac.uk (F.G.)
2   Nanoforce Technology Limited, London E1 4NS, UK
*   Correspondence: m.j.reece@qmul.ac.uk; Tel./Fax: +44-20-7882-2773

Received: 3 August 2018; Accepted: 21 August 2018; Published: 2 September 2018

Abstract: Wood-derived porous graphitic biocarbons with hierarchical structures were obtained by high-temperature (2200–2400 °C) non-catalytic graphitization, and their mechanical, electrical and thermal properties are reported for the first time. Compared to amorphous biocarbon produced at 1000 °C, the graphitized biocarbon-2200 °C and biocarbon-2400 °C exhibited increased compressive strength by ~38% (~36 MPa), increased electrical conductivity by ~8 fold (~29 S/cm), and increased thermal conductivity by ~5 fold (~9.5 W/(m·K) at 25 °C). The increase of duration time at 2200 °C contributed to increased thermal conductivity by ~12%, while the increase of temperature from 2200 to 2400 °C did not change their thermal conductivity, indicating that 2200 °C is sufficient for non-catalytic graphitization of wood-derived biocarbon.

Keywords: graphitization; wood-derived biocarbon; thermal conductivity

1. Introduction

Wood-derived biocarbon (biochar, charcoal) structures have gained much attention owing to the hierarchical architecture of their cellular pore structures and the ability to produce complex shapes [1–4]. The graphitization of carbon has a significant impact on its properties, i.e., the electronic, magnetic and thermal properties [5–7]. Graphitic porous biocarbon monoliths are promising because they combine good mechanical properties with low density (0.11–0.97 g/cm^3) with the properties of graphite (high degree of ordering, low thermal expansion coefficient, good thermal and electrical conductivities) [8]. Two main techniques have been used to graphitize wood-derived biocarbons, including non-catalytic high-temperature (up to 3000 °C) graphitization [5], and low-temperature (1300–1600 °C) catalytic graphitization with Fe, Co, Mn and Ni etc. [8–12]. During the catalytic graphitization process, the catalysts introduce impurities (i.e., carbides, metal particles) into the biocarbon structure, and the graphitic carbon surrounding the catalyst particles (i.e., Fe, Co, and Ni), can be formed at 1000–1600 °C [8,13]. Acid washing (i.e., HNO$_3$) is required to remove metal particles in order to achieve pure graphitic carbon. Byrne et al. [14] graphitized wood-derived biocarbon at 2500 °C without the use of a catalyst, however, they did not report their mechanical properties, or electrical and thermal conductivities. Until now, there are few reported works on the effect of temperature and duration time on the properties (especially thermal conductivity) of graphitized wood-derived biocarbon structures prepared by non-catalytic high-temperature (above 2000 °C) graphitization [5].

Porous carbon materials with high thermal conductivity are needed for thermal energy storage, such as thermal enhancers and containers for phase change materials [15,16]. Rico et al. [8] evaluated the thermal conductivity of Fe-catalyst graphitized wood-derived carbon, and found that the thermal diffusivity of graphitized carbon increased with increasing pyrolysis temperatures up to 800 °C, mainly resulting from an increased degree of graphitization. Johnson et al. [17] found that Ni-catalyst graphitized wood-derived carbon has similar properties, and they further infiltrated copper into the pore structures to increase the thermal conductivity.

In this work, graphitized porous biocarbon monoliths derived from beech wood were obtained by heating at high temperatures (2200–2400 °C) without the use of a catalyst. This heat treatment was performed in a Spark Plasma Sintering (SPS) furnace with high heating and cooling rates (up to 200 °C/min). Accordingly, we report for the first time the effects of temperature and duration time on the properties (compressive strength, electrical and thermal conductivity) of these samples prepared by non-catalytic high temperature graphitization.

2. Experimental Process

Cylindrical pieces of beech wood (DOW003100, Tilgear Ltd., Hertfordshire, UK) were chosen as the carbon source. Cylindrical biocarbon structures (Ø = ~6 mm, H = ~9 mm) were prepared by pyrolyzing the beech wood (DOW003100, Tilgear Ltd.) at 1000 °C for 4 h, as performed in our previous work [18]. The prepared biocarbon structures were then heated to higher temperatures (2200 °C and 2400 °C) in Ar for different duration times (2–15 min) in a SPS furnace. A heating rate of 200 °C/min and cooling rate of 100 °C/min were used during this thermal processing. A pressureless mode in SPS was used in order to retain the porous biomorphic structure derived from the wood. The bulk density (geometrical density, which includes pores) of the samples was estimated by dividing the weight by the geometrical volume. The solid density (which excludes the pores) of the samples was measured using the Archimedes' method.

An FEI Inspect-F scanning electron microscope (SEM, Hillsboro, OR, USA) was used to characterize the morphology of the samples. Transmission electron microscopy (TEM, JEOL 2010, JEOL, Akishima, Japan) and X-ray diffraction (XRD, Siemens Diffraktometer-D5000, Siemens, Berlin, Germany) analysis with Cu Kα radiation were used to detect the crystalline structures in the samples. Raman spectroscopy (Labspec 6, Horiba Jobin-Yvon, Kyoto, Japan) at room temperature was used to determine the degree of structural disorder in the carbons using an excitation of 514 nm. The degree of crystallinity (β) was calculated using the following Equation (1) [8]:

$$\beta = \frac{I_G}{I_G + I_D} \tag{1}$$

where I_G and I_D are the intensities (area under the peak) of the bands G (~1580 cm^{-1}) and D (~1350 cm^{-1}) in the Raman spectra, respectively.

The nitrogen absorption-desorption isotherm was measured using an Autosorb-IQ2-MP-C system (Quantachrome Instruments, Boynton Beach, FL, USA). The specific surface area and pore size distribution were calculated using the multipoint Brunauer–Emmett–Teller (BET, Quantachrome Instruments, Boynton Beach, FL, USA) method and Quenched Solid Density Function Theory (QSDFT), respectively.

The compressive strength of a set of six samples with nominal dimensions of Ø = 6 ± 0.1 mm and H = 9 ± 0.3 mm was measured in the axial direction at room temperature using a universal testing device (Model 4202, Instron, Canton, MA, USA). The displacement speed was set at 0.5 mm/min.

The room-temperature electrical conductivity of the samples was measured using a two-point conductivity measurement technique, using a picoameter (Keithley 6485, Keithley, Solon, OH, USA) and DC voltage source (Agilent 6614C, Agilent, Santa Clara, CA, USA).

The thermal diffusivity (α) was measured on cylinder samples (diameter: ~6 mm, thickness: ~1.5 mm) using a Netzsch LFA-457 thermal analyzer (Netzsch, Hamburg, Germany). Three measurements were carried out at each temperature in the range of 25–800 °C in a flowing Ar atmosphere. The thermal conductivity (κ) was calculated using the following equation: κ = C_p × D × α. In our work, the specific heat capacity (C_p) of samples was taken from the literature (0.25–2.0 J/(g·K) in the temperature range of 25 to 800 °C [19], and D was taken as the bulk density (geometric density).

3. Results and Discussion

Figure 1 shows the microstructures and pore size distributions of the wood-derived biocarbons after different heat treatments. The biocarbon-2400 °C exhibited uniform and nearly round macropores with diameters of ~50 μm and ~8 μm, as shown in Figure 1a,b. Dense struts (Figure 1c) were also observed, providing strong mechanical support for the structures. The biocarbon-1000 °C (Figure 1d) exhibited a relatively wide range of micropores (0–25 nm), while the graphitized biocarbon-2400 °C exhibited a micropore distribution mainly concentrated in the range of 0–10 nm (Figure 1e). This might result from the shrinkage of large nano-sized pores (10–50 nm) during the graphitization process. In addition, the specific pore volume and specific surface area of the biocarbon-2400 °C were two orders of magnitude smaller than that of the biocarbon-1000 °C, indicating the disappearance of micropores during the high temperature (2400 °C) treatment. This mainly resulted from the disappearance of small pores (≤50 nm) caused by the rearrangement of carbon structures at high temperatures up to 2400 °C. The shrinkage of nano-sized pores might limit the application of the graphitized biocarbon in the electrochemical energy storage applications.

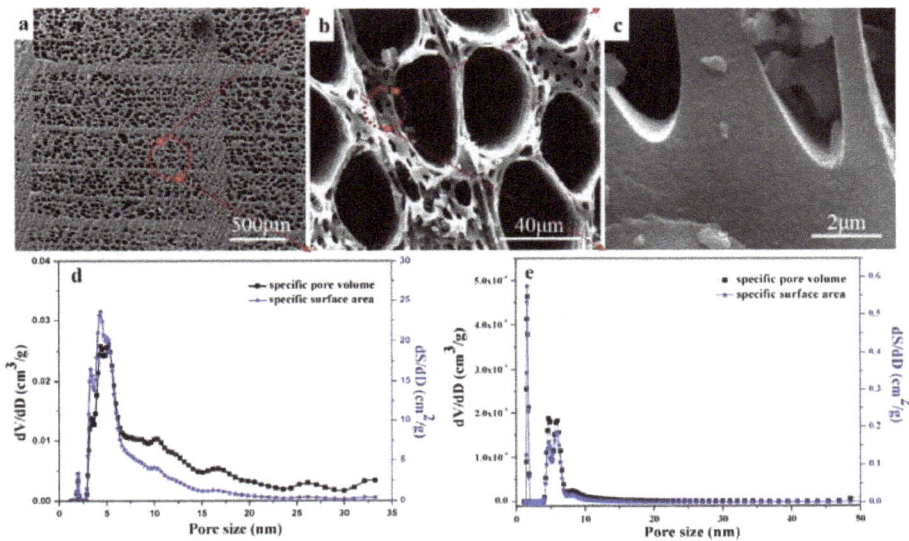

Figure 1. (**a–c**) SEM micrographs and (**d,e**) pore size distributions (based on BET analysis) of the biocarbon structures obtained at different heat treatment conditions. (**d**) is the sample prepared at 1000 °C for 4 h and (**e**) is the sample prepared at 2400 °C for 10 min.

The Raman spectra for the biocarbon-1000 °C exhibited a broad weak D peak at 1360 cm^{-1} and G peak at 1584 cm^{-1}, indicating that it contained little graphitic carbon (Figure 2). All of the biocarbons prepared at 2200 °C and 2400 °C exhibited both a sharp D and G peak, which are related to the defect structure of graphite and perfect graphite structure (in-plane stretching of graphite lattice, in-plane vibration of sp^2 carbon atoms), respectively. The G/D ratio increased in the graphitized biocarbon-2200 °C with the dwell time increasing from 2 to 15 min. This indicates a higher degree of graphitization in the biocarbon, which is further confirmed by the XRD patterns (Figure 3a) and TEM images (Figure 3b,c). The biocarbon-2400 °C exhibited a slightly higher G/D ratio compared to the biocarbon-2200 °C. Both biocarbon-2200 °C and biocarbon-2400 °C exhibited a smaller (~50%) full width at half maximum (FWHM) of their G band compared with biocarbon-1000 °C, indicating a high relative amount of graphitic carbon to amorphous carbon. The corresponding crystalline ratio of the samples was calculated based on the Equation (1), and is shown in Table 1.

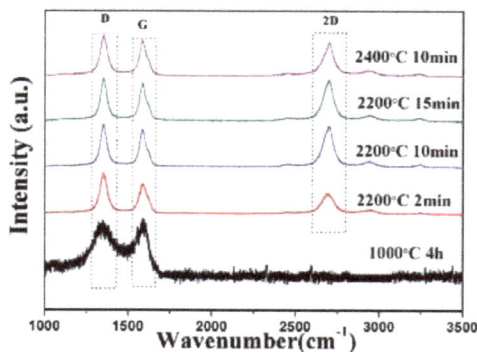

Figure 2. Raman spectra of wood-derived biocarbon prepared at different temperatures and dwell times.

The XRD and TEM analysis were also used to further investigate the graphitization of the biocarbons, as shown in Figure 3. The biocarbon-1000 °C exhibited two broad peaks at 2θ = 20–26° and 2θ = 41–46°, which are characteristic of amorphous carbon. Both the biocarbon-2200 °C and biocarbon-2400 °C showed a superposition of two peaks (a broad peak and a sharp peak) at 2θ = 20–28°. Both the biocarbon-2200 °C and biocarbon-2400 °C showed characteristic peaks at 2θ = 26° and 2θ = 43°, which correspond to the reflections of the (002) and (001) planes of graphitic carbon, respectively [20,21], indicating the formation of graphitic carbon, which is in good agreement with the Raman data (Figure 2). The biocarbon-1000 °C exhibited a typical HRTEM image for an amorphous structure (Figure 3b), while the biocarbon-2400 °C showed graphitic carbon layers (see red dashed circle) and some amorphous carbon regions (see red solid circle in Figure 3c). The SAED pattern (inset of Figure 3b) further confirmed the amorphous nature of biocarbon-1000 °C, which is consistent with the XRD data (Figure 3a). The SAED pattern (inset of Figure 3c) further confirmed the crystallinity of the biocarbon-2400 °C, consistent with the peaks in the XRD pattern.

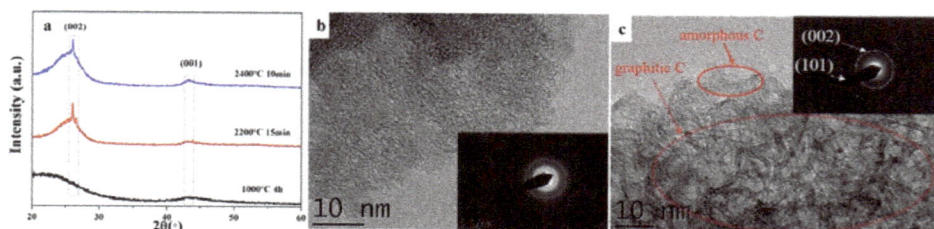

Figure 3. (a) XRD patterns of biocarbon structures obtained using different temperatures and dwell times; (b,c) are high resolution transmission electron microscope (HRTEM) images of the biocarbon structures prepared at 1000 °C and 2400 °C, respectively. The insets are the corresponding selected area electron diffraction (SAED) patterns.

X-ray photoelectron spectroscopy (XPS) was used to identify c, as shown in Supplementary Figure S1. The XPS survey spectra shown in Figure S1a indicates the presence of C and O in both the biocarbon-1000 °C and biocarbon-2400 °C. The biocarbon-2400 °C exhibited a smaller atomic percentage of O (3.6 at %) than the biocarbon-1000 °C (9.7 at. %). In the high-resolution C 1s spectra

(Figure S1b,c), the higher dominant peak at 285.6 eV indicates a higher volume of C=C/C-C in the biocarbon-2400 °C. Both biocarbon-1000 °C and biocarbon-2400 °C exhibited the peaks of C-O and C=O, which are further confirmed in the high resolution O 1s spectra (Supplementary Figure S2).

Table 1 shows the weight loss, density, specific surface area, electrical conductivity, thermal conductivity, crystallinity ratio and compressive strength of the biocarbons. The biocarbon-1000 °C exhibited a bulk density of 0.51 g/cm^3 and a solid density of 1.85 g/cm^3. The bulk density of the graphitized biocarbons-2200–2400 °C exhibited a slight decrease (from 0.51 to 0.48 g/cm^3), owing to a further weight loss of ~10 wt%, probably caused by a mild oxidation and evaporation of the carbon in the SPS chamber during the high temperature graphitization process. However, the solid density of the graphitized biocarbons-2200 and -2400 °C moderately increased to ~2.02 g/cm^3, owing to the disappearance of nanopores and rearrangement of carbon during the graphitization process at high temperatures (2200–2400 °C). The specific surface area and specific pore volume of graphitized biocarbon-2400 °C compared to the biocarbon-1000 °C decreased from 356 to 144 m^2/g and from 0.267 to 0.232 cm^3/g, respectively, owing to the disappearance of micropores (<50 nm) shown in Figure 1e. Compared to biocarbon-1000 °C (~2.8 S/cm), the electrical conductivity of the graphitized biocarbons increased by ten fold (~29 S/cm). This increase was produced by the formation of the graphitic carbon, which is confirmed by the increase of the calculated crystallinity ratio of the graphitized samples given in Table 1. In addition, the compressive strength (36 MPa) of the graphitized samples increased by ~38% compared to biocarbon-1000 °C, again probably resulting from the graphitic carbon formed at 2200–2400 °C. However, the increased duration time from 2 to 15 min and higher temperature from 2200 to 2400 °C, did not significantly increase their compressive strength.

Table 1. The weight loss, bulk density, specific surface area, electrical conductivity, thermal conductivity and compressive strength of wood-derived biocarbon prepared at different conditions.

Heat Treatment Condition	Weight Loss (wt %)	Bulk Density (g/cm³)	Solid Density (g/cm³)	Specific Surface Area (m²/g)	Specific Pore Volume (cm³/g)	RT Electrical Conductivity (S/cm)	RT Thermal Conductivity (W/(m·K))	Crystallinity Ratio β	Compressive Strength (MPa)
1000 °C, 4 h, Ar	76.6 ± 0.1	0.51 ± 0.02	1.85 ± 0.03	356	0.267	2.8 ± 0.8	1.8	0.25	26 ± 1
2200 °C, 2 min, Ar	86.7 ± 0.1	0.49 ± 0.03	2.03 ± 0.05	-	-	24 ± 0.7	7.8	0.48	35 ± 1
2200 °C, 10 min, Ar	87.5 ± 0.2	0.47 ± 0.02	2.04 ± 0.03	-	-	25 ± 1	9.2	0.48	34 ± 2
2200 °C, 15 min, Ar	85.1 ± 0.1	0.47 ± 0.02	2.03 ± 0.02	-	-	29 ± 0.8	9.4	0.52	36 ± 2
2400 °C, 10 min, Ar	85.2 ± 0.1	0.48 ± 0.04	2.02 ± 0.04	144	0.232	24 ± 0.5	9.5	0.49	36 ± 2

Note. Heat treatment conditions refer to the highest temperature and its corresponding duration time, and heating atmosphere. The weight loss is relative to the starting wood.

Figure 4 shows the thermal transport properties versus temperatures (25–800 °C) for the wood-derived biocarbon structures prepared at different temperatures (1000–2400 °C) and duration times (2–15 min). As shown in Figure 4a, the measured thermal diffusivity of biocarbon-1000 °C slightly increased with the measuring temperature increasing from 25 to 800 °C. On the contrary, the graphitized biocarbons exhibited decreasing thermal diffusivity with increasing temperature. These thermal diffusivity trends versus measuring temperature are consistent with the reported data for Fe-graphitized biocarbons in the literature [8]. Compared to amorphous biocarbon-1000 °C, the graphitized biocarbons-2200 °C and -2400 °C exhibited much higher thermal diffusivity (up to ~6 mm²/s). As shown in Figure 4b, the graphitized biocarbon-2400 °C exhibited similar thermal diffusivity during the heating and cooling process, indicating the stability of the samples during the high-temperature measurements (below 800 °C).

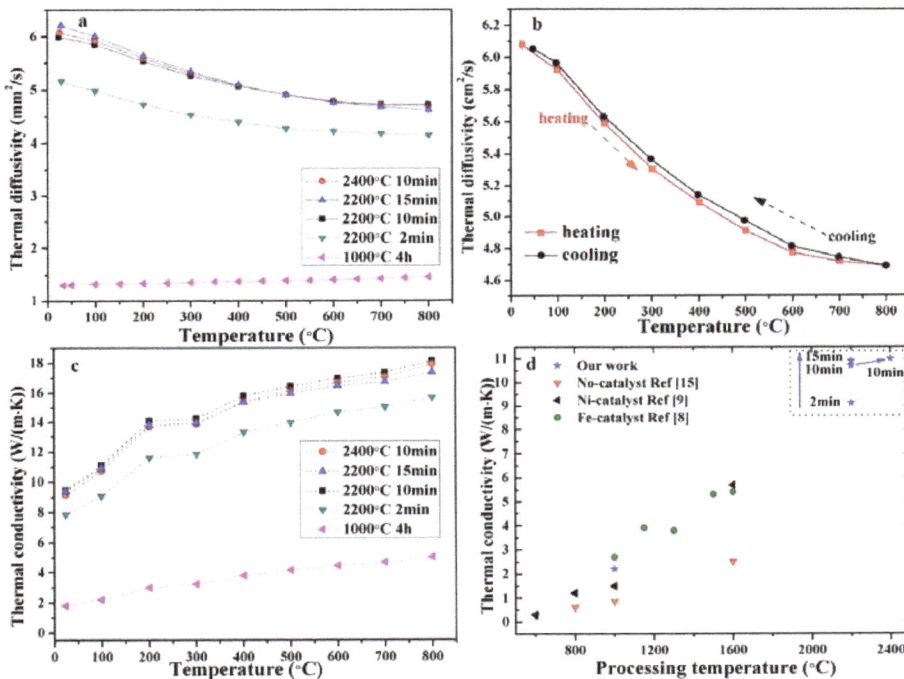

Figure 4. (a) Thermal diffusivity as a function of measuring temperatures for biocarbons obtained using different processing temperatures and dwell time; (b) Thermal diffusivity versus measuring temperatures during the heating and cooling process of the biocarbon prepared at 2400 °C for 10 min; (c) Thermal conductivity as a function of measuring temperatures for biocarbons; (d) Comparison of thermal conductivity (at 100 °C) of beech-derived biocarbons prepared using different techniques.

Figure 4c shows the corresponding thermal conductivity calculated based on the measured diffusivity (Figure 4a) and using values for the heat capacity reported in the literature [15]. All of the samples exhibited increasing thermal conductivity with increasing measuring temperature from 25 to 800 °C. This phenomenon is consistent with the reported results for Fe-graphitized biocarbon structures [8]. The total thermal transfer of the porous biocarbon was mainly through the pores by radiation and struts (pore walls) by electrons and phonons. The contribution of large pores (~1–500 μm) to heat loss by radiation plays a significant role in the thermal transport of porous ceramic foams [22], resulting in the increase of thermal conductivity of biocarbon with increasing measuring temperature.

The thermal conductivity of biocarbon-2400 °C is up to 5 times higher than the biocarbon-1000 °C at the same measurement temperature. As polycrystalline graphite has more than two orders higher thermal conductivity than amorphous carbon [15], this higher thermal conductivity of graphitized biocarbon mainly resulted from the formation of graphitic carbon. With the increase of duration time from 2 to 15 min at 2200 °C, the thermal conductivity of samples increased moderately by ~12%. This mainly resulted from the increased degree of crystallinity of the biocarbon. The electronic contribution of the thermal conductivity (estimated using Wiedemann-Franz law) in the biocarbon-2400 °C was estimated to be only < ~0.004 W/m/K [8], which is far smaller than the total thermal conductivity ($\geq$ ~2 W/m/K).

Figure 4d shows the comparison of our results with the reported thermal conductivity data for graphitized biocarbon structures derived from beech wood from the literature [8,17]. Our samples exhibited ~89% higher thermal conductivity than the highest result reported in the literature [8,9]. Since the crystallinity ratio of graphitized biocarbon in our work is similar to the reported one for Fe-graphitized biocarbon [8], the high thermal conductivity of our graphitized biocarbon might result from the increased phonon contributions produced by the massively reduced micropore and nanopore volumes (as shown in the BET data in Figure 1).

4. Conclusions

The wood-derived porous monolithic biocarbon structures were graphitized without the use of a catalyst at 2200–2400 °C with high heating and cooling rates (up to 200 °C/min) in a Spark Plasma Sintering (SPS) furnace. The effects of temperatures and duration time on the microstructures of the biocarbons were investigated in detail. Furthermore, the properties of the graphitized biocarbons were also investigated including their mechanical, electrical, and thermal properties. Compared to the un-graphitized biocarbon, the graphitized biocarbons at 2200 °C and 2400 °C exhibited an increased compressive strength of ~38%, and increased room-temperature electrical conductivity of ~8 fold. In addition, the graphitized biocarbons exhibited up to 5 times higher thermal conductivity than the ungraphitized biocarbon. With the increase of duration time from 2 to 15 min at 2200 °C, the graphitized biocarbon exhibited increased thermal conductivity by ~12%, while for the increase of temperature from 2200 to 2400 °C for 10 min, the graphitized biocarbon exhibited similar thermal conductivity. This indicates that 2200 °C might be the optimum temperature for non-catalytic graphitization of wood-derived biocarbon.

Supplementary Materials: The following are available online at http://www.mdpi.com/1996-1944/11/9/1588/s1, Figure S1: (a) Full-scan XPS of biocarbon structures obtained using different temperatures. (b) and (c) are the high resolution C 1s XPS scan of the biocarbon-1000 °C and biocarbon-2400 °C, Figure S2: The high resolution O 1s XPS scan of the biocarbon-1000 °C and biocarbon-2400 °C.

Author Contributions: Conceptualization, M.Y., M.J.R. and T.S. (Theo Saunders); Methodology, M.Y., M.J.R. and T.S.(Theo Saunders); Software, M.Y; Validation, M.Y. and T.S. (Theo Saunders); Formal Analysis, M.Y.; Investigation, M.Y. and T.S.(Theo Saunders); Thermal Conductivity Measurement and Analysis, M.Y., T.S. (Taicao Su), F.G. and T.S. (Theo Saunders); Resources, M.J.R.; Data Curation, M.Y.; Writing-Original Draft Preparation, M.Y.; Writing-Review & Editing, M.J.R.; Visualization, M.Y.; Supervision, M.J.R.; Project Administration, M.J.R.; Funding Acquisition, M.J.R.

Funding: "This research was funded by the European Union's Horizon 2020 Programme through a Marie Skłodowska-Curie Innovative Training Network ('CoACH-ETN", http://www.coach-etn.eu/, g.a. no. 642557), Sunchon National University, South Korea, through the BK21+ programme and EPSRC (EP/K008749/1, XMat) and EC FP7 2007-2013 (ADMACOM).

Conflicts of Interest: The authors declare no conflict of interest.

References

1. Vogli, E.; Sieber, H.; Greil, P. Biomorphic SiC-ceramic prepared by Si-vapor phaseinfiltration of wood. *J. Eur. Ceram. Soc.* **2002**, *22*, 2663–2668. [CrossRef]

2. Wilkes, T.E.; Young, M.L.; Sepulveda, R.E.; Dunand, D.C.; Faber, K.T. Composites by aluminum infiltration of porous silicon carbide derived from wood precursors. *Scr. Mater.* **2006**, *55*, 1083–1086. [CrossRef]

3. Rambo, C.R.; Cao, J.; Rusina, O.; Sieber, H. Manufacturing of biomorphic (Si, Ti, Zr)-carbide ceramics by sol–gel processing. *Carbon* **2005**, *43*, 1174–1183. [CrossRef]

4. Yukhymchuk, V.O.; Kiselov, V.S.; Valakh, M.Y.; Tryus, M.P.; Skoryk, M.A.; Rozhin, A.G.; Belyaev, A.E. Biomorphous SiC ceramics prepared from cork oak as precursor. *J. Phys. Chem. Solids* **2016**, *91*, 145–151. [CrossRef]

5. Cheng, H.M.; Endo, H.; Okabe, T.; Saito, K.; Zheng, G.B. Graphitization behavior of wood ceramics and bamboo ceramics as determined by X-ray diffraction. *J. Porous. Mater.* **1999**, *6*, 233–237. [CrossRef]

6. Steiner III, S.A.; Baumann, T.F.; Bayer, B.C.; Blume, R.; Worsley, M.A.; MoberlyChan, W.J.; Wardle, B.L. Nanoscale zirconia as a nonmetallic catalyst for graphitization of carbon and growth of single-and multiwall carbon nanotubes. *J. Am. Chem. Soc.* **2009**, *131*, 12144–12154. [CrossRef] [PubMed]

7. Maldonado-Hódar, F.J.; Moreno-Castilla, C.; Rivera-Utrilla, J.; Hanzawa, Y.; Yamada, Y. Catalytic graphitization of carbon aerogels by transition metals. *Langmuir* **2000**, *16*, 4367–4373. [CrossRef]

8. Ramirez-Rico, J.; Gutierrez-Pardo, A.; Martinez-Fernandez, J.; Popov, V.V.; Orlova, T.S. Thermal conductivity of Fe graphitized wood derived carbon. *Mater. Des.* **2016**, *99*, 528–534. [CrossRef]

9. Johnson, M.T.; Faber, K.T. Catalytic graphitization of three-dimensional wood-derived porous scaffolds. *J. Mater. Res.* **2011**, *26*, 18–25. [CrossRef]

10. Sevilla, M.; Fuertes, A.B. Catalytic graphitization of templated mesoporous carbons. *Carbon* **2006**, *44*, 468–474. [CrossRef]

11. Dudina, D.V.; Ukhina, A.V.; Bokhonov, B.B.; Korchagin, M.A.; Bulina, N.V.; Kato, H. The influence of the formation of Fe3C on graphitization in a carbon-rich iron-amorphous carbon mixture processed by Spark Plasma Sintering and annealing. *Ceram. Int.* **2017**, *43*, 11902–11906. [CrossRef]

12. Guo, H.; Song, Y.; Chen, P.; Lou, H. Effects of Graphitization of Carbon Nanosphere on Hydrodeoxygenation Activity of Molybdenum Carbide. *Catal. Sci. Technol.* **2018**, *8*, 4199–4208. [CrossRef]

13. Thambiliyagodage, C.J.; Ulrich, S.; Araujo, P.T.; Bakker, M.G. Catalytic graphitization in nanocast carbon monoliths by iron, cobalt and nickel nanoparticles. *Carbon* **2018**, *134*, 452–463. [CrossRef]

14. Byrne, C.E.; Nagle, D.C. Carbonized wood monoliths—Characterization. *Carbon* **1997**, *35*, 267–273. [CrossRef]

15. Balandin, A.A. Thermal properties of graphene and nanostructured carbon materials. *Nat. Mater.* **2011**, *10*, 569. [CrossRef] [PubMed]

16. Inagaki, M.; Qiu, J.; Guo, Q. Carbon foam: Preparation and application. *Carbon* **2015**, *87*, 128–152. [CrossRef]

17. Johnson, M.T.; Childers, A.S.; Ramirez-Rico, J.; Wang, H.; Faber, K.T. Thermal conductivity of wood-derived graphite and copper–graphite composites produced via electrodeposition. *Compos. Part A Appl. Sci. Manuf.* **2013**, *53*, 182–189. [CrossRef]

18. Yu, M.; Bernardo, E.; Colombo, P.; Romero, A.R.; Tatarko, P.; Kannuchamy, V.K.; Titirici, M.-M.; Castle, E.G.; Picot, O.T.; Reece, M.J. Preparation and properties of biomorphic potassium-based geopolymer (KGP)-biocarbon (CB) composite. *Ceram. Int.* **2018**, *44*, 12957–12964. [CrossRef]

19. Wiener, M.; Reichenauer, G.; Hemberger, F.; Ebert, H.P. Thermal conductivity of carbon aerogels as a function of pyrolysis temperature. *Int. J. Thermophys.* **2006**, *27*, 1826–1843. [CrossRef]

20. Shang, H.; Lu, Y.; Zhao, F.; Chao, C.; Zhang, B.; Zhang, H. Preparing high surface area porous carbon from biomass by carbonization in a molten salt medium. *RSC Adv.* **2015**, *5*, 75728–75734. [CrossRef]

21. Chia, C.H.; Joseph, S.D.; Rawal, A.; Linser, R.; Hook, J.M.; Munroe, P. Microstructural characterization of white charcoal. *J. Anal. Appl. Pyrolysis* **2014**, *109*, 215–221. [CrossRef]

22. Shimizu, T.; Matsuura, K.; Furue, H.; Matsuzak, K. Thermal conductivity of high porosity alumina refractory bricks made by a slurry gelation and foaming method. *J. Eur. Ceram. Soc.* **2013**, *33*, 3429–3435. [CrossRef]

materials

MDPI

Letter

Detrimental Effects of Doping Al and Ba on the Thermoelectric Performance of GeTe

Bhuvanesh Srinivasan *[iD], Alain Gellé, Jean-François Halet, Catherine Boussard-Pledel and Bruno Bureau

Univ. Rennes, ISCR UMR 6226, IPR UMR 6251, CNRS, 35000 Rennes, France; alain.gelle@univ-rennes1.fr (A.G.); jean-francois.halet@univ-rennes1.fr (J.-F.H.); catherine.boussard@univ-rennes1.fr (C.B.-P.); bruno.bureau@univ-rennes1.fr (B.B.)
* Correspondence: bhuvanesh.srinivasan@univ-rennes1.fr; Tel.: +33-223-233-688; Fax: +33-223-235-611

Received: 15 October 2018; Accepted: 9 November 2018; Published: 11 November 2018

Abstract: GeTe-based materials are emerging as viable alternatives to toxic PbTe-based thermoelectric materials. In order to evaluate the suitability of Al as dopant in thermoelectric GeTe, a systematic study of thermoelectric properties of $Ge_{1-x}Al_xTe$ (x = 0–0.08) alloys processed by Spark Plasma Sintering are presented here. Being isoelectronic to $Ge_{1-x}In_xTe$ and $Ge_{1-x}Ga_xTe$, which were reported with improved thermoelectric performances in the past, the $Ge_{1-x}Al_xTe$ system is particularly focused (studied both experimentally and theoretically). Our results indicate that doping of Al to GeTe causes multiple effects: (i) increase in p-type charge carrier concentration; (ii) decrease in carrier mobility; (iii) reduction in thermopower and power factor; and (iv) suppression of thermal conductivity only at room temperature and not much significant change at higher temperature. First principles calculations reveal that Al-doping increases the energy separation between the two valence bands (loss of band convergence) in GeTe. These factors contribute for $Ge_{1-x}Al_xTe$ to exhibit a reduced thermoelectric figure of merit, unlike its In and Ga congeners. Additionally, divalent Ba-doping [$Ge_{1-x}Ba_xTe$ (x = 0–0.06)] is also studied.

Keywords: Thermoelectrics; GeTe; Al-doping; Ba-doping; loss of band convergence; lowered zT

1. Introduction

The generation, storage and transport of energy are among the greatest challenges, if not the most formidable challenge of all, for years to come. In this regard, thermoelectric (TE) materials and devices have drawn increasing interest and attention due to their potential to reversibly convert waste heat into electricity [1]. The TE material's efficiency is quantified by a dimensionless figure of merit, $zT = S^2\sigma T/\kappa$ where S, σ, T and κ are the Seebeck coefficient, electrical conductivity, absolute temperature and total thermal conductivity (sum of the electronic part, κ_e, and the lattice part, κ_{latt}), respectively. Seebeck coefficient, electrical and thermal conductivity are inter-locked and there is a high degree of challenge to decouple the electronic and thermal transport [2]. To tackle these challenges, thermoelectric material research has recently flourished with the emergence of novel concepts of band engineering, nanostructuring and discoveries of various novel materials. Amongst the state-of-the-art TE materials, the extensively studied PbTe-based materials are limited by their toxicity for any practical applications, despite their high zT [3–6]. Recently, GeTe-based materials have emerged as a clear alternative choice, as they have proven to exhibit higher performance ($zT > 1$), if optimally doped with suitable elements [7–10]. Some of the strategies for GeTe-based materials to enhance the power factor ($S^2\sigma$) and/or to suppress κ_{latt} were adopted on compositions such as GeTe-AgSbTe$_2$ (TAGS) [11], GeTe-LiSbTe$_2$ [12], GeTe-AgInTe$_2$ [13],GeTe-AgSbSe$_2$ [14], (GeTe)$_n$Sb$_2$Te$_3$ [15], $Ge_{1-x}Pb_xTe$ [16], $Ge_{1-x}Bi_xTe$ [17], (Bi$_2$Te$_3$)$_n$$Ge_{1-x}Pb_xTe$ [18], $Ge_{1-x}In_xTe$ [19], GeTe$_{1-x}$Se [20], $Ge_{1-x}Sb_xTe$ [21], $Ge_{1-x}Ag_xTe$ [7], $Ge_{1-x}Mn_xTe$ [22], $Ge_{1-x-y}Sn_xPb_yTe$ [23], $Ge_{1-x}Sb_xTe_{1-y}Se_y$ [24],

GeTe-GeSe-GeS [25], $Ge_{1-x-y}Bi_xSb_yTe$ [26], $Ge_{1-x-y}Bi_xIn_yTe$ [9], $Ge_{0.9-y}Pb_{0.1}Bi_yTe$ [27], and more recently $Ge_{1-x-y}Ga_xSb_yTe$ [8]. The crystal structure of these GeTe-based compounds undergoes a second-order ferroelectric structural transition from rhombohedral symmetry (low temperature phase) to cubic symmetry (high temperature phase) at around 700 K [10].

This work tries to explore the suitability of trivalent Al and divalent Ba as dopants for improving the thermoelectric performance of GeTe. The choice of Al is particularly interesting, as its isoelectronic group-13 counterparts In and Ga, if doped in optimum concentration, have proven to strongly enhance the thermoelectric performance of GeTe [8,19].

2. Materials and Methods

The samples $Ge_{1-x}Al_xTe$ (x = 0–0.08) and $Ge_{1-x}Ba_xTe$ (x = 0–0.06) were synthesized by vacuum sealed-tube melt processing. The obtained ingots were crushed into powder and consolidated by Spark Plasma sintering, SPS (FCT Systeme GmbH) at 773 K for 5 min under an axial pressure of 60 MPa. Details pertaining to experimental procedures, and materials characterization including electrical and thermal transport property measurements were discussed in detail in our previous works [6–9,28–30].

Density Functional Theory (DFT) calculations were performed to understand the effect of doping on the electronic states of GeTe. We used the projector-augmented-wave (PAW) approach [31] implemented in the Vienna ab initio simulation package (VASP) [32]. Calculations were performed using the generalized gradient approximation (GGA) for the exchange-correlation term parametrized by J. P. Perdew et al. [33] Similar to our previous work on Ga-doped GeTe [8], spin orbit coupling was included in the computations. As we were interested in high temperature behavior of doped GeTe, calculations were performed on cubic structural models. Impurities were substituted to Ge atom in a 4 × 4 × 4 super-cell. In order to understand the effect of Al, the calculations were performed on $Al_2Ge_{62}Te_{64}$ model (which is close to the experimental $Ge_{0.97}Al_{0.03}Te$ composition). The distance between two Al atoms was 12.02 Å. For the irreducible cell, the Brillouin-zone integration was performed using a 25 × 25 × 25 Monkhorst−Pack k-mesh. For the super-cell, we used a 3 × 3 × 3 k-mesh for the atomic relaxation and a 7 × 7 × 7 k-mesh for the electronic density of states (DOS) calculations.

3. Results and Discussion

The sharp reflections observed from X-ray diffraction (XRD) patterns for Al and Ba doped GeTe (Figure 1a,b, respectively) indicate the crystalline nature of the phases. The main reflections in both cases could be indexed to the rhombohedral (*R3m*) GeTe phase (PDF# 00-047-1079). The rhombohedral phase was further confirmed by the presence of double reflections [(0 2 4) and (2 2 0)] in the range of 2θ values between 41° and 44°. A minor proportion of Ge-rich secondary phase was also present, as in agreement with the previous studies [7–9]. Based on the evolution of lattice parameters, the solubility limit for Al in GeTe was estimated to be 4 mol%. At higher content of Al (for x > 0.04), Al_2Te_3 secondary phase started to appear and the GeTe main phase was not rhombohedral anymore (change of symmetry). The solubility limit for Ba in GeTe was found to be minimum (< 2 mol%), as $Ba_2Ge_2Te_5$ phase existed in all the samples. This is unsurprising given the larger atomic radii of Ba compared to that of Ge.

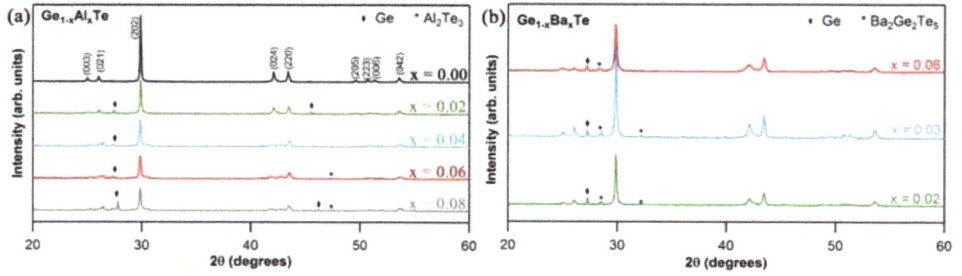

Figure 1. XRD patterns for $Ge_{1-x}Al_xTe$ (**a**) and $Ge_{1-x}Ba_xTe$ (**b**) systems.

Results from Hall measurements tabulating carrier concentration (n) and mobility (μ) are presented in Table 1. Holes are the major charge carriers (p-type), as the Hall voltage was found to be positive (p-type) in both $Ge_{1-x}Al_xTe$ and $Ge_{1-x}Ba_xTe$ systems. Doping Al to GeTe provides extra holes to the system, which is reflected in the enhancement in charge carrier density. This is in contrast to the effect observed in In and Ga (isoelectronic with Al) doped GeTe, where In and Ga decreased the hole concentration by filling up Ge vacancies [8,19]. On the other hand, the mobility reduction can be attributed to the alloy scattering mechanism arising from the doping of Al and Ba to GeTe [34]. Due to decreased mobility, the electrical conductivity at room temperature was decreased for both $Ge_{1-x}Al_xTe$ (Figure 2a) and $Ge_{1-x}Ba_xTe$ (Figure 3a) systems with respect to that of GeTe. However, this trend was reversed at higher temperatures (cross over point at ~450 K) for Al-doped GeTe (Figure 2a). Such similar cases were reported for SnTe and PbTe-based materials, and those cross over effects were attributed to the changes in the electronic band structure with increasing temperature [35,36]. The electrical conductivity of all the samples decreased with increasing temperature, which suggests a degenerate semi-conducting behavior. The positive Seebeck coefficient confirmed the p-type charge carriers in Al and Ba doped GeTe (Figures 2 and 3), which was consistent with the Hall measurement results. The thermopowers of $Ge_{1-x}Al_xTe$ and $Ge_{1-x}Ba_xTe$ monotonically increased with temperature. With increasing Al and Ba content, the change in S values at room temperature were not much evident, but they decreased significantly with increasing temperature, when compared to pristine GeTe. Doping of Al and Ba to GeTe decreased the S values, as it drastically inflated the hole carrier concentration. Consequently, the reduction of Seebeck coefficient with Al and Ba content also considerably affected the thermoelectric power factor (Figures 2 and 3). Finally, with Al and Ba doping, the total thermal conductivity decreased considerably at room temperature (Figures 2 and 3). However, it remained almost constant with temperature for Al-doped GeTe. The decreased thermopower significantly affected the thermoelectric figure of merit, zT (Figures 2 and 3), which plunged with dopant concentration.

Table 1. Hall measurement results (at ~300 K) of carrier concentration (n) and mobility (μ) for $Ge_{1-x}Al_xTe$ (x = 0.00–0.08) and $Ge_{1-x}Ba_xTe$ (x = 0.00–0.06) samples.

Sample	Carrier Concentration, n (cm^{-3})	Mobility, μ (cm^{2}V^{-1}s^{-1})
GeTe	9.08×10^{20}	57.0
$Ge_{0.98}Al_{0.02}Te$	1.75×10^{21}	21.8
$Ge_{0.96}Al_{0.04}Te$	2.88×10^{21}	10.6
$Ge_{0.94}Al_{0.06}Te$	2.17×10^{21}	12.5
$Ge_{0.92}Al_{0.08}Te$	3.01×10^{21}	8.8
$Ge_{0.98}Ba_{0.02}Te$	9.78×10^{20}	28.2
$Ge_{0.97}Ba_{0.03}Te$	9.06×10^{20}	33.6
$Ge_{0.94}Ba_{0.06}Te$	1.62×10^{21}	16.2

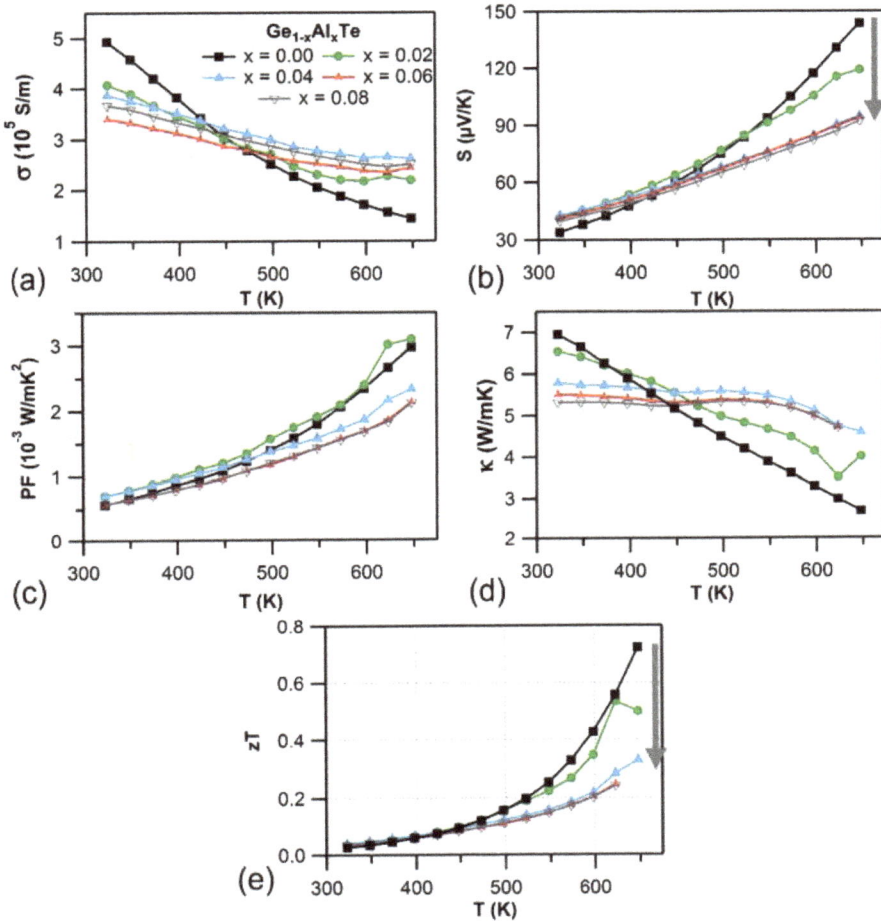

Figure 2. Temperature-dependent (**a**) electrical conductivity (σ), (**b**) Seebeck coefficient (S), and (**c**) power factor (PF = $S^2\sigma$), (**d**) total thermal conductivity (κ), and (**e**) figure of merit (zT) for $Ge_{1-x}Al_xTe$ (x = 0.00–0.08) samples.

To have a more cogent understanding on the detrimental effects of these dopants in GeTe, DFT calculations were performed. As we were interested in the high temperature domain for thermoelectric application, these DFT calculations were carried out on 4 × 4 × 4 super-cells derived from the cubic structural arrangement of GeTe. The electronic density of states (DOS) computed for the cubic models of pristine and Al-doped GeTe ($Al_2Ge_{62}Te_{64} \approx Ge_{0.97}Al_{0.03}Te$) are shown and compared in Figure 4a. The Al-induced resonant states (distinctly indicated by a sharp hump) are present around the Fermi level (E_F), just like its isoelectronic counterparts In and Ga [8,19]. In such a situation, the Seebeck coefficient is expected to increase, which is not the case with Al (Figure 2b).

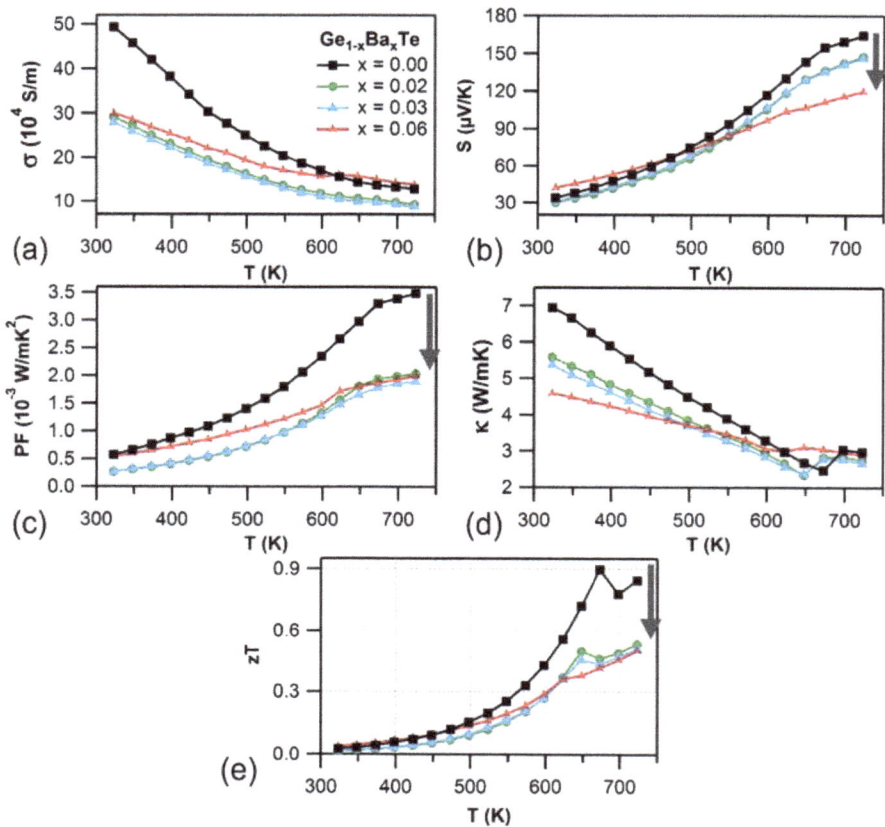

Figure 3. Temperature-dependent (**a**) electrical conductivity (σ), (**b**) Seebeck coefficient (S), and (**c**) power factor (PF = $S^2\sigma$), (**d**) total thermal conductivity (κ), and (**e**) figure of merit (zT) for $Ge_{1-x}Ba_xTe$ ($x = 0.00$–0.06) samples.

Since the DOS calculations yielded inconclusive evidence, electronic band structures were computed to decipher the role of Al in GeTe. The band structures are plotted in Figure 4 along some high symmetry lines of the cubic Brillouin zone (BZ). The energy difference between light and heavy hole valence bands ($\Delta E_{L\Sigma}$) for undoped cubic $Ge_{64}Te_{64}$ was found to be 64 meV, consistent with a recent report [37]. The flat and localized Al bands are located within the principal band bap (Figure 4d). Al-doping in GeTe increased the energy separation between the light hole and heavy hole valence bands to 179 meV (180% increment in $\Delta E_{L\Sigma}$ when compared to pristine c-GeTe), thus disfavoring the band convergence. According to Mott's relationship, Seebeck coefficient strongly depends on the total DOS effective mass, which in fact is directly proportional to the product of $N_v^{2/3}$ and the average DOS effective mass for each pocket (N_v is the number of degenerate carrier pockets) [38]. For GeTe, N_v is 4 for the L band and it increases to 12 for the Σ band [21]. Hence, by increasing the energy separation between L and Σ bands by doping of Al to GeTe, the contributions from the additional carriers (from Σ valence band) to electrical transport are lost, thus resulting in a significant reduction in the Seebeck coefficient.

For composition at $x = 0.02$, Al-doped GeTe exhibits a thermopower of ~110 µV/K at 623 K. For the same level of doping, the isoelectronic In-doped GeTe is known to exhibit a much higher thermopower of ~200 µV/K at the same temperature [19]. $\Delta E_{L\Sigma}$ for the In-doped GeTe (for $InGe_{63}Te_{64}$

$\approx Ge_{0.98}In_{0.02}Te$ model) is calculated to be 95 meV, which is almost two times lower than the $\Delta E_{L\Sigma}$ for Al-doped GeTe. It must be noted that, even though the $\Delta E_{L\Sigma}$ for In-doped GeTe is marginally higher than that of pristine GeTe, the presence of In-induced resonant states near the E_F has helped it to exhibit a superior thermopower compared to pristine GeTe. However, for the Al-doped GeTe, the beneficial effect of the presence of Al-induced resonant states near the E_F is nullified and severely affected by the extremely large energy separation between the light hole and heavy hole valence bands (179 meV). This explains the juxtaposition of high thermoelectric performance of $Ge_{0.98}In_{0.02}Te$ and low thermoelectric performance of $Ge_{0.98}Al_{0.02}Te$ (isoelectronic) compounds.

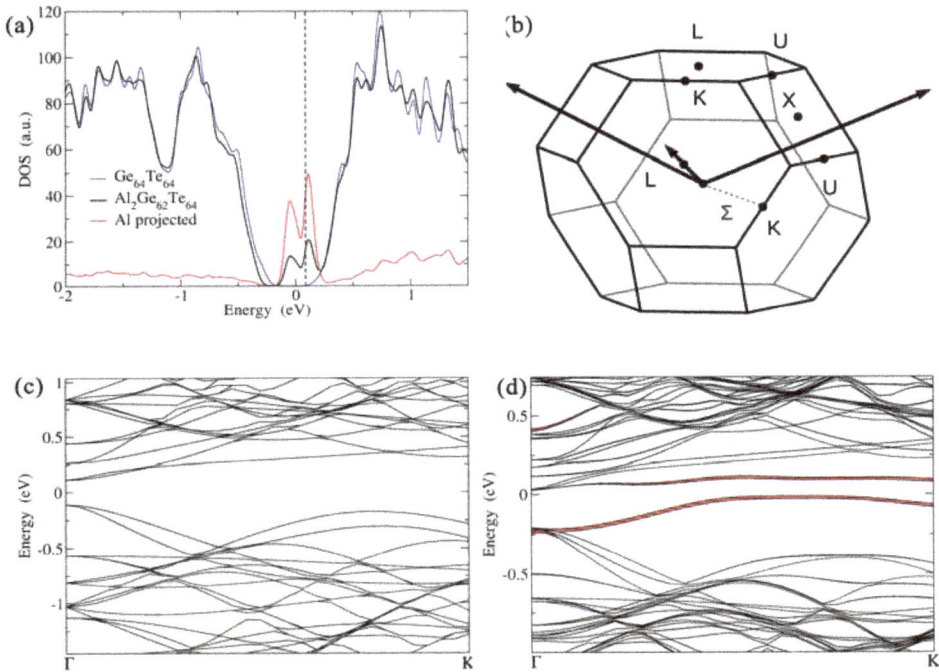

Figure 4. (**a**) Calculated DOS for $Al_2Ge_{62}Te_{64}$ ($Ge_{0.97}Al_{0.03}Te$) model, which is compared with that of the pristine cubic phase $Ge_{64}Te_{64}$ (c-GeTe). The Fermi level (E_F) of pristine c-GeTe is set arbitrarily at 0 eV. The dashed line represents the shifted E_F for the doped compositions. Additional Gaussian smearing of 25 meV was applied and the Al projected DOS is magnified for a better readability of the curves. (**b**) Brillouin zone of c-GeTe. Band structures for (**c**) c-$Ge_{64}Te_{64}$ using a 4 × 4 × 4 supercell showing band folding in the $\Gamma \rightarrow K$ (Σ) direction, and (**d**) $Al_2Ge_{62}Te_{64}$ ($Ge_{0.97}Al_{0.03}Te$) highlighting Al projections. Line thickness is proportional to the projection of the wave function on the Al (in red) orbitals.

For the case of Ba-doped GeTe, though the DFT results were inconclusive in portraying a clearer picture to explain the reduction in thermopower, it can be attributed to the presence of the secondary phase ($Ba_2Ge_2Te_5$). More in-depth studies, like experiments to synthesize this $Ba_2Ge_2Te_5$ phase and measure its transport properties (to estimate the role of contribution of that secondary phase to the overall thermoelectric behavior of the $Ge_{1-x}Ba_xTe$ compound) are required to understand the causes for Ba-doped GeTe to exhibit lower TE performance.

4. Conclusions

The crystalline ingots of $Ge_{1-x}Al_xTe$ ($x = 0$–0.08) and $Ge_{1-x}Ba_xTe$ ($x = 0$–0.06) were prepared by the vacuum-sealed tube melting route, followed by Spark Plasma Sintering processing. Al and Ba are found to not be the best choice of dopants for GeTe, as they subside its thermoelectric performance. Al-doping, unlike other isoelectronic group-13 elements (In and Ga), inflates the hole concentration and drastically increases the energy separation between light and heavy hole bands in GeTe, thus resulting in a reduced thermopower.

Author Contributions: B.S. designed and performed experiments, analyzed the results, and wrote the manuscript. A.G. performed DFT calculations and helped in understanding those results along with J.-F.H., C.B.-P. and B.B. supervised the work. All authors discussed the results and contributed to the final manuscript.

Funding: This research and the article processing charges were funded by European Commission's Horizon 2020 research and innovation program under the Marie Skłodowska-Curie grant (GA. 642557, CoACH-ETN).

Acknowledgments: Prof. Mike Reece is gratefully acknowledged for providing B.S. a secondment opportunity in Queen Mary University of London, where most of the thermoelectric measurements were performed. Régis Gautier and François Chevré from ISCR Rennes are acknowledged for their constant support.

Conflicts of Interest: The authors declare no conflict of interest.

References

1. He, J.; Tritt, T.M. Advances in thermoelectric materials research: Looking back and moving forward. *Science* **2017**, *357*, eaak9997. [CrossRef] [PubMed]
2. Chen, Z.; Jian, Z.; Li, W.; Chang, Y.; Ge, B.; Hanus, R.; Yang, J.; Chen, Y.; Huang, M.; Snyder, G.J.; et al. Lattice Dislocations Enhancing Thermoelectric PbTe in Addition to Band Convergence. *Adv. Mater.* **2017**, *29*, 1606768. [CrossRef] [PubMed]
3. Snyder, G.J.; Toberer, E.S. Complex thermoelectric materials. *Nat. Mater.* **2008**, *7*, 105–114. [CrossRef] [PubMed]
4. He, J.; Kanatzidis, M.G.; Dravid, V.P. High performance bulk thermoelectrics via a panoscopic approach. *Mater. Today* **2013**, *16*, 166–176. [CrossRef]
5. Zhao, L.-D.; Dravid, V.P.; Kanatzidis, M.G. The panoscopic approach to high performance thermoelectrics. *Energy Environ. Sci.* **2013**, *7*, 251–268. [CrossRef]
6. Srinivasan, B.; Gucci, F.; Boussard-Pledel, C.; Chevré, F.; Reece, M.J.; Tricot, S.; Calvez, L.; Bureau, B. Enhancement in thermoelectric performance of n-type Pb-deficit Pb-Sb-Te alloys. *J. Alloys Compd.* **2017**, *729*, 198–202. [CrossRef]
7. Srinivasan, B.; Gautier, R.; Gucci, F.; Fontaine, B.; Halet, J.-F.; Chevré, F.; Boussard-Pledel, C.; Reece, M.J.; Bureau, B. Impact of Coinage Metal Insertion on the Thermoelectric Properties of GeTe Solid-State Solutions. *J. Phys. Chem. C* **2018**, *122*, 227–235. [CrossRef]
8. Srinivasan, B.; Gellé, A.; Gucci, F.; Boussard-Pledel, C.; Fontaine, B.; Gautier, R.; Halet, J.-F.; Reece, M.; Bureau, B. Realizing a Stable High Thermoelectric zT ~2 over a Broad Temperature Range in $Ge_{1-x-y}Ga_xSb_yTe$ via Band Engineering and Hybrid Flash-SPS Processing. *Inorg. Chem. Front.* **2018**, in press. [CrossRef]
9. Srinivasan, B.; Boussard-Pledel, C.; Bureau, B. Thermoelectric performance of codoped (Bi, In)-GeTe and (Ag, In, Sb)-SnTe materials processed by Spark Plasma Sintering. *Mater. Lett.* **2018**, *230*, 191–194. [CrossRef]
10. Perumal, S.; Roychowdhury, S.; Biswas, K. High performance thermoelectric materials and devices based on GeTe. *J. Mater. Chem. C* **2016**, *4*, 7520–7536. [CrossRef]
11. Yang, S.H.; Zhu, T.J.; Sun, T.; He, J.; Zhang, S.N.; Zhao, X.B. Nanostructures in high-performance $(GeTe)_x(AgSbTe_2)_{100-x}$ thermoelectric materials. *Nanotechnology* **2008**, *19*, 245707. [CrossRef] [PubMed]
12. Schröder, T.; Schwarzmüller, S.; Stiewe, C.; de Boor, J.; Hölzel, M.; Oeckler, O. The Solid Solution Series $(GeTe)_x(LiSbTe_2)_2$ ($1 \leq x \leq 11$) and the Thermoelectric Properties of $(GeTe)_{11}(LiSbTe_2)_2$. *Inorg. Chem.* **2013**, *52*, 11288–11294. [CrossRef] [PubMed]
13. Schröder, T.; Rosenthal, T.; Giesbrecht, N.; Maier, S.; Scheidt, E.-W.; Scherer, W.; Snyder, G.J.; Schnick, W.; Oeckler, O. TAGS-related indium compounds and their thermoelectric properties—The solid solution series $(GeTe)_xAgIn_ySb_{1-y}Te_2$ ($x = 1$–$12; y = 0.5$ and 1). *J. Mater. Chem. A* **2014**, *2*, 6384–6395. [CrossRef]

14. Samanta, M.; Roychowdhury, S.; Ghatak, J.; Perumal, S.; Biswas, K. Ultrahigh Average Thermoelectric Figure of Merit, Low Lattice Thermal Conductivity and Enhanced Microhardness in Nanostructured $(GeTe)_x(AgSbSe_2)_{100-x}$. *Chem. Eur. J.* **2017**, *23*, 7438–7443. [CrossRef] [PubMed]

15. Fahrnbauer, F.; Souchay, D.; Wagner, G.; Oeckler, O. High Thermoelectric Figure of Merit Values of Germanium Antimony Tellurides with Kinetically Stable Cobalt Germanide Precipitates. *J. Am. Chem. Soc.* **2015**, *137*, 12633–12638. [CrossRef] [PubMed]

16. Gelbstein, Y.; Davidow, J. Highly efficient functional $Ge_xPb_{1-x}Te$ based thermoelectric alloys. *Phys. Chem. Chem. Phys.* **2014**, *16*, 20120–20126. [CrossRef] [PubMed]

17. Perumal, S.; Roychowdhury, S.; Biswas, K. Reduction of thermal conductivity through nanostructuring enhances the thermoelectric figure of merit in $Ge_{1-x}Bi_xTe$. *Inorg. Chem. Front.* **2016**, *3*, 125–132. [CrossRef]

18. Wu, D.; Zhao, L.-D.; Hao, S.; Jiang, Q.; Zheng, F.; Doak, J.W.; Wu, H.; Chi, H.; Gelbstein, Y.; Uher, C.; et al. Origin of the High Performance in GeTe-Based Thermoelectric Materials upon Bi_2Te_3 Doping. *J. Am. Chem. Soc.* **2014**, *136*, 11412–11419. [CrossRef] [PubMed]

19. Wu, L.; Li, X.; Wang, S.; Zhang, T.; Yang, J.; Zhang, W.; Chen, L.; Yang, J. Resonant level-induced high thermoelectric response in indium-doped GeTe. *NPG Asia Mater.* **2017**, *9*, e343. [CrossRef]

20. Yang, L.; Li, J.Q.; Chen, R.; Li, Y.; Liu, F.S.; Ao, W.Q. Influence of Se Substitution in GeTe on Phase and Thermoelectric Properties. *J. Electron. Mater.* **2016**, *45*, 5533–5539. [CrossRef]

21. Perumal, S.; Roychowdhury, S.; Negi, D.S.; Datta, R.; Biswas, K. High Thermoelectric Performance and Enhanced Mechanical Stability of p-type $Ge_{1-x}Sb_xTe$. *Chem. Mater.* **2015**, *27*, 7171–7178. [CrossRef]

22. Lee, J.K.; Oh, M.W.; Kim, B.S.; Min, B.K.; Lee, H.W.; Park, S.D. Influence of Mn on crystal structure and thermoelectric properties of GeTe compounds. *Electron. Mater. Lett.* **2014**, *10*, 813–817. [CrossRef]

23. Rosenberg, Y.; Gelbstein, Y.; Dariel, M.P. Phase separation and thermoelectric properties of the $Pb_{0.25}Sn_{0.25}Ge_{0.5}Te$ compound. *J. Alloys Compd.* **2012**, *526*, 31–38. [CrossRef]

24. Li, J.; Zhang, X.; Lin, S.; Chen, Z.; Pei, Y. Realizing the High Thermoelectric Performance of GeTe by Sb-Doping and Se-Alloying. *Chem. Mater.* **2017**, *29*, 605–611. [CrossRef]

25. Samanta, M.; Biswas, K. Low Thermal Conductivity and High Thermoelectric Performance in $(GeTe)_{1-2x}(GeSe)_x(GeS)_x$: Competition between Solid Solution and Phase Separation. *J. Am. Chem. Soc.* **2017**, *139*, 9382–9391. [CrossRef] [PubMed]

26. Perumal, S.; Bellare, P.; Shenoy, U.S.; Waghmare, U.V.; Biswas, K. Low Thermal Conductivity and High Thermoelectric Performance in Sb and Bi Codoped GeTe: Complementary Effect of Band Convergence and Nanostructuring. *Chem. Mater.* **2017**, *29*, 10426–10435. [CrossRef]

27. Li, J.; Zhang, X.; Chen, Z.; Lin, S.; Li, W.; Shen, J.; Witting, I.T.; Faghaninia, A.; Chen, Y.; Jain, A.; et al. Low-Symmetry Rhombohedral GeTe Thermoelectrics. *Joule* **2018**, *2*, 976–987. [CrossRef]

28. Srinivasan, B.; Boussard-Pledel, C.; Dorcet, V.; Samanta, M.; Biswas, K.; Lefèvre, R.; Gascoin, F.; Cheviré, F.; Tricot, S.; Reece, M.; et al. Thermoelectric Properties of Highly-Crystallized Ge-Te-Se Glasses Doped with Cu/Bi. *Materials* **2017**, *10*, 328. [CrossRef] [PubMed]

29. Srinivasan, B.; Fontaine, B.; Gucci, F.; Dorcet, V.; Saunders, T.G.; Yu, M.; Cheviré, F.; Boussard-Pledel, C.; Halet, J.-F.; Gautier, R.; et al. Effect of the Processing Route on the Thermoelectric Performance of Nanostructured $CuPb_{18}SbTe_{20}$. *Inorg. Chem.* **2018**, *57*, 12976–12986. [CrossRef] [PubMed]

30. Srinivasan, B.; Cui, S.; Prestipino, C.; Gellé, A.; Boussard-Pledel, C.; Ababou-Girard, S.; Trapananti, A.; Bureau, B.; Di Matteo, S. Possible Mechanism for Hole Conductivity in Cu–As–Te Thermoelectric Glasses: A XANES and EXAFS Study. *J. Phys. Chem. C* **2017**, *121*, 14045–14050. [CrossRef]

31. Kresse, G.; Joubert, D. From ultrasoft pseudopotentials to the projector augmented-wave method. *Phys. Rev. B* **1999**, *59*, 1758–1775. [CrossRef]

32. Kresse, G.; Furthmüller, J. Efficient iterative schemes for ab initio total-energy calculations using a plane-wave basis set. *Phys. Rev. B* **1996**, *54*, 11169–11186. [CrossRef]

33. Perdew, J.P.; Burke, K.; Ernzerhof, M. Generalized Gradient Approximation Made Simple. *Phys. Rev. Lett.* **1996**, *77*, 3865–3868. [CrossRef] [PubMed]

34. Wang, H.; LaLonde, A.D.; Pei, Y.; Snyder, G.J. The Criteria for Beneficial Disorder in Thermoelectric Solid Solutions. *Adv. Funct. Mater.* **2013**, *23*, 1586–1596. [CrossRef]

35. Roychowdhury, S.; Shenoy, U.S.; Waghmare, U.V.; Biswas, K. Tailoring of Electronic Structure and Thermoelectric Properties of a Topological Crystalline Insulator by Chemical Doping. *Angew. Chem. Int. Ed.* **2015**, *54*, 15241–15245. [CrossRef] [PubMed]

36. Zhang, L.J.; Qin, P.; Han, C.; Wang, J.L.; Ge, Z.H.; Sun, Q.; Cheng, Z.X.; Li, Z.; Dou, S.X. Enhanced thermoelectric performance through synergy of resonance levels and valence band convergence via Q/In (Q = Mg, Ag, Bi) co-doping. *J. Mater. Chem. A* **2018**, *6*, 2507–2516. [CrossRef]

37. Hong, M.; Chen, Z.-G.; Yang, L.; Zou, Y.-C.; Dargusch, M.S.; Wang, H.; Zou, J. Realizing zT of 2.3 in $Ge_{1-x-y}Sb_xIn_yTe$ via Reducing the Phase-Transition Temperature and Introducing Resonant Energy Doping. *Adv. Mater.* **2018**, *30*, 1705942. [CrossRef] [PubMed]

38. Pei, Y.; Shi, X.; LaLonde, A.; Wang, H.; Chen, L.; Snyder, G.J. Convergence of electronic bands for high performance bulk thermoelectrics. *Nature* **2011**, *473*, 66–69. [CrossRef] [PubMed]

materials

MDPI

Article

Mechanical Performance of Glass-Based Geopolymer Matrix Composites Reinforced with Cellulose Fibers

Gianmarco Taveri [1,*], Enrico Bernardo [2] and Ivo Dlouhy [1]

[1]   Institute of Physics of Materials, Czech Academy of Science, Žižkova 22, 61662 Brno, Czech Republic; idlouhy@ipm.cz

[2]   Department of Industrial Engineering, University of Padova, 35131 Padova, Italy; enrico.bernardo@unipd.it

*   Correspondence: taveri@ipm.cz or gianmarco.taveri@outlook.com; Tel.: +420-777-542-929

Received: 1 November 2018; Accepted: 26 November 2018; Published: 28 November 2018

Abstract: Glass-based geopolymers, incorporating fly ash and borosilicate glass, were processed in conditions of high alkalinity (NaOH 10–13 M). Different formulations (fly ash and borosilicate in mixtures of 70–30 wt% and 30–70 wt%, respectively) and physical conditions (soaking time and relative humidity) were adopted. Flexural strength and fracture toughness were assessed for samples processed in optimized conditions by three-point bending and chevron notch testing, respectively. SEM was used to evaluate the fracture micromechanisms. Results showed that the geopolymerization efficiency is strongly influenced by the SiO_2/Al_2O_3 ratio and the curing conditions, especially the air humidity. The mechanical performances of the geopolymer samples were compared with those of cellulose fiber–geopolymer matrix composites with different fiber contents (1 wt%, 2 wt%, and 3 wt%). The composites exhibited higher strength and fracture resilience, with the maximum effect observed for the fiber content of 2 wt%. A chemical modification of the cellulose fiber surface was also observed.

Keywords: geopolymer composite; wastes incorporation; cellulose fibers; cellulose modification

---

1. Introduction

Geopolymers and alkali-activated materials (AAMs) are considered as the cementitious materials of the future [1], to be applied mainly in building and civil infrastructures [2–4]. What makes these materials widely attractive is the low CO_2 emission process of production, coupled with mechanical properties at least comparable to Portland cement (OPC) [5–7]. To date, however, extensive market diffusion has failed due to several reasons, including the cost of production, upscaling, and standardization of the process [8]. Nevertheless, so far, no other materials have been found to be more suitable than geopolymers and AAMs for facing the constantly increasing concern regarding climate change due to greenhouse gas emissions in the atmosphere, with 8% of the annual CO_2 emissions being accounted for by OPC production [1,7,9].

To decrease the cost of production, a fundamental solution consists of the incorporation of aluminosilicate waste, such as fly ash (a byproduct of coal combustion in thermal power plants) [10–16], as raw materials. The alkali-activation of fly ash induces the formation of a Ca-modified sodium aluminosilicate hydrates (N-A-S-H) gel through the polycondensation of aluminosilicate–lime species in a semiamorphous network composed of long-chain molecules [9,17,18]. In contrast to the benefits of its low cost and versatility of production, the extensive presence of hydrate groups in the chemistry of fly ash-based AAMs does not make this material more durable than OPC, unlike geopolymers [19–22]. The latter materials are characterized by a low Ca content (which normally favors the formation of a calcium silicate hydrates gel, C–S–H) and a silica-to-alumina ratio (SiO_2/Al_2O_3) exceeding 2, and thus they yield a semiamorphous three-dimensional and highly cross-linked aluminosilicate microstructure with a much lower amount of hydrate groups than in AAMs [21,23,24]. The additional supply of reactive silica could be provided by introducing additional waste-derived raw materials, such as recycled glass (from urban and industrial waste collection), with the obvious advantages of cost reduction and waste management [16,25–28]. Among all the possible variants of recycled glass utilized in geopolymerization, borosilicate glass (BSG) cullet from dismantled or discarded pharmaceutical vials is an intriguing alternative [29], since it was demonstrated that it also provides reactive borates in polycondensation, replacing alumina in its role in geopolymerization [30].

Despite all these recent developments, geopolymers still suffer from sudden unstable fractures due to their extreme brittleness [31,32]. The excessive low resistance to crack initiation/propagation is not due only to the fragility of the geopolymer product of reaction, but also to the extent of porosity and crack production during hardening. The amount of macrodefects generated by the process could be controlled to a limited extent through chemical (e.g., alkalinity, liquid-to-solid ratio) and physical conditions (e.g., humidity) in curing [33–35]. Alternatively, the production of composites from a geopolymeric matrix was extensively investigated as an effective solution to increase the fracture toughness, due to the mechanisms of pull-out and crack bridging of dispersed fibers [36–40]. Above all, cellulose seems to be suitable for geopolymer composites due to its chemical stability and specific tensile strength [41–43].

Here, a comparative study carried out on the effects of chemical (alkalinity and silica-to-alumina ratio) and physical parameters (relative humidity and soaking time) on the mechanical and microstructural properties of geopolymers is reported. Geopolymer composites were also produced by dispersing cellulose fibers, and the effect of fiber content was assessed in terms of bending strength, fracture toughness (chevron notch test), and fracture micromechanisms.

2. Materials and Methods

2.1. Manufacturing of Geopolymer Samples

Fly ash, a coal combustion byproduct, from a Bohemian thermal power plant (Počerady power plant, North Bohemia, Czech Republic) was used as a primary aluminosilicate source. The fly ash was dry-mixed with borosilicate glass from recycled pharmaceutical vial cullet (Kimble/Kontes, Vineland, NJ, USA) and activated with a caustic soda solution, prepared by dissolving NaOH pellets (American Ceramic Society reagent, 97%, pellets) purchased from Sigma Aldrich (Saint Louis, MO, USA) in distilled water. The geopolymer samples were manufactured according to 6 different formulations, in which the powder mixture (SiO_2/Al_2O_3 ratio), the molarity of the activator, and the curing time were modulated in order to investigate the influence of these parameters on geopolymerization. A schematic summary of the methodology is reported in Table 1. In batches 'Mix-1' to 'Mix-4', samples were prepared by activating dry mixes of fly ash from 70 wt% to 30 wt% and BSG from 30 wt% to 70 wt% in a 13 M NaOH solution and cured for 1 day. The samples processed through the 'Mix-5' batch were based on the 'Mix-2' formulation and cured for 3 days, and finally, 'Mix-6' samples were activated using a 10 M NaOH solution. The alkali solution was added in a sufficient amount to ensure workability to the slurry. In all the mixes, the liquid-to-solid ratio ranged between 0.4 and 0.5. The obtained slurry was cast in rubber molds and cured at 85 °C. The humidity in the samples was retained in two different ways [35]:

- Method 1: Molds were sealed in latex bags.
- Method 2: Molds were closed in an air-tightened jar with some water.

Table 1. Summary of the methodology of preparation of the fly ash (FA) and borosilicate (BSG) based geopolymer samples and related parameters.

Sample	Powder Mixture (FA%–BSG%)	Molarity (M)	Curing Time (day)	SiO$_2$/Al$_2$O$_3$	α Coefficient
Mix-1	70–30	13	1	2.7	2.2
Mix-2	55–45	13	1	3.3	2.5
Mix-3	40–60	13	1	4.2	2.8
Mix-4	30–70	13	1	5.0	3.0
Mix-5	55–45	13	3	3.3	2.5
Mix-6	55–45	10	1	3.3	2.5

Irrespective of the methodologies of processing, after curing, the samples were demolded and exposed to air for one week prior to testing in order to complete the geopolymerization.

The relative density of the samples in all the cases was calculated to be around 70%, according to the Archimedes method (theoretical density of 2.08 g/cm^3 and real density of 1.44 g/cm^3 on average), using high-purity water as a buoyant. The weighting of the specimens was carried out using a Denver analytical balance ($\pm$0.0001 g precision).

2.2. Manufacturing of Geopolymer Composite Samples

Mix-1 was properly modified in order to combine the dry mix with cellulose fibers in different percentages (from 1 to 3 wt%); the according weight content substitute was fly ash (from 69 wt% to 67 wt%). The reduction of the fly ash content was not considered sufficient to effect significant change in the properties of the material. The cellulose fibers were provided by CIUR A.s. (Brandýs nad Labem, Czech Republic). The dry mix was then diluted in distilled water and sonicated for one hour to guarantee homogenization of the mix and to unravel the bundles of cellulose fibers. The suspension was dried overnight and the dry mix was activated in a 13 M NaOH solution, forming a slurry whose liquid-to-solid ratio was between 0.8 and 0.9. The slurry was cast according to method 1 and cured at 85 °C for one day. As mentioned previously, the samples were demolded after curing and kept in air for one week prior to testing.

2.3. Mechanical Testing

Flexural strength and fracture toughness were determined through a 3-point bending test and chevron notch (CVN) test, respectively, conducted on 3 × 4 × 16 mm specimens using a ZWICK Z50 screw-driven machine (Ulm, Germany). The CVN test was performed in a 3-point bending configuration on specimens with a notch depth of 1 mm. The value of the fracture toughness was calculated from the ultimate bending load according to the following formula:

$$K_{IC} = \frac{Y^*_{min} \cdot F}{B \cdot W^{\frac{1}{2}}},\tag{1}$$

where F is the measured bending load corresponding to unstable crack development, Y^*_{min} is the minimum of the geometry function Y^* which is dependent on the notch depth a_0 and the geometry, B is the thickness, and W is the width of the specimen. The reliability of the chevron notch technique for fracture toughness determination of fiber-reinforced brittle matrix composites has been probed elsewhere [44]. The data elaboration was carried out using interquartile range (IQR) () statistics with a whisker factor of 1.5 [45,46].

2.4. Microstructural Investigation

The microstructure was investigated through SEM microscopy with a Tescan LYRA 3 XMH FEG equipped with an X-Max80 Energy-dispersive X-ray spectroscopy (EDS) detector for X-ray chemical analysis. The chemical composition of the raw materials was analyzed through X-ray fluorescence

(XRF) analysis using a RIGAKU (Tokio, Japan) ZSX100e model operating at 60 kV and 150 mA and equipped with Wavelength-dispersion X-ray spectroscopy (WDS), an X-ray Rh tube working at 3 kW, a scintillation counter for heavy element detection, and a gas-flow proportional counter (Ar–methane 10 %) for the detection of light elements.

3. Results and Discussion

3.1. Geopolymer Characterization

The geopolymerization process was tuned by considering the influence of the SiO_2/Al_2O_3 ratio, the molarity of the alkali solution, and the soaking time and humidity during curing, although the parameters influencing the process also include the temperature of curing, nature of the alkali solution, silica-to-alkali ratio, liquid-to-solid ratio, and so forth. The calculation of the SiO_2/Al_2O_3 ratio was based on the chemical composition of the raw materials, that is, fly ash and BSG glass. Table 2 reports the chemical composition of both the raw materials.

Table 2. Chemical compositions of raw fly ash and recycled BSG glass powders.

	SiO_2	Al_2O_3	B_2O_3	Fe_2O_3	CaO	K_2O	Na_2O	LOI [1]	Remainder
Fly ash (wt%)	46.3	26	/	13.9	3.5	3.9	0.2	0.7	5.5
BSG (wt%)	72	7	12	/	1	2	6	/	/

[1] Loss of ignition (LOI)—calculated through weight loss after fly ash firing.

The four fly ash/BSG formulations (from Mix-1 to Mix-4), as described in the experimental section, are characterized by a silica-to-alumina ratio of 2.7, 3.3, 4.2, and 5.0, respectively (Table 1). Considering other ratios as well (silica-to-soda and water-to-soda), all the formulations fall in the chemical ranges giving a stoichiometric geopolymer [23,24]. Interestingly, it was demonstrated in previous studies that the utilization of BSG glass as a silicate supplier of the geopolymerization also provides, during the dissolution stage, borate species, which eventually become part of the final microstructure as additional building blocks, giving rise to a B–Al–Si network [30]. Boron oxide, similarly to alumina, can be found in nature as diboron trioxide (B_2O_3), and dissolved borates can display a trigonal planar (BO_3) or a tetrahedral (BO_4) configuration (3-fold and 4-fold configurations, respectively). During geopolymerization, aluminates and borates concurrently rearrange in a 4-fold configuration (with charge compensation provided by alkali ions) to condense with tetrahedral silicates and to form the boron–alumino–silicate chains [30,47]. Due to this reason, a more refined formula of the previous ratio is provided, also taking into account the borates' contribution:

$$\alpha = \frac{\%SiO_2}{\%(Al_2O_3 + B_2O_3)},$$

(2)

For the calculation of the α-coefficient, we employed the weight percentages given by XRF analysis (Table 2); the α-coefficients for Mix-1, Mix-2, Mix-3, and Mix-4 were calculated to be 2.2, 2.5, 2.8, and 3.0, respectively (see Table 1).

Figure 1a reports the flexural strength values of the whole set of the tested specimens in three-point bending. The results of the three-point bending test revealed an increase of the flexural resistance of the geopolymers up to 50% if the α-coefficient is increased from 2.2 to 2.5. The higher amount of silicates in the formulation, indeed, may help the formation of the more cross-linked three-dimensional network, providing higher strength to the geopolymer. The maximum flexural strength values were observed to decrease if the α-coefficient was increased from 2.5 to 3.0, whereas the median value of the IQR whiskers box experienced a continuous increase and the data dispersion became narrower (Mix-3 and Mix-4 in Figure 1a). The reason of this trend was attributed to a deterioration of the microstructure, giving rise to a lowering of the maximum flexural strength value and to a decrease of the porosity and crack formation, inducing a refinement of the statistic and the average flexural strength values.

The macrodefects' formation in the geopolymer samples could be unrelated to the changes in the chemistry of the system.

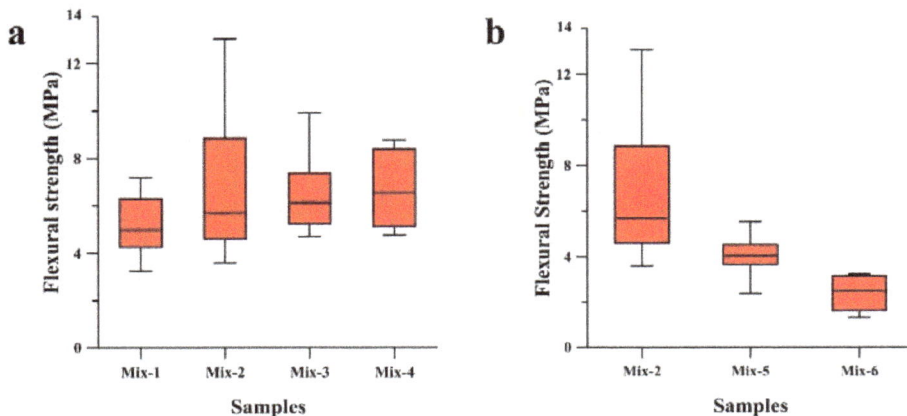

Figure 1. Comparative study of the flexural strength of geopolymer samples in terms of (**a**) silica-to-alumina ratio of the mixtures from Mix-1 to Mix-4 (2.7, 3.3, 4.2, 5.0 respectively) and (**b**) molarity of the activator and curing time used for Mix-2 (13 M for 1 day), Mix-5 (13 M for 3 days) and Mix-6 (10 M for 1 day).

A similar trend in the flexural strength is also associated with the higher molarity of the alkali solution: Figure 1b demonstrates that an increase in molarity of the NaOH solution from 10 M to 13 M leads to an increase of the flexural strength of geopolymer samples from 4 MPa (Mix-5) to almost 14 MPa (Mix-2), and to a higher data dispersion. This result is discordant with those of several prior studies, asserting that a higher amount of alkali cations hinders the condensation of long polymeric chains, which generally gives less microstructural stability to the geopolymeric compound [48]. The microstructural refinement is not always synonymous with the mechanical optimization of the material, as it has to be appropriately weighted to the efficiency of other stages of the geopolymerization, which in turn depends on other factors. One of these is the initial stage of dissolution, where the building blocks of the polymeric structure, namely borates, aluminates, and silicates, are provided from. The dissolution of the ionic species is strictly related to the raw materials' nature and the concentration of the alkali solvent. Specifically, a higher concentration of the alkali activator yields to a better dissolution of the boron–alumino–silicate sources, improving the degree of reaction [49,50].

The results from the three-point bending test conducted on Mix-6 samples showed that an increased duration of exposure to the curing temperature decreases the flexural strength of the material by up to half of the average flexural strength (Figure 1b). This drop in the bending properties could be attributed to multiple reasons, but in general, an increase in the temperature exposure can induce a larger shrinkage and more substantial crack formation. A comparative study between the microstructures of the geopolymer samples tested in bending is not reported in this work, as the samples did not show any significant differences. Evidence of the microstructure is reported in Figure 2a, in which porosity and cracks are homogenously present everywhere, reporting a characteristic morphology associated with hardening and shrinkage during the curing of geopolymers featuring a high SiO_2/Al_2O_3 ratio [12,17,51]. The amount of porosity and defects is inferable from the calculated relative density (see the Materials and Methods section).

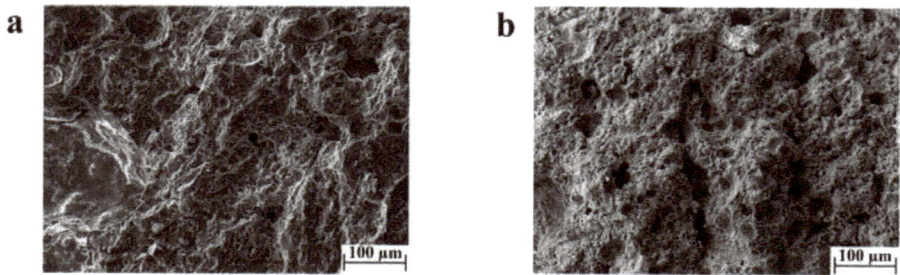

Figure 2. SEM images of (**a**) the 'Mix-2' sample sealed according to method 1 and (**b**) the 'Mix-2' sample sealed according to method 2.

In all the cases, the Mix-2 samples provide the higher bending strength, but at the same time, larger data scattering. This is also attributed to porosity and crack formation, which are strictly related to the amount of water in the slurry and to the methodology of sealing and amount of humidity in the bag. For instance, a few percentages of humidity change can induce a significant variation in the mechanical properties of the geopolymers [33].

The parametrical study of the geopolymerization as a function of humidity goes beyond the aim of this work, but two sealing methodologies of the samples in the curing phase were tested. Method 1 involves humidity only by water evaporation of the slurry, while method 2 provides more water to the system. The flexural results related to Figure 1 are attributed to samples cured according to method 1. The data testing of the samples processed according to method 2 are not reported, as they do not show any appreciable bending resistance. The microstructure of these samples was observed by SEM microscopy and compared to the Mix-2 sample. Figure 2b revealed a deficiency in the degree of reaction and a much lower amount of geopolymeric product in samples processed by method 2 (Figure 2b) as compared to method 1 (Figure 2a), confirming that the supply of extra water in the curing stage hinders the geopolymerization reaction [33,35].

3.2. Geopolymer Composites

The comparative study of the fracture performance of geopolymer samples, albeit tested, were not reported, as the CVN test gave values of fracture toughness all below 0.3 $MPa \cdot m^{1/2}$. The resistance to the crack propagation was significantly improved with the addition of 1 wt%, 2 wt%, and 3 wt% cellulose fibers into the geopolymeric matrix (see Materials and Methods section). Besides the fracture resistance, the flexural strength similarly improved, as depicted in Figure 3. In all the three composite samples, the flexural strength improved as compared to the plain geopolymer samples. The highest flexural strength was observed with 2 wt% cellulose dispersed throughout the sample, with an average value of 18 MPa. This value is three-fold higher than the 'Mix-1' value (see Figure 1a), and higher than the flexural strength reported in the literature for an analogous fly ash-based geopolymer matrix composite with dispersed cellulose fibers [41]. Geopolymeric composite samples with a cellulose content exceeding 2 wt% induce a reduction of the flexural strength (Figure 3a), attributed to excessive Na^+ absorption by the cellulose fibers. The same trend was also observed for the fracture toughness, reported in Figure 3b, in which the highest fracture resistance (0.6 $MPa \cdot m^{1/2}$) is displayed by the 2 wt% sample, with an average increase of the fracture toughness of up to 300% as compared to 'Mix-1' sample. The value of the fracture toughness is in line with the values reported in the literature [41].

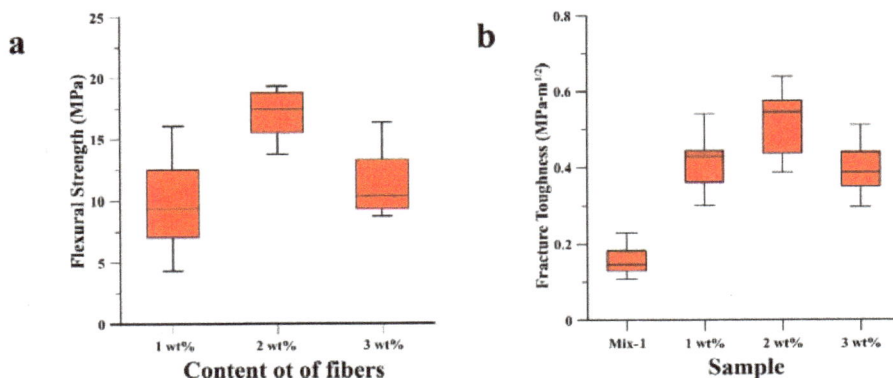

Figure 3. Comparative study of the mechanical and fracture performances of the cellulose fibers geopolymer matrix composites in terms of (**a**) Flexural strength of samples with 1 wt%, 2 wt% and 3 wt% of dispersed cellulose fibers and (**b**) fracture toughness of the geopolymer sample (Mix-1 formulation) and geopolymer matrix composite samples (1 wt%, 2 wt% and 3 wt%).

The absorption of alkali cations (especially Na^+) by the cellulose fibers is a widely known process known as 'mercerization'. The cellulose swells due to the interference of alkali cations between the polymeric chains, creating electrostatic layers of positively charged Na^+ ions and negatively charged polymeric chains [52]. What is not equally reported is the absorption of other chemical elements, which creates, in cellulose-based composites, a sort of transitional interface between the inner part of the fiber and the matrix. Tonoli et al. [53] demonstrated the formation of a modified surface on cellulose fibers dispersed in the cement matrix, due to Ca^{2+} absorption from the C–S–H network of the cement. Previously, Merrill et al. [54] published a study reporting the sodium silicate sorption in cellulose fibers, in which it was proved that the cellulose can also absorb silicon ions from the water glass, although to a lower extent than sodium, involving a polycondensation with the polymeric chains.

In our geopolymer matrix samples, irrespective of the weight amount of the dispersed fibers, cellulose was found to undertake a surface modification and a new phase formation at the interface with the geopolymeric matrix. Figure 4a shows detail of a cellulose fiber, whose surface is clearly modified. It is evident from the figure that the affected polymeric fibrils at the fiber surface are rearranged in an orthogonal configuration to the geopolymer interface (see insert in Figure 4a), whereas at the core, the fibrils are parallel to the interface line. EDS analysis was conducted to detect the chemical species absorbed by the new phase, and the results are reported in Figure 4b. The chemical analysis was carried out in three different zones in order to draw a conclusion regarding the elements' absorption: zone 1—the cellulose fiber core, zone 2—the cellulose/geopolymer interface, and zone 3— the geopolymeric matrix. In all zones, the amount of sodium was the same, whereas the silicon was observed to be more abundant in zone 3 than zone 2, as expected. What was, however, unexpected was a deficiency of oxygen in the modified interface (zone 2), suggesting a condensation between the hydroxyl groups of the polymeric chains of the cellulose and geopolymer with corresponding water release. However, this hypothesis should be supported by further spectroscopic analysis, unequivocally demonstrating the formation of a new phase formed by organic–inorganic polymeric chains.

The presence of the new phase between the two organic–inorganic compounds can influence the mechanisms of fracture of the composite samples. The observation of the fracture surface (Figure 5a,b) and the polished surface (Figure 5c,d) of the geopolymer matrix samples revealed the presence of both the mechanisms of pull-out and crack bridging. For instance, the investigation of the fracture surface clearly evidenced that during fracture, the fibers were exposed to pull-out mechanisms from the geopolymeric matrix (Figure 5a). This was especially highlighted by observations of delamination sites, such as the one reported in Figure 5b. The mechanisms of crack bridging were rather evident on the polished surface (Figure 5c). Due to the presence of the fiber, the crack was both hindered and deflected from the direction of propagation. However, the formation of the new phase induces also mechanisms of fiber tearing, as shown in Figure 5d. Indeed, in some cases, the interface strength may be higher than the resistance of the single compound, causing fiber failure instead of fiber delamination. Figure 5b shows the remainder of a cellulose fiber in the delaminated site, suggesting an inhomogeneous nature of this phase. Therefore, to understand the whole dynamics of failure in the geopolymer matrix composite, a detailed investigation of the chemical nature and mechanical performance of the new phase is required and will be the object of future work.

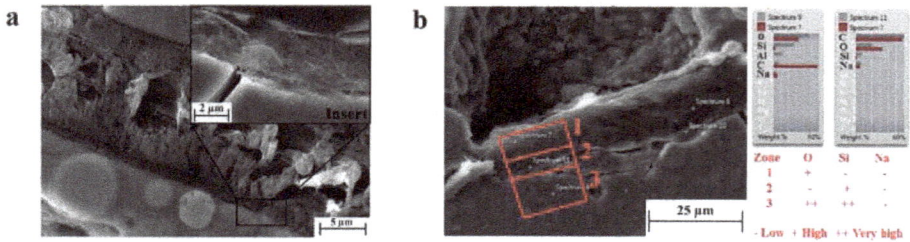

Figure 4. (**a**) SEM image of a cellulose–geopolymer interface, including a close-up of the new phase (insert), and (**b**) Energy-dispersed X-ray spectroscopy (EDS) analysis of the three zones of interest.

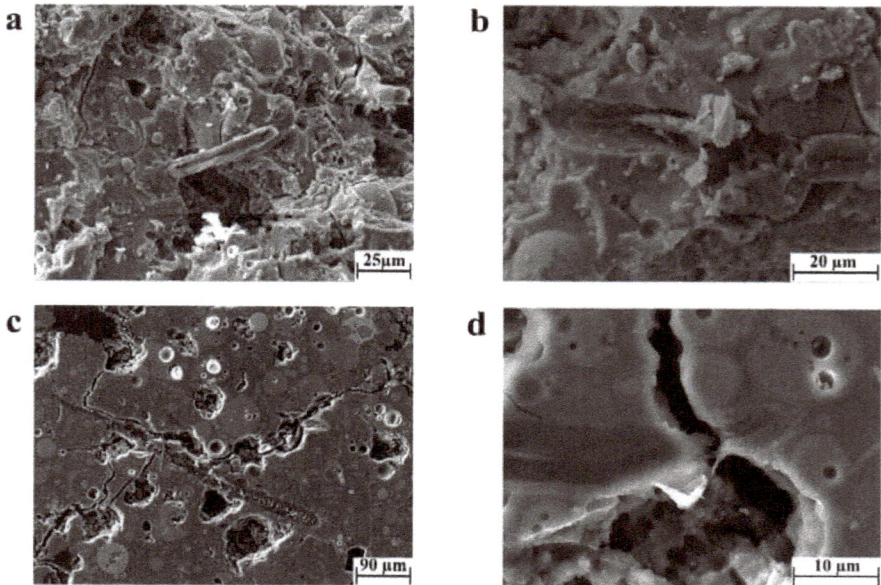

Figure 5. Fracture and polished surface observations of the microstructure of the geopolymer matrix composite through SEM images of (**a**) fiber pull-out and (**b**) negative site of a delaminated fiber, (**c**) crack-bridging, and (**d**) fiber failure.

4. Conclusions

In summary:

- Geopolymer samples incorporating fly ash and borosilicate wastes were processed using four different formulations (Mix-1, Mix-2, Mix-3, and Mix-4), two alkali solution molarities (10–13 M, Mix-5), and curing times (1–3 h, Mix-6) in order to evaluate the influence of these chemical/physical parameters on geopolymerization.
- The rate of influence was assessed in terms of flexural strength. The results of the bending test revealed that the higher flexural strength (13 MPa) was associated with the Mix-2 sample, which was activated with a 13 M NaOH solution and cured for 1 day.
- The fracture toughness of the geopolymer was enhanced by up to 4 times when geopolymer matrix samples including 2 wt% of cellulose fiber were processed.
- A mercerization and a surface modification of the cellulose fibers were also observed, and EDS analysis suggested a possible chemical interaction between the organic and inorganic polymeric chains.
- The dynamics of failure of the geopolymer matrix composite includes crack bridging, fiber pull-out, and fiber-tearing mechanisms.

Author Contributions: Methodology, G.T.; Investigation, G.T.; Resources, E.B. and I.D.; Data Curation, G.T.; Writing—Original Draft Preparation, G.T.; Writing—Review & Editing, E. B. and I.D.; Supervision, I.D.; Funding Acquisition, E.B. and I.D.

Funding: This work has received funding from the European Union's Horizon 2020 research and innovation program under the Marie Sklodowska-Curie grant agreement No. 642557.

Acknowledgments: The authors are giving thanks to Nuova OMPI (Piombino Dese, Padova, Italy) for the supply of the recycled BSG glass and to CIUR a.s. (Brandýs nad Labem, Czechia) for fiber supply. The authors also thank SASIL S.p.a. (Brusnengo, Biella, Italy) for the valuable technical support and Carla Celozzi (DISAT Dpt., Polytechnic of Turin) for the XRF measurements.

Conflicts of Interest: The authors declare no conflict of interest.

References

1. Palomo, A.; Grutzeck, M.W.; Blanco, M.T. Alkali-activated fly ashes: A cement for the future. *Cem. Concr. Res.* **1999**, *29*, 1323–1329. [CrossRef]
2. Provis, J.L.; Bernal, S.A. Alkali Activated Materials: State of the Art Report. *Taylor Fr.* **2014**, *13*, 125–144. [CrossRef]
3. Provis, J.L.; van Deventer, J.S.J. *Geopolymers: Structures, Processing, Properties and Industrial Applications*; Elsevier: Amsterdam, The Netherlands, 2009; ISBN 9781845694494.
4. Duxson, P.; Fernandez-Jimenez, A.; Provis, J.L.; Lukey, G.C.; Palomo, A.; Van Deventer, J.S.J. Geopolymer technology: The current state of the art. *J. Mater. Sci.* **2007**, *42*, 2917–2933. [CrossRef]
5. Živica, V.; Palou, M.T.; Križma, M. Geopolymer Cements and Their Properties: A Review. *Build. Res. J.* **2015**, *61*. [CrossRef]
6. Pan, Z.; Sanjayan, J.G.; Vijay, R.B. Fracture properties of geopolymer paste and concrete. *Mag. Concr. Res.* **2011**, *63*, 763–777. [CrossRef]
7. Shi, C.; Jiménez, A.F.; Palomo, A. New cements for the 21st century: The pursuit of an alternative to Portland cement. *Cem. Concr. Res.* **2011**, *41*, 750–763. [CrossRef]
8. Hooton, R.D. Bridging the gap between research and standards. *Cem. Concr. Res.* **2008**, *38*, 247–258. [CrossRef]
9. Palomo, A.; Krivenko, P.; Garcia-Lodeiro, I.; Kavalerova, E.; Maltseva, O.; Fernandez-Jimenez, A. A review on alkaline activation: New analytical perspectives. *Mater. Constr.* **2014**, *64*, 1–23. [CrossRef]
10. Toniolo, N.; Taveri, G.; Hurle, K.; Roether, J.A.A.; Ercole, P.; Dlouhý, I.; Boccaccini, A.R.R. Fly-ash-based geopolymers: How the addition of recycled glass or red mud waste influences the structural and mechanical properties. *J. Ceram. Sci. Technol.* **2017**, *8*, 411–419. [CrossRef]

11. Fernández Jiménez, A.; Palomo, A. Factors affecting early compressive strength of alkali activated fly ash (OPC-free) concrete. *Mater. Constr.* **2007**, *57*, 7–22.
12. Fernández-Jiménez, A.; Palomo, A. Composition and microstructure of alkali activated fly ash binder: Effect of the activator. *Cem. Concr. Res.* **2005**, *35*, 1984–1992. [CrossRef]
13. Zhang, Z.; Provis, J.L.; Reid, A.; Wang, H. Fly ash-based geopolymers: The relationship between composition, pore structure and efflorescence. *Cem. Concr. Res.* **2014**, *64*, 30–41. [CrossRef]
14. Provis, J.L.; Yong, C.Z.; Duxson, P.; van Deventer, J.S.J. Correlating mechanical and thermal properties of sodium silicate-fly ash geopolymers. *Colloids Surf. A Physicochem. Eng. Asp.* **2009**, *336*, 57–63. [CrossRef]
15. Fernández-Jiménez, A.; Palomo, A. Characterisation of fly ashes. Potential reactivity as alkaline cements. *Fuel* **2003**, *82*, 2259–2265. [CrossRef]
16. Toniolo, N.; Boccaccini, A.R. Fly ash-based geopolymers containing added silicate waste. A review. *Ceram. Int.* **2017**, *43*, 14545–14551. [CrossRef]
17. Fernández-Jiménez, A.; Palomo, A.; Criado, M. Microstructure development of alkali-activated fly ash cement: A descriptive model. *Cem. Concr. Res.* **2005**, *35*, 1204–1209. [CrossRef]
18. Martin, A.; Pastor, J.Y.; Palomo, A.; Fernández Jiménez, A. Mechanical behaviour at high temperature of alkali-activated aluminosilicates (geopolymers). *Constr. Build. Mater.* **2015**, *93*, 1188–1196. [CrossRef]
19. Pacheco-Torgal, F.; Abdollahnejad, Z.; Camões, A.F.; Jamshidi, M.; Ding, Y. Durability of alkali-activated binders: A clear advantage over Portland cement or an unproven issue? *Constr. Build. Mater.* **2012**, *30*, 400–405. [CrossRef]
20. Davidovits, J. *The Pyramids*; Dorset Press: New York, NY, USA, 1990; ISBN 290293307X.
21. Davidovits, J. Properties of Geopolymer Cements. In Proceedings of the First International Conference on Alkaline Cements and Concretes, Kiev, Ukraine, 11–14 October 1994; pp. 131–149.
22. Duxson, P.; Provis, J.L.; Lukey, G.C.; Mallicoat, S.W.; Kriven, W.M.; Van Deventer, J.S.J. Understanding the relationship between geopolymer composition, microstructure and mechanical properties. *Colloids Surf. A Physicochem. Eng. Asp.* **2005**, *269*, 47–58. [CrossRef]
23. Davidovits, J. *Geopolymer: Chemistry & Applications*; Institut Géopolymère: Saint-Quentin, France, 2008; ISBN 9782951482098.
24. Klinowski, J. Nuclear magnetic resonance studies of zeolites. *Prog. Nucl. Magn. Reson. Spectrosc.* **1984**, *16*, 237–309. [CrossRef]
25. Torres-Carrasco, M.; Puertas, F. Waste glass in the geopolymer preparation. Mechanical and microstructural characterisation. *J. Clean. Prod.* **2015**, *90*, 397–408. [CrossRef]
26. Torres-Carrasco, M.; Puertas, F.; Blanco-Varela, M. Preparación de cementos alcalinos a partir de residuos vítreos. Solubilidad de residuos vítreos en medios fuertemente básicos. In *XII Congreso Nacional de Materiales and XII Iberomat: Conference Proceedings*; Instituto Universitario de Materiales: Alicante, Spain, 2012.
27. Torres-Carrasco, M.; Rodríguez-Puertas, C.; Del Mar Alonso, M.; Puertas, F. Alkali activated slag cements using waste glass as alternative activators. Rheological behaviour. *Boletín de la Sociedad Española de Cerámica y Vidrio* **2015**, *54*, 45–57. [CrossRef]
28. Puertas, F.; Torres-Carrasco, M. Use of glass waste as an activator in the preparation of alkali-activated slag. Mechanical strength and paste characterisation. *Cem. Concr. Res.* **2014**, *57*, 95–104. [CrossRef]
29. Chinnam, R.K.; Francis, A.A.; Will, J.; Bernardo, E.; Boccaccini, A.R. Review. Functional glasses and glass-ceramics derived from iron rich waste and combination of industrial residues. *J. Non-Cryst. Solids* **2013**, *365*, 63–74. [CrossRef]
30. Taveri, G.; Tousek, J.; Bernardo, E.; Toniolo, N.; Boccaccini, A.R.R.; Dlouhy, I. Proving the role of boron in the structure of fly-ash/borosilicate glass based geopolymers. *Mater. Lett.* **2017**, *200*, 105–108. [CrossRef]
31. Côrtes Pires, E.F.; Campinho de Azevedo, C.M.; Pimenta, A.R.; da Silva, F.J.; Darwish, F.A.I. Fracture Properties of Geopolymer Concrete Based on Metakaolin, Fly Ash and Rice Rusk Ash. *Mater. Res.* **2017**, *20*, 630–636. [CrossRef]
32. Dias, D.P.; Thaumaturgo, C. Fracture toughness of geopolymeric concretes reinforced with basalt fibers. *Cem. Concr. Compos.* **2005**, *27*, 49–54. [CrossRef]
33. Oderji, S.Y.; Chen, B.; Taseer, S.; Jaffar, A. Effects of relative humidity on the properties of fly ash-based geopolymers. *Constr. Build. Mater.* **2017**, *153*, 268–273. [CrossRef]

34. García-Mejía, T.A.; de Lourdes Chávez-García, M. Compressive Strength of Metakaolin-Based Geopolymers: Influence of KOH Concentration, Temperature, Time and Relative Humidity. *Mater. Sci. Appl.* **2016**, *7*, 772–791. [CrossRef]

35. Criado, M.; Fernández Jiménez, A.; Sobrados, I.; Palomo, A.; Sanz, J. Effect of relative humidity on the reaction products of alkali activated fly ash. *J. Eur. Ceram. Soc.* **2012**, *32*, 2799–2807. [CrossRef]

36. Yan, S.; He, P.; Jia, D.; Yang, Z.; Duan, X.; Wang, S.; Zhou, Y. Effect of fiber content on the microstructure and mechanical properties of carbon fiber felt reinforced geopolymer composites. *Ceram. Int.* **2016**, *42*, 7837–7843. [CrossRef]

37. Assaaedi, H.; Alomayri, T.; Shaikh, A.; Low, I. Characterisation of mechanical and thermal properties in flax fabric reinforced geopolymer composites. *J. Adv. Cermics* **2015**, *4*, 272–281. [CrossRef]

38. Shaikh, F.U.A. Deflection hardening behaviour of short fibre reinforced fly ash based geopolymer composites. *Mater. Des.* **2013**, *50*, 674–682. [CrossRef]

39. Yuan, J.; He, P.; Jia, D.; Yan, S.; Cai, D.; Xu, L.; Yang, Z.; Duan, X.; Wang, S.; Zhou, Y. SiC fiber reinforced geopolymer composites, part 1: Short SiC fiber. *Ceram. Int.* **2016**, *42*, 5345–5352. [CrossRef]

40. Sá Ribeiro, R.A.; Sá Ribeiro, M.G.; Kriven, W.M. Review of particle- and fiber-reinforced metakaolin-based geopolymer composites. *J. Ceram. Sci. Technol.* **2017**, *8*, 307–321. [CrossRef]

41. Alomayri, T.; Shaikh, F.U.A.; Low, I.M. Characterisation of cotton fibre-reinforced geopolymer composites. *Compos. Part B Eng.* **2013**, *50*, 1–6. [CrossRef]

42. Alomayri, T.; Low, I.M. Synthesis and characterization of mechanical properties in cotton fiber-reinforced geopolymer composites. *J. Asian Ceram. Soc.* **2013**, *1*, 30–34. [CrossRef]

43. Alomayri, T.; Shaikh, F.U.A.; Low, I.M. Thermal and mechanical properties of cotton fabric-reinforced geopolymer composites. *J. Mater. Sci.* **2013**, *48*, 6746–6752. [CrossRef]

44. Dlouhý, I.; Boccaccini, A.R. Reliability of the chevron notch technique for fracture toughness determination in glass composites reinforced by continuous fibres. *Scr. Mater.* **2001**, *44*, 531–537. [CrossRef]

45. Abubakar, M.; Mat Yajid, M.A.; Ahmad, N. Comparison of Weibull and normal probability distribution of flexural strength of dense and porous fired clay. *J. Teknol.* **2016**, *78*, 73–78. [CrossRef]

46. Sahin, Z.; Ergun, G.; Author, C. The Assessment of Some Physical and Mechanical Properties of PMMA Added Different Forms of Nano-ZrO 2. *J. Dent. Oral. Heal.* **2017**, *3*, 64–74.

47. Nicholson, C.; Murray, B.; Fletcher, R.A.; Brew, D.R.M.; MacKenzie, K.J.D.; Schmuker, M. Novel geopolymer materials containing borate structural units. In *Geopolymer, Green Chemistry and Sustainable Development Solutions*; Geopolymer Institute: Saint-Quentin, France, 2005.

48. Xu, H.; van Deventer, J.S.J. Geopolymerisation of alumino-silicate minerals. *Int. J. Miner. Process.* **2000**, *59*, 247–266. [CrossRef]

49. Weng, L.; Sagoe-Crentsil, K. Dissolution processes, hydrolysis and condensation reactions during geopolymer synthesis: Part I-Low Si/Al ratio systems. *J. Mater. Sci.* **2007**, *42*, 2997–3006. [CrossRef]

50. Weng, L.; Sagoe-Crentsil, K. Dissolution processes, hydrolysis and condensation reactions during geopolymer synthesis: Part II-High Si/Al ratio systems. *J. Mater. Sci.* **2007**, *42*, 3007–3014. [CrossRef]

51. Trochez, J.J.; De Gutiérrez, R.M.; Rivera, J.; Bernal, S.A. Synthesis of Geopolymer from Spent FCC: Effect of SiO_2/Al_2O_3 and Na_2O/SiO_2 Molar Ratios. *Mater. Constr.* **2015**, *65*, 1–11. [CrossRef]

52. Wang, Y. Cellulose Fiber Dissolution in Sodium Hydroxide Solution at Low Temperature: Dissolution Kinetics and Solubility Improvement. Ph.D. Thesis, Georgia Institute of Technology, Atlanta, GA, USA, 2008.

53. Tonoli, G.H.D.; Rodrigues Filho, U.P.; Savastano, H.; Bras, J.; Belgacem, M.N.; Rocco Lahr, F.A. Cellulose modified fibres in cement based composites. *Compos. Part A Appl. Sci. Manuf.* **2009**, *40*, 2046–2053. [CrossRef]

54. Merrill, R.; Spencer, R. Sorption of Sodium Silicates and Silica Sols by Cellulose Fibers. *Ind. Eng. Chem.* **1950**, *42*, 744–747. [CrossRef]

materials

MDPI

Article

Solid-Liquid Interdiffusion (SLID) Bonding of p-Type Skutterudite Thermoelectric Material Using Al-Ni Interlayers

Katarzyna Placha [1,2,*], Richard S. Tuley [1], Milena Salvo [2], Valentina Casalegno [2] and Kevin Simpson [1]

[1] European Thermodynamics Ltd., 8 Priory Business Park, Leicester LE8 0RX, UK; richard.tuley@etdyn.com (R.S.T.); kevin@etdyn.com (K.S.)
[2] Politecnico di Torino, Department of Applied Science and Technology, Corso Duca degli Abruzzi, 10129 Turin, Italy; milena.salvo@polito.it (M.S.); valentina.casalegno@polito.it (V.C.)
* Correspondence: katarzyna@etdyn.com; Tel.: +44-7922-810-492

Received: 30 October 2018; Accepted: 4 December 2018; Published: 6 December 2018

Abstract: Over the past few years, significant progress towards implementation of environmentally sustainable and cost-effective thermoelectric power generation has been made. However, the reliability and high-temperature stability challenges of incorporating thermoelectric materials into modules still represent a key bottleneck. Here, we demonstrate an implementation of the Solid-Liquid Interdiffusion technique used for bonding $Mm_y(Fe,Co)_4Sb_{12}$ p-type thermoelectric material to metallic interconnect using a novel aluminium–nickel multi-layered system. It was found that the diffusion reaction-controlled process leads to the formation of two distinct intermetallic compounds (IMCs), Al_3Ni and Al_3Ni_2, with a theoretical melting point higher than the initial bonding temperature. Different manufacturing parameters have also been investigated and their influence on electrical, mechanical and microstructural features of bonded components are reported here. The resulting electrical contact resistances and apparent shear strengths for components with residual aluminium were measured to be $(2.8 \pm 0.4) \times 10^{-5}$ $\Omega \cdot cm^2$ and 5.1 ± 0.5 MPa and with aluminium completely transformed into Al_3Ni and Al_3Ni_2 IMCs were $(4.8 \pm 0.3) \times 10^{-5}$ $\Omega \cdot cm^2$ and 4.5 ± 0.5 MPa respectively. The behaviour and microstructural changes in the joining material have been evaluated through isothermal annealing at hot-leg working temperature to investigate the stability and evolution of the contact.

Keywords: solid-liquid interdiffusion (SLID) bonding; transient-liquid phase bonding (TLPB); skutterudite; high-temperature thermoelectric material; joining

1. Introduction

Skutterudite materials with excellent high-temperature thermoelectric performance and reasonably good mechanical properties are being extensively investigated for power generation use, including automotive waste heat recovery [1,2] and deep space energy generation [3]. However, a major limitation to fully realizing the potential of skutterudites and many other medium-high temperature range thermoelectric materials is the lack of reliable, low-cost assembly technology for robust thermoelectric generators (TEGs). The conventional thermoelectric device manufacturing comprises of labour-intensive methodology including soldering, brazing and adhesive bonding [4] which are limited by the thermal stability (as the braze melting temperature is normally its re-melting temperature) and the extensive growth of reaction layers at the thermoelectric–interconnect interfaces, causing mechanical or electrical deterioration of the entire assembly. In the conventional skutterudite devices, metal interconnects are joined to the metallized semiconductor elements to provide permanent interconnection either by the mechanical (i.e., using compression springs [5]) or chemical (i.e., solid-state diffusion bonding [6,7] or brazing [8]) methods of joining. Designing a suitable metallization and bonding technique for the skutterudite family of compounds is particularly difficult with the complex reactivity of their individual components, since antimony (Sb) can react with most of the transition metals to form brittle antimonide intermetallic compounds (IMCs). Among the many different metallization and interconnects reported, pure elements including those of transition and refractory metals [9,10], nickel-based alloys [11,12] and metal transition silicides [13] are the most commonly explored in a variety of skutterudite-based thermoelectric systems. Single transition metals are proven to not be a suitable metallization for skutterudites as they usually suffer from the intensive thermal stresses induced during the high-temperature service including a volumetric change, associated with the phase transformation at the skutterudite–metallization interface or components thermal expansion coefficient (CTE) mismatch. Although the recently reported Fe-Ni low-expansion alloys exhibit exceptionally good electrical contact resistance of ~0.4 $\mu\Omega\cdot cm^2$ on $(Ce_{1-z}Nd_z)_y$ $Fe_{4-x}Co_xSb_{12}$ p-type material, a higher cost two-step consolidation process can be needed due to the low density of Fe-Ni powders sintered at relatively low temperatures (~85 % of the theoretical value) [13]. Consequently, there is an urgent need for joining processes that could maximize thermoelectric module performance and are transferable to low-cost, high-volume manufacturing.

One promising interconnect technology for high-temperature microelectronic packaging is solid–liquid interdiffusion (SLID) bonding [14], also called transient liquid-phase bonding (TLPB) [15], and diffusion brazing [16]. This technique was developed in the 1970s by Duvall, et al. [17] and constitutes a suitable alternative to high-temperature furnace brazing operations while being commonly considered as the best alternative for solder bonding e.g., for wafer/chip stacked bonding for advanced micro-electro-mechanical systems (MEMS) [18,19] and in sensors manufacturing [20]. The SLID bonding technique is based on binary interlayer systems comprising a high-melting-point material, $T_{M\,HIGH}$, and a low-melting-point material, $T_{M\,LOW}$, which is heated to its molten state, resulting in the formation of intermetallic compounds (IMCs) through a diffusion-reaction mechanism. SLID bonding has been commonly explored using Sn and In as $T_{M\,LOW}$ filler systems [21–25] with several advantages over standard solder interconnections such as outstanding thermal stability due to increased re-melting temperature and power handling capability with current densities exceeding the capability of solders [26]. However, SLID technology faces many limitations before industrial implementation due to having a relatively slow and time-consuming diffusion controlled reaction along with inherent Kirkendall voiding and shrink holes created within the bonding material [18]. While, Cu-Sn and Ni-Sn bilayers have been found to form robust IMCs with relatively high melting points, i.e. T_M (Cu_3Sn) = 676 °C and T_M (Ni_3Sn_4) = 794 °C, the possible excess of residual Sn within a joint reduces its attractiveness in high-temperature electronic packaging. To that end, development of new joining materials with high-temperature chemical stabilities and significant oxidation resistances that can be utilized in thermoelectric modules manufacturing are essentially needed.

In the present work, we assess Al-Ni diffusion couples as a potential joining material for high-temperature thermoelectric modules assembly application. Among various intermetallic compounds, Al-Ni IMCs are associated with unique properties such as high thermal stability, tremendous oxidation and corrosion resistance along with excellent performance in creep strength [27]. The aluminium-nickel bi-layer system has already been classified as a particularly suitable structural energetic material used in reactive, multilayer brazing foils due to its exceptional exothermic properties [28,29]. Encouraged by the remarkable properties of Al-Ni IMCs and the recent findings on flux-less solid-liquid interdiffusion (SLID) bonding technique in thermoelectric manufacturing [30,31], a new joining technique utilizing the aluminium–nickel system has been developed and is reported here.

2. Materials and Methods

Polycrystalline $Mm_y(Fe,Co)_4Sb_{12}$ (*Mm*—Mischmetal, i.e., 50 at.% Ce, 25 at.% La) p-type thermoelectric material was synthesized by placing commercial pre-alloyed powder (99.995% purity) inside a graphite die and then sealed with graphite punches. To uniformly distribute heat and forces along sintered material, a graphite sheet (Erodex, Halesowen, UK) was placed between the graphite punch and thermoelectric material. An in-situ synthesis was carried out in Spark Plasma Sintering (SPS) apparatus (FCT HPD25, Rauenstein, Frankenblick, Germany) by applying uniaxial pressure of 50 MPa at the peak temperature of 600 °C for 5 min. The process was performed under a vacuum with heating and cooling rates of 50 °C/min and as-resulted bulk material was 98% dense as determined by the Archimedes principle. Afterwards, the thermoelectric material was ground to 3-mm thick discs using a 320-grit diamond wheel and was cleaned using isopropanol and ultrasonic bath.

Subsequently, 0.3-mm thick copper plate along with as-sintered thermoelectric material were subjected to an electroless nickel plating process with a phosphorous content of 10 wt.% resulting in the nickel–phosphorous Ni(P) coating layer. Both substrates required different surface pre-treatments prior to the electroless nickel plating process due to the poor adherence of the Ni(P) layer on the un-treated skutterudite surface and the presence of a native oxide layer on copper substrates. The thermoelectric material used an electrolytic nickel Wood's strike comprising $NiSO_4 \cdot 6H_2O$ and 12M HCl electrolyte solution to deposit a seed layer on the surface of the thermoelectric material by passing a current with a density of 5 A/dm^2 for 2 min. Copper plate was subjected to native oxides removal process including acid cleaning, chemical etching, pre-activation dipping in 5% H_2SO_4, and Pd-activation process. Afterwards, both the copper plate and thermoelectric disc were immersed in an electroless nickel plating bath comprising of 45 wt.% $NiSO_4 \cdot 6H_2O$, 25 wt.% $NaPO_2H_2 \cdot H_2O$ and 1 wt.% $NH_3 \cdot H_2O$. The bath temperature was kept at 95 °C during the deposition process. As-prepared thermoelectric material was cut to $2.5 \times 2.5 \times 3$ mm rectangular-shaped legs using a diamond cut-off wheel (M1D15, Struers, Ballerup, Denmark) and soaked in acetone to remove any organic residues. A standard scotch tape test was performed on an as-coated thermoelectric surface to evaluate the adhesion quality according to ASTM D3359 standard [32].

A 17 μm-thick Al foil (Goodfellow, Cambridge, UK, 99.0% purity) was incorporated between the Ni(P) coatings on the thermoelectric pellet and the copper plate and supportive uniaxial pressure of ~0.5 MPa was applied. To provide mechanical stability within brazing jig, two symmetrical joints were fabricated by placing a copper plate between two thermoelectric pellets at the same time, as presented in Figure 1. Afterwards, the brazing jig was placed inside a tube furnace (Carbolite, Hope Valley, UK) and the assembly was joined at 585 °C and 660 °C in Ar (4 L/min) with a heating rate of 7 °C/min. All joined assemblies were dwelled at the peak temperature for 4.6 min and afterwards the furnace was switched off and cooled down (at approximately 3 °C/min cooling rate) until the samples reached an ambient temperature. The behaviour and microstructural changes in the joining material were evaluated through isothermal annealing at 450 °C for 48 h and 96 h in flowing argon to investigate the stability and evolution of the joint.

For the microstructural observation, specimens were mounted in conductive resin, mechanically ground with abrasive papers and polished using 1/4 μm diamond suspension. The chemical composition of the reaction products was characterized by high-resolution field emission gun scanning electron microscope (FEGSEM, 1530VP Zeiss GmbH, Oberkochen, Germany) with 80 mm^2 energy dispersive X-Ray spectroscopy detector (Oxford Instruments, High Wycombe, UK) and analyzed by Aztec EDS/EBSD microanalysis software.

The electrical contact resistance (R_C) of contacting interfaces was periodically measured during the isothermal ageing process using in-house developed four-point probe resistance measurement. The set-up uses a Keithley 2400 (Keithley Instruments, Inc., Cleveland, OH, US) and DPP205 probe positioner (Cascade Microtech, Inc., Beaverton, OR, USA) to measure the voltage drop across two probes as a function of applied short current pulses. The electrical contact resistance (R_C) is given by:

$$R_C = (R_1 - R_0) \cdot A \tag{1}$$

where R_1, R_0 are measured resistance at the metallic electrode and the thermoelectric material, respectively, and A is a cross-sectional area of this interface. At least three measurements were made for each specimen prepared at different joining conditions.

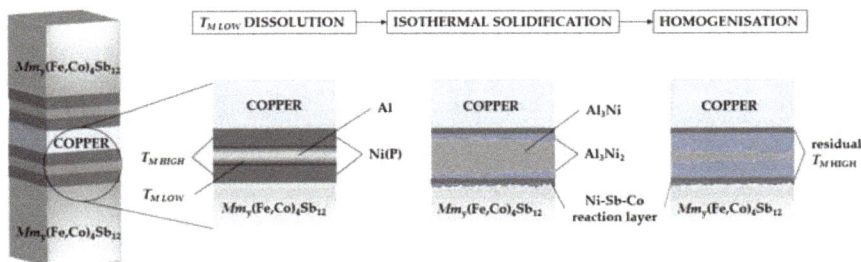

Figure 1. A schematic illustration of the Solid-Liquid Interdiffusion (SLID) bonding process demonstrating the interlayer configuration and joining mechanism when specimens where joined at 660 °C for 4.6 min (described as "$T_{M\,LOW}$ dissolution" and "isothermal solidification") and when isothermally aged at 450 °C for 96 h in Ar (shown as "homogenization").

The mechanical strength of joints was assessed using specimens with the cross-sectional size of 2.5 mm × 2.5 mm by measuring apparent shear strength using MTS Criterion 43 (MTS Systems Corporation, Eden Prairie, MN, USA) at room temperature. The shear configuration was adapted according to ASTM D905 standard [33] which is designed to expose the assembly to direct contact with a shearing blade at the thermoelectric/joint interface, as depicted in Figure 2a,b. A shear load was applied by moving the blade perpendicularly to the longitudinal axis of the specimen with a speed of 0.2 mm/min. The apparent shear strength was calculated by dividing the maximum force applied by the joining area of at least two specimens of different joining conditions.

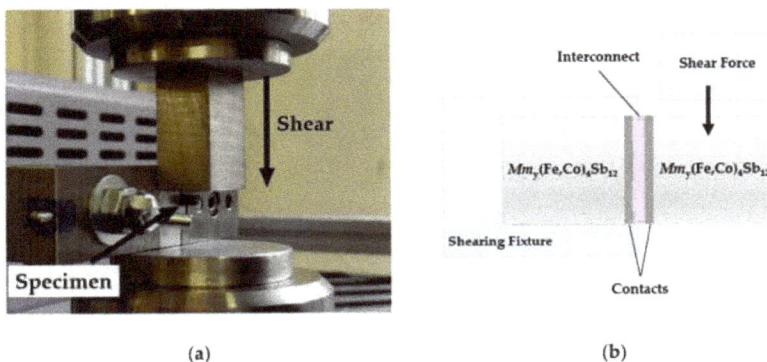

(a)   (b)

Figure 2. Configuration used for mechanical strength evaluation showing: (**a**) Photograph of sample clamped in the shear test fixture and (**b**) schematic illustration of the specimen and shear test configuration adapted from ASTM D905 standard [24]; the thermoelectric leg length is 3 mm, the copper Ni(P) interconnect is 0.3 mm thick and the contact thickness are ~30 μm. The shearing blade makes contact with the thermoelectric material ≤ 0.3mm from the contact.

3. Results and Discussion

3.1. Evolution of Microstructure Morphologies

Off-the-shelf $Mm_y(Fe,Co)_4Sb_{12}$ p-type thermoelectric material (with $zT_{AVG} = 0.2$ measured between 30 °C and 450 °C) was used in this study, with thermoelectric performance significantly

Materials **2018**, *11*, 2483

lower in comparison to similar compositions previously reported [34]. Commercially available material was used as it is available to be reproduced in the high volume needed for joining trials and this bonding technique is expected to be transferable to other similar skutterudite material compositions. In order to test the feasibility of solid–liquid interdiffusion bonding, Ni(P) coating on p-type skutterudite and copper interconnect was introduced. Phosphorous content in the coating was 18 at.% which, as reported in the literature [35], results in the mixture of amorphous and microcrystalline microstructure. The as-deposited 14 μm-thick Ni(P) layer shows good adherence to the thermoelectric material (4B-level according to ASTM D3359 standard [32]) and is commonly used as a diffusion barrier on copper substrates in semiconductors manufacturing. It has been reported, that due to the lack of high-temperature thermal stability of skutterudites and metal interconnects, implementing an additional diffusion barrier is needed [9,36]. Fundamental research on high-temperature stability between $CoSb_3$-based thermoelectric material and Ni interconnect conducted by Chen, et al. [37], revealed extensive growth of reaction layers, i.e., (Co,Ni)Sb and Ni_5Sb_2 phases at the skutterudite–nickel interface. In spite of the above-mentioned issues arising from the $Ni/CoSb_3$ high-temperature instabilities, no diffusion barrier was implemented as it is beyond the scope of this paper. In this research, the bonding process is designed so that layers of Ni(P)-Al-Ni(P) at the temperature of molten aluminium ($T_{M\,LOW}$) undergo solid–liquid interdiffusion reaction leading to the formation of intermetallic compounds (IMCs) through a diffusion-reaction mechanism (Figure 1). Based on the phase diagram [38], there are five thermodynamically stable intermetallic compounds existing in the Al-Ni binary system but Al_3Ni and Al_3Ni_2 phases are the only two expected to be found in the reactive diffusion zone in both solid–solid and solid–liquid, nickel-aluminium interfaces [39,40]. To achieve the subject of this study for maintaining a single Al_3Ni_2 phase between two Ni layers, we calculated, based on the following equation, the initial thickness of Ni needed for 17 μm-thick Al foil:

$$\frac{t_{Ni}}{t_{Al}} = \frac{2M_{Ni}/\rho_{Ni}}{3M_{Al}/\rho_{Al}} \qquad (2)$$

where t are the thicknesses of both nickel and aluminium layers, M are the atomic weights and ρ are elemental densities, i.e., 8.9 g/cm^3 and 2.7 g/cm^3 of Ni and Al, respectively. However, deviations from the theoretical model were expected as Ni(P) instead of pure nickel was used in this study. According to Equation (2), a symmetrical joint consisting of two 3.75 μm-thick Ni(P) layers and 17 μm-thick Al foil (Figure 1) should be enough to maintain a single Al_3Ni_2 phase. While it would be advantageous to completely consume the nickel layer to avoid any subsequent reaction with the skutterudite material, initial experiments indicated that an excess of nickel is required, as any excess aluminium can react with the thermoelectric to form the wider band gap [41] aluminium antimonide, increasing the contact resistance. Moreover, elemental nickel is also consumed by the formation of intermetallic compounds (IMCs) at the skutterudite–Ni(P) interface and its consumption at the interconnection side caused by the significant Cu solubility in Ni [42] at all concentration ranges. Therefore, thicker than calculated Ni(P) layers were deposited.

Based on the cross-sections of the joints, a 6 μm-thick Co-Ni-Sb reaction layer at the Ni(P)–skutterudite interface can be observed at different bonding conditions (Figure 3a,b) and stays unchanged after isothermal ageing at 450 °C for 96 h (Figure 3c). Based on EDS analysis, the reaction layer is composed of Co—44 at.% Ni—52 at.% Sb which is consistent with results of Ref. [37]. The joining process was performed at two different temperatures, i.e., 585 °C ($\leq T_{M\,LOW}$) and 660 °C (=$T_{M\,LOW}$) and the resulting joints' microstructure can be seen in Figure 3a,b. It is clear that joining at 585 °C is not sufficient for complete phase transformation, as some residual aluminium at the joined interface can be found, and the only formed IMC is Al_3Ni_2 (40 at.% Ni and 60 at.% Al) as identified by EDS analysis (Figure 3a). It is believed, that during the heating of the assembly, connected metallic interfaces form a solid solution until its saturation and, as expected, the formation of IMCs through the solid-state diffusion process occurs slower compared to that of the solid–liquid kinetic.

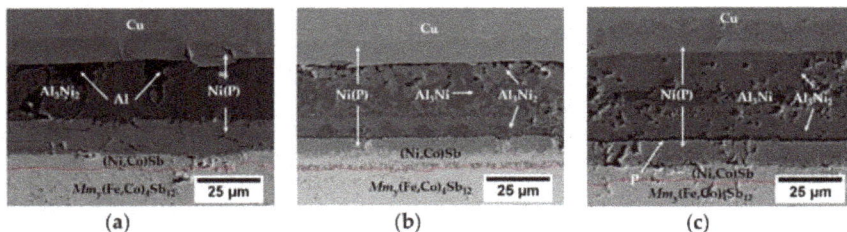

Figure 3. Cross-sectional microstructure of the skutterudite–interconnect interfaces bonded at (a) 585 °C; (b) 660 °C and (c) 660 °C and isothermally aged at 450 °C for 96 h, under flowing Ar, with the Co-Ni-Sb reaction layer marked in red. The high magnification image of Ni(P)–Al$_3$Ni$_2$ interface can be found in the Supplementary Materials (Figure S1).

A complete transformation of aluminium filler is observed at a joining temperature of 660 °C as two distinct IMCs, i.e., Al$_3$Ni2 (38 at.% Ni and 62 at.% Al) near the Ni(P) coating and Al$_3$Ni (24 at.% Ni and 76 at.% Al) in the centre line can be found (Figure 3b), which is consistent with other studies performed on non-thermoelectric, aluminium–nickel diffusion couples [39,43].

A kinetic model was used to explain Al$_3$Ni$_2$ IMCs growth behaviour and to determine the process heating profile. Considering the joint design based on two nickel-based layers to be contacted with 17 μm-thick aluminium foil, the transition time was calculated based on the formation of Al$_3$Ni$_2$ at the expense of 7.5 μm of Ni(P) deposit. The thickness of the resulting bonding layer was calculated from the following equation:

$$t_{Al_3Ni_2} = \frac{M_{Al_3Ni_2}}{\rho_{Al_3Ni_2}} \times \frac{t_{Al}\rho_{Al}}{3M_{Al}} \tag{3}$$

where t are the thicknesses of both aluminium and the resulting Al$_3$Ni$_2$ IMCs layer, M are the atomic weights and ρ are the elemental densities, i.e., 4.7 g/cm^3 Al$_3$Ni$_2$ phase. According to Equations (2) and (3), from the consumption of 7.5 μm Ni (P) from both exposed joined surfaces and 17 μm-thick Al foil, a final joint with a 24 μm-thick Al$_3$Ni$_2$ layer is expected. As shown in Figure 3b, the 28.6 μm-thick joint is thicker than its theoretical estimation which might be induced by the formation of a two-phase bonding region (Al$_3$Ni + Al$_3$Ni$_2$). The parabolic equation based on Fick's diffusion law was used to determine the bonding conditions as follows:

$$X(t, T) = k \times t^n \tag{4}$$

where X is the average thickness of the intermetallic compound, t is the joining process time, T is the bonding temperature, k is the constant rate, and n is a time exponent. In addition, the temperature dependence of reaction rate k can be expressed by the following Arrhenius relationship:

$$k = k_0 exp\left(-\frac{Q}{RT}\right) \tag{5}$$

where k_0 is the frequency factor, R is the Boltzmann constant, and Q is the activation energy for the growth of the designed Al$_3$Ni$_2$ phase. According to Tumminello, et al. [44], the two empirical parameters attributed to Al$_3$Ni$_2$ phase growth, k and n are 8.5×10^{-8} m/s and 0.844, respectively. Based on Equations (4) and (5), complete transition to the Al$_3$Ni$_2$ phase can be achieved within 4.6 min, which was used in the experiment. One of the attributes of the SLID bonding process is that both solid–liquid and solid–solid diffusion mechanism play the key roles in isothermal solidification, which makes a process more time consuming than standard soldering [45]. However, due to the high diffusivity of solid nickel in molten aluminium [46], a combination of Al-Ni diffusion couple is a good choice, making the joining process as fast as 4.6 min. Additionally, the resulting IMCs, Al$_3$Ni and

Al$_3$Ni$_2$ are characterized with higher than initial re-melting temperature with values of 854 °C and 1133 °C, respectively [38].

In order to investigate the microstructural evolution of the fully transformed contact, isothermal ageing on the assembly joined at 660 °C was performed and the cross-sectional view of that microstructure can be seen in Figure 3c. Isothermal ageing was proven to promote homogenisation within the bond line as a reduction of the Al$_3$Ni phase in favour of growing Al$_3$Ni$_2$ was observed. It is believed that during the homogenisation process, excess nickel from Ni(P) is causing unstable Al$_3$Ni phase continual conversion into Al$_3$Ni$_2$ IMCs through grain boundary diffusion, which may influence the joint's thermal stability, as the Al$_3$Ni$_2$ phase has a higher melting point than Al$_3$Ni IMC. The residual phosphorous was accumulated at the Al$_3$Ni$_2$ / Ni(P) interface as a result of Ni depletion from the Ni-P coating due to the formation of desirable Al$_3$Ni$_2$ IMCs. The high magnification image of the Ni(P)–Al$_3$Ni$_2$ interface can be found in the Supplementary Materials (Figure S1). Additionally, Kirkendall voids are found within the two-phase region (Al$_3$Ni + Al$_3$Ni$_2$), representing a surface fraction of 4.4%, although some of these are believed to be induced by pulling out the brittle intermetallic grains during the polishing process and might not affect the joining process itself.

3.2. Electrical Contact Resistance (R$_C$) Evaluation

According to Ref. [47], in order to maintain >80% thermoelectric material efficiency in the working device, the electrical contact resistance (R$_c$) at the thermoelectric–metal contacts needs to be at least less than 30% of the total thermoelectric leg resistance (assuming no thermal contact resistances). By considering one thermoelectric pellet dimension of 2.5 × 2.5 × 3 mm and the electrical resistivity of Mm_y(Fe,Co)$_4$Sb$_{12}$ p-type thermoelectric material of 7.59 μΩ·m (measured by four-point probe resistivity at room temperature), the electrical contact resistance should be < 6.8 × 10^{-5} Ω·cm^2 in order to achieve the high module performance promised by the material measurements. The contact resistance (R$_c$) of assemblies joined at 585 °C and 660 °C was measured to be (2.8 ± 0.4) × ·10^{-5} Ω·cm^2 and (4.8 ± 0.3) × 10^{-5} Ω·cm^2 respectively, which contribute approximately 12% and 21% to the total Mm_y(Fe,Co)$_4$Sb$_{12}$ p-type thermoelectric pellet resistivity. Moreover, the electrical resistivity of thermoelectric leg was measured using the same technique and showed negligible changes in the material electrical performance after the SLID bonding process. The evolution of the electrical performance during isothermal ageing at 450 °C in flowing Ar, as depicted in Figure 4c, shows that high-temperature degradation leads to the R$_c$ increase by 370% and 68% within 96 h for assemblies joined at 585 °C and 660 °C respectively, which is likely caused by the Al-Ni phase transformation within the joint and a partial Ni(P) delamination from the skutterudite material as observed in SEM images (Figure 3c).

3.3. Mechanical Strength Evaluation

The mechanical strength of bonded specimens was evaluated by measuring apparent shear strength at room temperature adapted to the ASTM D905 standard [33]. As shown in Figure 5a, the apparent shear strength of 5.1 ± 0.5 MPa and 4.5 ± 0.5 MPa was achieved for specimens joined at 585 °C ($\leq T_{M\,LOW}$) and 660 °C (= $T_{M\,LOW}$) respectively. Taking into account that the mechanical properties of skutterudite–metallic interconnects are rarely reported in the literature, and the mechanical strength depends on several variables, i.e., measurement set-up configuration, it is challenging to quantitatively compare to other high-temperature thermoelectric junctions previously reported. Nonetheless, according to Ref. [48], the maximum bonding strength of 13.2 MPa was achieved in low-temperature (Pb, Sn) Te/Cu layers bonded by solid–liquid interdiffusion (SLID) using In-Ag system, suggesting that low mechanical reliability of IMCs-based joints might cause failure in long-term operations, thus requiring further improvement.

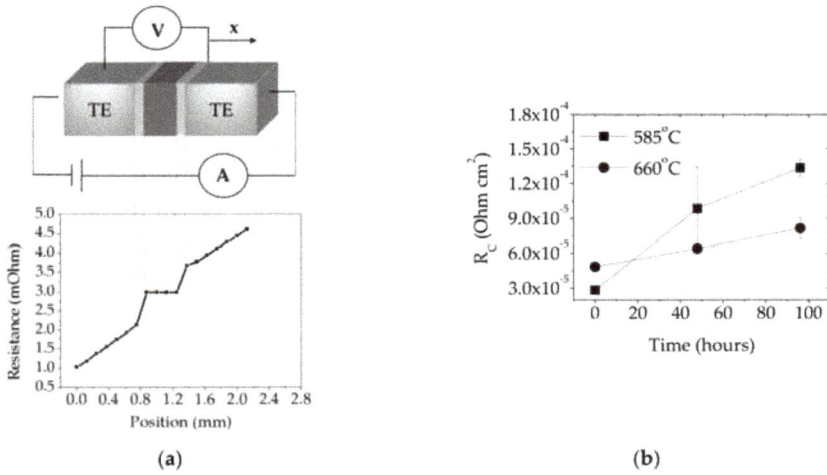

Figure 4. (**a**) Schematic illustration of four-point probe resistance set-up equipped with a scanning probe moving across the surface (x) where the contact resistance (R_c) is defined as the ratio of voltage drop across the contact to the current pulse applied to a pair of contacts; the graph below shows R_c measurement of the skutterudite specimen bonded at 660 °C for 4.6 min and (**b**) R_c evolution of metal interconnect–Mm_y(Fe,Co)$_4$Sb$_{12}$ p-type contacts as a function of isothermal ageing time. Note: As-joined contacts are denoted as '0 hours' specimens.

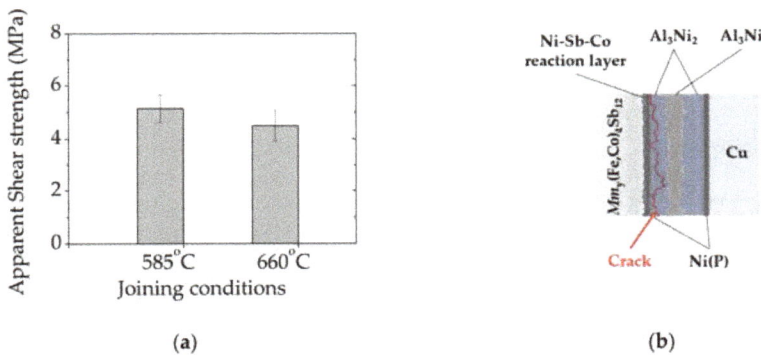

Figure 5. Mechanical performance of bonded specimens: (**a**) Apparent shear strength of interconnect–Mm_y(Fe,Co)$_4$Sb$_{12}$ interfaces formed at different joining temperature and (**b**) the schematic illustration with fracture mode of sheared assembly marked in red.

Scanning electron microscope along with EDS analysis was performed to observe the fractured interface and to understand the possible failure mode occurring within contacts, as presented in Figure 6a,b. The fracture seems to propagate in the mixed mode, along or very close to the Ni(P)–Al$_3$Ni$_2$ intermetallic interface and within the joining area (as schematically shown in Figure 5b) as the three distinct regions of Ni(P), Al$_3$Ni$_2$ and Al$_3$Ni can be detected by the EDS elemental mapping images.

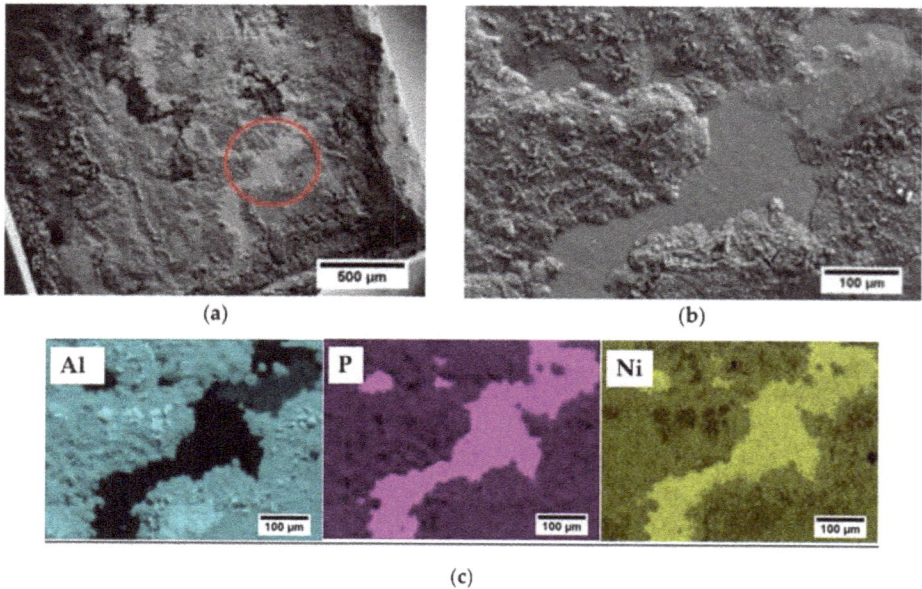

Figure 6. (**a**) Top-view of fractured $Mm_y(\text{Fe,Co})_4\text{Sb}_{12}$ surface after apparent shear strength testing; (**b**) red circle presents the enlarged area shown in (**a**), along with elemental mapping of Al, P and Ni obtained by EDS analysis (**c**).

4. Conclusions

It has been shown that Solid-Liquid Interdiffusion (SLID) is a possible technique for bonding metallic interconnects with $Mm_y(\text{Fe,Co})_4\text{Sb}_{12}$ p-type thermoelectric material using aluminium–nickel multi-layered system. Bonding parameters were found to create contacts with a final microstructure falling in the two-phase region comprised of Al_3Ni and Al_3Ni_2 intermetallic compounds. High-temperature reliability and microstructural changes in the joining material have been evaluated in terms of isothermal ageing at 450 °C in flowing Ar, showing that the homogenisation process leads to the reduction of the Al_3Ni phase in favour of growing Al_3Ni_2. In addition, it can be observed that avoiding residual aluminium in the joint by using a higher process temperature is advantageous for improved stability. Although joined components keep their integrity during the high-temperature isothermal ageing, further work is needed to test long term thermal stability including analysis of the potential influence of CTE mismatch of the constituent elements and thermal cycling of prototype modules. Moreover, as the formation of the (Ni, Co) Sb reaction layer at the metal interconnect–skutterudite interface cannot be avoided, implementation of an additional diffusion barrier is still needed.

Supplementary Materials: The following are available online at http://www.mdpi.com/1996-1944/11/12/2483/s1, Figure S1: (**a**) Cross-sectional microstructure of the skutterudite–interconnect interfaces bonded at 660 °C and isothermally aged at 450 °C for 96 h, under flowing Ar; (**b**) red circle presents the enlarged area shown in (**a**) with a high magnification Scanning electron microscope (SEM) image of the Ni(P)–Al_3Ni_2 interface.

Author Contributions: K.P. performed the experiment and wrote the paper. R.S.T., M.S. and V.C. reviewed and edited the paper. K.S. assisted with the project funding and administration.

Funding: This research was funded by the European Union's Horizon 2020 Programme through a Marie Skłodowska-Curie Innovative Training Network (CoACH, grant number 642557).

Acknowledgments: Particular thanks to Francesco Gucci and Kan Chen from Nanoforce Technology Ltd. for thermoelectric material synthesis and Chris Stuttle for assistance with electroless nickel plating process. K.P. would like to acknowledge Cevriye Koz and Theo G. Saunders for helpful discussions.

Conflicts of Interest: The authors declare no conflict of interest.

References

1. Rogl, G.; Rogl, P. Skutterudites, a most promising group of thermoelectric materials. *Curr. Opin. Green Sustain. Chem.* **2017**, *4*, 50–57. [CrossRef]
2. Chiwanga, S.; Tuley, R.; Placha, K.; Robbins, M.; Gilchrist, B.; Simpson, K. Automotive power harvesting/thermoelectric applications. In *Thermoelectric Materials and Devices*; RSC Publishing: Cambridge, UK, 2017; Chapter 9; pp. 230–251.
3. Holgate, T.C.; Bennett, R.; Hammel, T.; Caillat, T.; Keyser, S.; Sievers, B. Increasing the Efficiency of the Multi-mission Radioisotope Thermoelectric Generator. *J. Electron. Mater.* **2015**, *44*, 1814–1821. [CrossRef]
4. Aswal, D.K.; Basu, R.; Singh, A. Key issues in development of thermoelectric power generators: High figure-of-merit materials and their highly conducting interfaces with metallic interconnects. *Energy Convers. Manag.* **2016**, *114*, 50–67. [CrossRef]
5. Caillat, T.; Firdosy, S.; Li, B.C.-Y.; Huang, K.; Cheng, B.; Paik, J.; Chase, J.; Arakelian, T.; Lara, L. Progress status of the development of high-efficiency segmented thermoelectric couples. *Nucl. Emerg. Technol. Space* **2012**, *2*.
6. Mo/Ti Diffusion Bonding for Making Thermoelectric Devices. Available online: https://www.techbriefs.com/component/content/article/tb/techbriefs/electronics-and-computers/2014 (accessed on 23 July 2018).
7. Fleurial, J.-P.; Caillat, T.; Chi, S.C. Electrical contacts for skutterudite thermoelectric materials. U.S. Patent 13,161,156, 12 January 2012.
8. Salvador, J.R.; Cho, J.Y.; Ye, Z.; Moczygemba, J.E.; Thompson, A.J.; Sharp, J.W.; Koenig, J.D.; Maloney, R.; Thompson, T.; Sakamoto, J.; et al. Conversion efficiency of skutterudite-based thermoelectric modules. *Phys. Chem. Chem. Phys.* **2014**, *16*, 12510–12520. [CrossRef] [PubMed]
9. Zhao, D.; Li, X.; He, L.; Jiang, W.; Chen, L. Interfacial evolution behavior and reliability evaluation of $CoSb_3$/Ti/Mo–Cu thermoelectric joints during accelerated thermal aging. *J. Alloys Compd.* **2009**, *477*, 425–431. [CrossRef]
10. Zybala, R.; Wojciechowski, K.; Schmidt, M.; Mania, R. Junctions and diffusion barriers for high temperature thermoelectric modules. *Mater. Ceram. Ceram. Mater.* **2010**, *62*, 481–485.
11. Park, S.H.; Jin, Y.; Cha, J.; Hong, K.; Kim, Y.; Yoon, H.; Yoo, C.-Y.; Chung, I. High-Power-Density Skutterudite-Based Thermoelectric Modules with Ultralow Contact Resistivity Using Fe–Ni Metallization Layers. *ACS Appl. Energy Mater.* **2018**, *1*, 1603–1611. [CrossRef]
12. Rao, A.; Bosak, G.; Joshi, B.; Keane, J.; Nally, L.; Peng, A.; Perera, S.; Waring, A.; Poudel, B. A TiAlCu Metallization for 'n' Type CoSb x Skutterudites with Improved Performance for High-Temperature Energy Harvesting Applications. *J. Electron. Mater.* **2017**, *46*, 2419–2431. [CrossRef]
13. Jie, Q.; Ren, Z.; Chen, G. Fabrication of stable electrode/diffusion barrier layers for thermoelectric filled skutterudite devices. U.S. Patent 15,627,593, 5 October 2017.
14. Bernstein, L. Semiconductor Joining by the Solid-Liquid-Interdiffusion (SLID) Process: I. The Systems Ag-In, Au-In, and Cu-In. *J. Electrochem. Soc.* **1966**, *113*, 1282–1288. [CrossRef]
15. MacDonald, W.D.; Eagar, T.W. Transient liquid phase bonding. *Annu. Rev. Mater. Sci.* **1992**, *22*, 23–46. [CrossRef]
16. Jacobson, D.M.; Humpston, G. *Principles of Brazing*; ASM International: Russel, OH, USA, 2005.
17. Owczarski, W.A.; King, W.H.; Duvall, D.S. Diffusion welding of the nickel-base superalloys. U.S. Patent 3,530,568, 29 September 1970.
18. Chu, K.; Sohn, Y.; Moon, C. A comparative study of Cn/Sn/Cu and Ni/Sn/Ni solder joints for low temperature stable transient liquid phase bonding. *Scr. Mater.* **2015**, *109*, 113–117. [CrossRef]
19. Xu, H.; Suni, T.; Vuorinen, V.; Li, J.; Heikkinen, H.; Monnoyer, P.; Paulasto-Kröckel, M. Wafer-level SLID bonding for MEMS encapsulation. *Adv. Manuf.* **2013**, *1*, 226–235. [CrossRef]
20. Mao, X.; Lv, X.-D.; Wei, W.-W.; Zhang, Z.; Yang, J.-L.; Qi, Z.-M.; Yang, F.-H. A wafer-level Sn-rich Au-Sn bonding technique and its application in surface plasmon resonance sensors. *Chin. Phys. Lett.* **2014**, *31*, 056803. [CrossRef]

21. Made, R.I.; Gan, C.L.; Yan, L.L.; Yu, A.; Yoon, S.W.; Lau, J.H.; Lee, C. Study of Low-Temperature Thermocompression Bonding in Ag-In Solder for Packaging Applications. *J. Electron. Mater.* **2009**, *38*, 365. [CrossRef]

22. Fukumoto, S.; Miyake, K.; Tatara, S.; Matsushima, M.; Fujimoto, K. Solid-Liquid Interdiffusion Bonding of Copper Using Ag-Sn Layered Films. *Mater. Trans.* **2015**, *56*, 1019–1024. [CrossRef]

23. Deillon, L.; Hessler-Wyser, A.; Hessler, T.; Rappaz, M. Solid-liquid interdiffusion (SLID) bonding in the Au-In system: Experimental study and 1D modelling. *J. Micromech. Microeng.* **2015**, *25*, 125016. [CrossRef]

24. Tollefsen, T.A.; Larsson, A.; Løvvik, O.M.; Aasmundtveit, K. Au-Sn SLID Bonding—Properties and Possibilities. *Metall. Mater. Trans. B* **2012**, *43*, 397–405. [CrossRef]

25. Larsson, A.; Tollefsen, T.A.; Martin, O.; Aasmundtveit, K.E. Ni-Sn solid-liquid interdiffusion (SLID) bonding for thermo-electric elements in extreme environments—FEA of the joint stress. In Proceedings of the Microelectronics Packaging Conference (EMPC), Friedrichshafen, Germany, 14 September 2015; pp. 1–6.

26. Huang, T.-C.; Smet, V.; Kawamoto, S.; Pulugurtha, M.R.; Tummala, R.R. Accelerated Metastable Solid—Liquid Interdiffusion Bonding with High Thermal Stability and Power Handling. *J. Electron. Mater.* **2018**, *47*, 368–377. [CrossRef]

27. Kumar, K.G. A Novel Intermetallic Nickel Aluminide (Ni_3Al) as an Alternative Automotive Body Material. *Int. J. Eng.* **2011**, *11*, 8.

28. Kanetsuki, S.; Kuwahara, K.; Egawa, S.; Miyake, S.; Namazu, T. Effect of thickening outermost layers in Al/Ni multilayer film on thermal resistance of reactively bonded solder joints. *Jpn. J. Appl. Phys.* **2017**, *56*, 06GN16. [CrossRef]

29. Kun, Y.; Hanqing, X.; Yilong, D.; Fei, T.; Sufeng, F.; Xueyan, Q.; Li, W. Bonding Process and Application Properties of an Al-Ni Layer Composite Sheet for Lithium-ion Battery Packaging. *Rare Met. Mater. Eng.* **2016**, *45*, 1100–1105. [CrossRef]

30. Chuang, T.H.; Lin, H.J.; Chuang, C.H.; Yeh, W.T.; Hwang, J.D.; Chu, H.S. Solid Liquid Interdiffusion Bonding of (Pb, Sn)Te Thermoelectric Modules with Cu Electrodes Using a Thin-Film Sn Interlayer. *J. Electron. Mater.* **2014**, *43*, 4610–4618. [CrossRef]

31. Lin, Y.C.; Lee, K.T.; Hwang, J.D.; Chu, H.S.; Hsu, C.C.; Chen, S.C.; Chuang, T.H. Solid Liquid Interdiffusion Bonding of Zn_4Sb_3 Thermoelectric Material with Cu Electrode. *J. Electron. Mater.* **2016**, *45*, 4935–4942. [CrossRef]

32. ASTM D3359-17. *Standard Test Methods for Rating Adhesion by Tape Test*; ASTM International: West Conshohocken, PA, USA, 2017.

33. ASTM D905-08. Standard Test Method for Strength Properties of Adhesive Bonds in Shear by Compression Loading. ASTM International: West Conshohocken, PA, USA, 2013.

34. Dahal, T.; Kim, H.S.; Gahlawat, S.; Dahal, K.; Jie, Q.; Liu, W.; Lan, Y.; White, K.; Ren, Z. Transport and mechanical properties of the double-filled p-type skutterudites $La_{0.68}Ce_{0.22}Fe_{4-x}Co_xSb_{12}$. *Acta Mater.* **2016**, *117*, 13–22. [CrossRef]

35. Mallory, G.O.; Hajdu, J.B. *Electroless Plating: Fundamentals and Applications*; William Andrew: New York, NY, USA, 1990.

36. Zhao, D.; Geng, H.; Teng, X. Fabrication and reliability evaluation of $CoSb_3$/W–Cu thermoelectric element. *J. Alloys Compd.* **2012**, *517*, 198–203. [CrossRef]

37. Chen, W.; Chen, S.; Tseng, S.; Hsiao, H.; Chen, Y.; Snyder, G.J.; Tang, Y. Interfacial reactions in Ni/$CoSb_3$ couples at 450 °C. *J. Alloys Compd.* **2015**, *632*, 500–504. [CrossRef]

38. Okamoto, H. Al-Ni (aluminum-nickel). *J. Phase Equilibria* **1993**, *14*, 257–259. [CrossRef]

39. Ding, Z.; Hu, Q.; Lu, W.; Sun, S.; Xia, M.; Li, J. In situ observation on the formation of intermetallics compounds at the interface of liquid Al/solid Ni. *Scr. Mater.* **2017**, *130*, 214–218. [CrossRef]

40. Edelstein, A.S.; Everett, R.K.; Richardson, G.R.; Qadri, S.B.; Foley, J.C.; Perepezko, J.H. Reaction kinetics and biasing in Al/Ni multilayers. *Mater. Sci. Eng. A* **1995**, *195*, 13–19. [CrossRef]

41. Dhakal, R.; Huh, Y.; Galipeau, D.; Y, X. AlSb Compound Semiconductor as Absorber Layer in Thin Film Solar Cells. In *Solar Cells—New Aspects and Solutions*; Kosyachenko, L.A., Ed.; InTech: London, UK, 2011.

42. Seith, W.; Heumann, T. *Diffusion of Metals: Exchange Reactions*; Springer Press: Berlin, Germany, 1955.

43. Wołczyński, W.; Okane, T.; Senderowski, C.; Kania, B.; Zasada, D.; Janczak-Rusch, J. Meta-Stable Conditions of Diffusion Brazing. *Arch. Metall. Mater.* **2011**, *56*, 311–323. [CrossRef]

44. Tumminello, S.; Sommadossi, S. Growth kinetics of intermetallic phases in transient liquid phase bonding process (TLPB) in Al/Ni system. *Defect Diffus. Forum* **2012**, *323*, 465–470. [CrossRef]

45. MacDonald, W.D.; Eagar, T.W. Isothermal solidification kinetics of diffusion brazing. *Metall. Mater. Trans. A* **1998**, *29*, 315–325. [CrossRef]

46. Eremenko, V.N.; Natanzon, Y.V.; Titov, V.P. Dissolution kinetics and diffusion coefficients of iron, cobalt, and nickel in molten aluminum. *Mater. Sci.* **1978**, *14*, 579–584. [CrossRef]

47. Bjørk, R. The Universal Influence of Contact Resistance on the Efficiency of a Thermoelectric Generator. *J. Electron. Mater.* **2015**, *44*, 2869–2876. [CrossRef]

48. Chuang, T.-H.; Yeh, W.-T.; Chuang, C.-H.; Hwang, J.-D. Improvement of bonding strength of a (Pb, Sn)Te-Cu contact manufactured in a low temperature SLID-bonding process. *J. Alloys Compd.* **2014**, *613*, 46–54. [CrossRef]

Article

Extension of the 'Inorganic Gel Casting' Process to the Manufacturing of Boro-Alumino-Silicate Glass Foams

Acacio Rincon Romero [1], **Sergio Tamburini** [2], **Gianmarco Taveri** [3], **Jaromír Toušek** [4], **Ivo Dlouhy** [3] and **Enrico Bernardo** [1,*]

1   Department of Industrial Engineering, University of Padova, Via Marzolo 9, 35131 Padova, Italy; acacio.rinconromero@unipd.it
2   National Research Council of Italy (CNR), Institute of Condensed Matter Chemistry and Technologies for Energy (ICMATE), Corso Stati Uniti 4, 35127 Padova, Italy; sergio.tamburini@cnr.it
3   Institute of Physics of Materials (IPM), Žižkova 22, 61662 Brno, Czech Republic; taveri@drs.ipm.cz (G.T.); idlouhy@ipm.cz (I.D.)
4   CEITEC—Central European Institute of Technology, Masaryk University, Kamenice 5, 62500 Brno, Czech Republic; jaromir.tousek@ceitec.muni.cz
*   Correspondence: enrico.bernardo@unipd.it; Tel.: +39-049-827-5510; Fax: +39-049-827-5505

Received: 26 November 2018; Accepted: 11 December 2018; Published: 14 December 2018

Abstract: A new technique for the production of glass foams, based on alkali activation and gel casting, previously applied to soda-lime glass, was successfully extended to boro-alumino-silicate glass, recovered from the recycling of pharmaceutical vials. A weak alkali activation (2.5 M NaOH or NaOH/KOH aqueous solutions) of fine glass powders (below 70 μm) allowed for the obtainment of well-dispersed concentrated aqueous suspensions, undergoing gelation by treatment at low temperature (75 °C). Unlike soda-lime glass, the progressive hardening could not be attributed to the formation of calcium-rich silicate hydrates. The gelation was provided considering the chemical formulation of pharmaceutical glass (CaO-free) to the formation of hydrated sodium alumino-silicate (N-A-S-H) gel. An extensive direct foaming was achieved by vigorous mechanical stirring of partially gelified suspensions, comprising also a surfactant. A sintering treatment at 700 °C, was finally applied to stabilize the cellular structures.

Keywords: glass recycling; alkali activation; gel casting; glass foams

---

1. Introduction

The recovery of glass from differentiated urban waste collection is undoubtedly favourable, for the significant savings in raw materials and energy consumption upon melting (pre-formed glass act as a flux for the reaction of mineral raw materials) [1] but it must face important limitations. In fact, an 'ideal' recycling, corresponding to a complete reuse of glass cullet in the manufacturing of the original glass articles, technically known as 'closed loop recycling', is far from being feasible. It was estimated that the saving in embodied energy (energy to be committed to create a mass of usable material) using recycled material instead of 'virgin' raw materials is about 20% for glass, whereas it approaches 90% for aluminium [2].

The controversy of glass recycling basically concerns the quality of glass articles which may be significantly degraded, when employing not properly purified glass cullet [3,4]. Crushed glass, from municipal waste collection, is typically subjected to an expensive and difficult sorting step, aimed at separating glass pieces with different colours and removing metal, plastic or ceramic impurities, before being considered as a real alternative to minerals. Glass fractions, in which these impurities are concentrated, are still landfilled [5]. Even in the case of limited impurities, some glass may be discarded, if the original glass articles are no longer produced, as in the case of glasses from dismantled cathode ray tubes (TV and PC screens abandoned CRT technology more than a decade ago) [6]. Additional difficulties arise with glass articles deriving from a quite complex processing chain, like pharmaceutical vials: highly chemically stable boro-alumino-silicate containers are typically manufactured by thermo-mechanical processing of an intermediate product (glass tube), produced elsewhere [7].

The difficulties in the direct recycling of glass open the way to 'open loop recycling' applications, when glass is considered as a raw material for articles different from the original ones [6]. In this context, a fundamental challenge concerns the value of the new articles: the use of glass as simple sintering aid, in limited quantities, in several types of ceramics, although successful (waste-derived ceramics actually benefit from the chemical stability of pharmaceutical glass [6]), may configure just a 'down-cycling' option. Products maximizing the content of recycled glass, placed in an advantageous market 'niche', like glass foams [5,6], may configure, on the contrary, as an 'up-cycling' option, a truly profitable way of reusing discarded materials.

The value of glass foams (or cellular glasses) is expressed by the distinctive set of properties, since they exhibit high strength-to-density ratios, high surface area, high permeability, low specific heat, high thermal and acoustic insulation combined with high chemical and thermal resistance (e.g., unlike polymer foams, glass foams are non-flammable and flame resistant) [8]. The value could be even maximized by optimization of the manufacturing process: in fact, while an energy-demanding melting process is avoided, exploiting the viscous flow sintering of glass powders, at much lower temperatures, the cost and the environmental impact of additives ('foaming agents'), aimed at releasing gas (by oxidation or decomposition reactions) during sintering, are still disputable [9].

A substantial change in the approach to glass foams is offered by the separation of foaming and sintering steps, obtainable by gel casting technology [10], that may be applied to solutions (from sol-gel processing) [11], as well as to suspensions of glass powders [12,13]. Air bubbles are incorporated by intensive mechanical stirring, with the help of surfactants, forming a cellular structure, stabilized firstly by the progressive hardening ('gelation') of the starting slurry and secondly by the sintering treatment. Unlike conventional glass foams, foams from gel casting may keep the open-celled morphology achieved in the foaming step, of fundamental importance in high value applications, such as bone tissue engineering [12,13].

Recent investigations were focused at adjusting the gel casting technology to foams from recycled glasses in an 'up-cycling' perspective. Instead of deriving from expensive organic polymerization (due to the addition of monomers, cross-linkers and catalysts), used for biomaterials [12,13] gelation of glass slurries was achieved simply by alkali activation. As shown by Rincon et al. [5] for soda-lime glass, glass slurries in alkaline media (KOH aqueous solutions) exhibited a marked pseudoplastic behaviour, due to the formation of C-S-H (calcium silicate hydrated) gel at the surface of glass particles. At low shear rate, the gel could 'glue' the particles, trapping many air bubbles (incorporated upon intensive mechanical stirring, at high shear rate), at low temperature. Sintering could be performed at 700–800 °C, far below the temperatures used for soda-lime glass-based foams, produced with foaming agents.

The proposed 'inorganic gel casting' process was successfully applied to other glasses, also for the obtainment of glass-ceramic foams [14–16]. More precisely, the alkali activation of glass powders was followed by sinter-crystallization, consisting of viscous flow sintering of glass with concurrent crystallization. The precipitation of crystals, by enhancing the apparent viscosity of the glass mass during sintering, was significant in 'freezing' the open-celled structure, developed at low temperature. Soda-lime glass, on the contrary, experienced a 'reshaping' of pores, owing to the decomposition of the same hydrated compounds [5]. In all cases, the hardening could be attributed to the formation of a C-S-H gel (clearly visible by FTIR spectroscopy).

The investigation here presented was essentially aimed at extending the inorganic gel casting approach to glass foams with the above mentioned boro-alumino-silicate pharmaceutical glass known to feature a very limited CaO content and thus expected to lead to different gels [17]. In particular, the alkali activation was expected to lead, instead of a 'tobermorite-like' (C-S-H) gel, a truly 'zeolite-like' gel, determined by the bridging of $[SiO_4]$ and $[AlO_4]$ tetrahedra, in analogy with the usual alkali-activated materials generally known as 'geopolymers' [18]. The significant content of boron oxide was reputed as favourable, since it is well known to contribute to 'zeolite-like' networks in the form of $[BO_4]$ units [19]. Although still of not completely clear origin, the gelation effectively occurred with the manufacturing of open-celled glass foams possessing good strength-to-density ratios and a good 'tunability' of the microstructure according to the processing conditions.

2. Materials and Methods

Pharmaceutical boro-alumino-silicated glass (later referred to as 'BSG'; chemical composition [17]: SiO_2 = 72 wt %, Al_2O_3 = 7 wt %, B_2O_3 = 12 wt %, Na_2O = 6 wt %, K_2O = 2 wt %, CaO = 1 wt %, BaO < 0.1 wt %) from crushed discarded vials, provided by the company Stevanato Group (Piombino Dese, Padova, Italy), was used as starting material.

Fine powders, after preliminary dry ball milling (Pulverisette 7 planetary ball mill, Fritsch, Idar-Oberstein, Germany) and manual sieving (<75 μm), were cast in aqueous solutions containing 2.5 M NaOH or 2.5 M NaOH/KOH (reagent grade, Sigma-Aldrich, Gillingham, UK) for a solid loading of 68 wt %. The glass powders were subjected to alkaline attack for 4 h under low speed mechanical stirring (500 rpm). After alkaline activation, the obtained suspensions of partially dissolved glass powders were poured in closed polystyrene cylindrical moulds (60 mm diameter) and dried at 75 °C for 4 h.

Partially dried suspensions were added with 4 wt % Triton X-100 (polyoxyethylene octyl phenyl ether—$C_{14}H_{22}O(C_2H_4O)_n$, n = 9–10, Sigma-Aldrich, Gillingham, UK) and then foamed by vigorous mechanical mixing (2000 rpm). 'Green' foams were demoulded after a second drying step (at 75 °C, for 24 h) and fired at 700 °C for 1 h with a heating rate of 10 °C/min. For comparison purposes, some samples were produced by replacing Triton X-100 with Tween 80, a polyoxyethylene sorbitan monooleate—$C_{64}H_{124}O_{26}$ (VWR BDH Prolabo, Milan, Italy) or an inexpensive ionic surfactant sodium lauryl sulphate (SLS; $CH_3(CH_2)_{11}OSO_3Na$ (Carlo Erba, Cornaredo, Milan, Italy) in an aqueous solution 1/10 in weight.

Selected samples, in the hardened and in the fired state, were subjected to Fourier-transform infrared spectroscopy (FTIR, FTIR model 2000, Perkin Elmer Waltham, MA, USA), nuclear magnetic resonance (NMR) spectroscopy and X-ray diffraction (XRD) analysis. NMR analysis consisted of ^{29}Si, ^{27}Al and ^{11}B studies: ^{29}Si and ^{27}Al spectra were collected on a Bruker AVANCE III spectrometer 300 (Bruker, Karlsruhe, Germany, magnetic field of 7.0 T corresponding to ^{29}Si and ^{27}Al Larmor frequencies of 59.623 and 78.066 MHz respectively) equipped for solid-state analysis in 4 mm diameter zirconia rotors. The magic angle was accurately adjusted prior to data acquisition using KBr. ^{29}Si chemical shifts were externally referenced to solid tetrakis(trimetylsilyl)silane at –9.8 ppm (in relation to TMS) and ^{27}Al chemical shifts were externally referenced to $AlCl_3 \cdot 6H_2O$ (0 ppm). The quantitative ^{29}Si single-pulse experiments were collected at a spinning frequency of 6 kHz, a recycling delay of 100 s and 2000 transients. ^{27}Al experiments were collected at a spinning frequency of 13 kHz with a recycle time of 2 s. About 4000 scans were needed using a single pulse experiment.

^{11}B NMR spectra were obtained using a Bruker Avance-500 spectrometer (Bruker, Karlsruhe, Germany), magnetic field of 11.7 T corresponding to ^{11}B, Larmor frequencies of 160.462 MHz equipped for solid-state analysis in 4 mm diameter zirconia rotors. ^{11}B chemical shifts were externally referenced to $B(OAc)_3$ (3.0 ppm). The quantitative single-pulse experiments were collected at a spinning frequency of 14 kHz, a recycling delay of 5 s and 400 transients.

The mineralogical analysis was conducted on powdered samples by means of X-ray diffraction (XRD) (Bruker D8 Advance, Karlsruhe, Germany), using CuKα radiation, 40 kV, 40 mA, 2θ = 10–70°, a step size of 0.02° and a counting time of 2 s, with a position sensitive detector (LinxEye, Bruker AXS, Karlsruhe, Germany). The phase identification was completed using the Match! programme package (Version 1.11, Crystal Impact GbR, Bonn, Germany), supported by data from the PDF-2 database (ICDD-International Centre for Diffraction Data, Newtown Square, PA, USA).

The morphological and microstructural characterizations were performed by scanning electron microscopy (FEI Quanta 200 ESEM, Eindhoven, The Netherlands).

The geometric density of both hardened foamed gels and fired glass foams was evaluated by considering the mass to volume ratio. The apparent and the true density were measured by using a helium pycnometer (Micromeritics AccuPyc 1330, Norcross, GA, USA), operating on bulk or on finely crushed samples, respectively. The three density values were used to compute the amounts of open and closed porosity.

Finally selected foams were subjected to compression and bending tests, employing samples of about $10 \times 10 \times 10$ mm^3 and of about $43 \times 4 \times 3$ mm^3, respectively, cut from larger specimens. Compression tests were applied by using an Instron 1121 UTS (Danvers, MA, USA) machine, with a crosshead speed of 1 mm/min (each data point corresponding to 10–12 samples). Bending tests were done on 3×4 mm^2 specimens section by using Zwick/Roell Z50 machine (Ulm, Germany), operating in 3–4 point configuration (span = 16 mm and 40 mm, respectively), with a crosshead speed of 10 μm/min (each data point corresponding to 10 samples).

3. Results and Discussion

As previously mentioned, the alkali activation determined the gelation of glass suspensions exploited for low temperature foaming, shown in Figure 1.

FTIR spectroscopy was considered as the first step in investigating the nature of the developed gel. Compared to the patterns for the as received glass and for the foam after firing at 700 °C, the pattern for alkali activated BSG features limited differences, as illustrated by Figure 2 (although similar in behaviour, we report data only for the foams produced with an activating solution of 2.5 M NaOH/KOH, exhibiting the clearest FTIR pattern). In all spectra, the most intense band is centred around 1000–1050 cm^{-1} corresponding to tetrahedral stretching modes of two types of bridges, such as Si–O (Si) and Si–O (Al) [20]. The main band is accompanied by two broad shoulders, one at around 1200 cm^{-1} (S1), attributed to asymmetric stretching of B–O bonds in trigonal BO$_3$, the other at around 900 cm^{-1} (S2), assigned to stretching vibrations of B–O bonds in BO$_4$ tetrahedral unit [21,22].

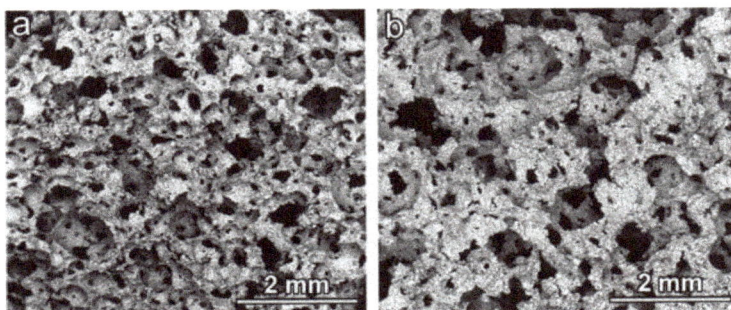

Figure 1. Morphology of foams in the 'green' state (before firing): (**a**) NaOH/KOH activation; (**b**) KOH activation.

The alkali activation, determining some broadening and downshifting of the main band, is interpreted as effects of the formation of a gel with mixed ions (in geopolymers, the band displacement is correlated with an increase of Al or B in the network structure [23]). The band at 900 cm^{-1}, ascribable to the amount of boron in tetragonal configuration, visibly increased in intensity passing from the as received state to the activated and fired states. The bands around 800 and 1400 cm^{-1}, attributed to the B–O bond bending vibration and asymmetric stretching vibration of the BO$_3$ trigonal units, respectively [24–26], slightly decreased in intensity in the hardened foam.

Exclusively in the hardened foams, some additional bands were ascribed to the formation of silanol groups (SiOH) and incorporation of molecular water (SiO-H stretching at around 1600 cm^{-1}, bending vibration of H_2O at 1650 cm^{-1}, to stretching vibration of H_2O at 3430 cm^{-1} [21]). Finally, the band at about 2800 cm^{-1} was attributed to C-H_2 stretch vibrations from the surfactant (undergoing complete decomposition upon heating) [5].

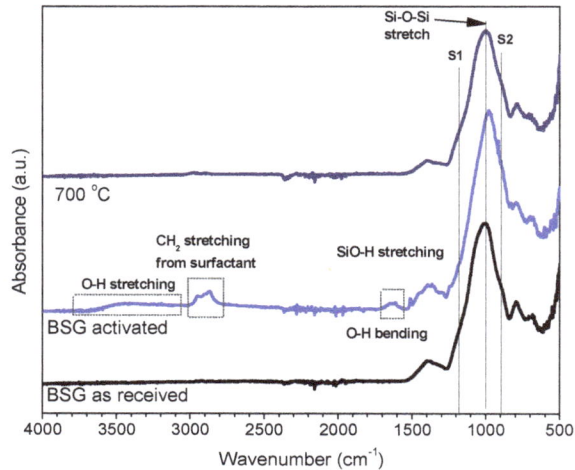

Figure 2. Fourier transform infrared (FTIR) analysis of boro-alumino-silicated glass (BSG) in the as received state, after alkali activation (in 'green' foams, NaOH/KOH activation) and after firing (in final foams)

The X-ray diffraction analysis (although performed with a position sensitive detector, yielding a distinctive high signal-to-noise ratio), shown in Figure 3, clarified only partially the gel nature. Considering that the patterns are nearly identical independently from the state (as received, hardened and fired), we can conclude that the gel was mostly amorphous and in limited quantities, unlike the case of soda-lime glass and other CaO rich glasses [5,14–16], for which the shape of typical amorphous 'halo' of glasses had significant modifications (i.e., clearly visible 2θ displacements). The pattern of the green sample after NaOH activation actually features some weak peaks consistent with the formation of crystalline N-A-S-H (sodium alumino-silicate hydrated) phases, such as gmelinite (PDF#38-0435), fajausite (PDF#38-0238) and paragonite (PDF#42-0602), although some uncertainties remain, testified by the arrows in Figure 3. Whereas the not perfect match in the position of fajausite (2θ ~ 27°) main peak could be justified by formation of solid solutions, no phase was compatible with the peak at 2θ ~ 39.5°. Interestingly, both gmelinite and fajausite are zeolites [27].

Samples from NaOH/KOH, for which the formation of a zeolitic gel could not be evaluated even as 'probable' (from mineralogical analysis), were subjected to NMR studies, which led to interesting observations on the coordination of Si, Al and B ions, caused by the alkali activation and kept in the material after firing. In particular, from the ^{29}Si spectra in Figure 4a, we can observe that BSG in the initial state presented a peak centered at about −105 ppm, composed by two equivalent peaks at −107.4 ppm and −101.7 ppm consistent with a mixed glass network (pure silica glass exhibits a peak centered at about −110 ppm, attributed to Q_4 silicons, SiO_4 [28,29]). The formation of silanol groups, observed with FTIR in the alkali activated BSG, is confirmed by the study of the deconvoluted broadened spectrum (Figure 4b) that show the presence of three peaks at −109.3, −103.5 and −90.6 ppm. The presence of the peak at higher chemical shift can be attributed to the new hydrated specie $SiO_2(OH)_2$ that is known to provide signals at ca −89 ppm [29]). The quantitative analysis shows that the peak intensity at −109.3 ppm (Q_4 silicons, SiO_4) decreases almost the same amount of the peak at −90.6 ppm while the peak at 103.5 remains constant. The peak returned quite symmetrical after firing, which caused dehydration; it is composed by two again equal peaks at −105.63 ppm and −99.66 ppm but with increased upshifting respect the initial state, compared to pure silica glass, attributed to the incorporation of AlO_4 and BO_4 units (as observed in geopolymers, the chemical shifts decrease with increasing Al/Si ratio with the formation of Q_4 (1Al) species [30]).

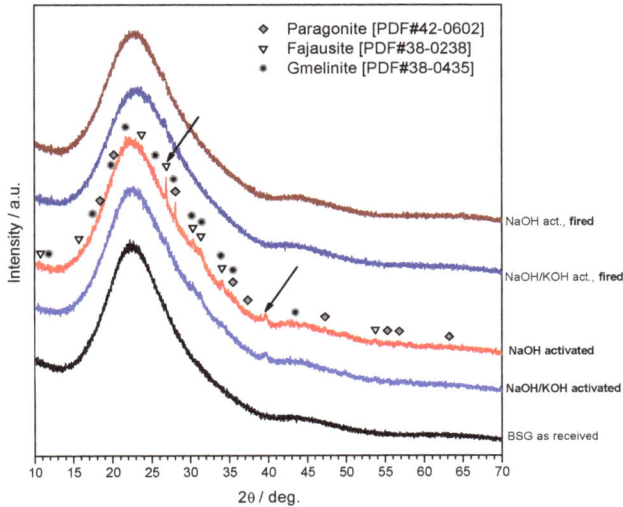

Figure 3. Mineralogical analysis of the studied materials.

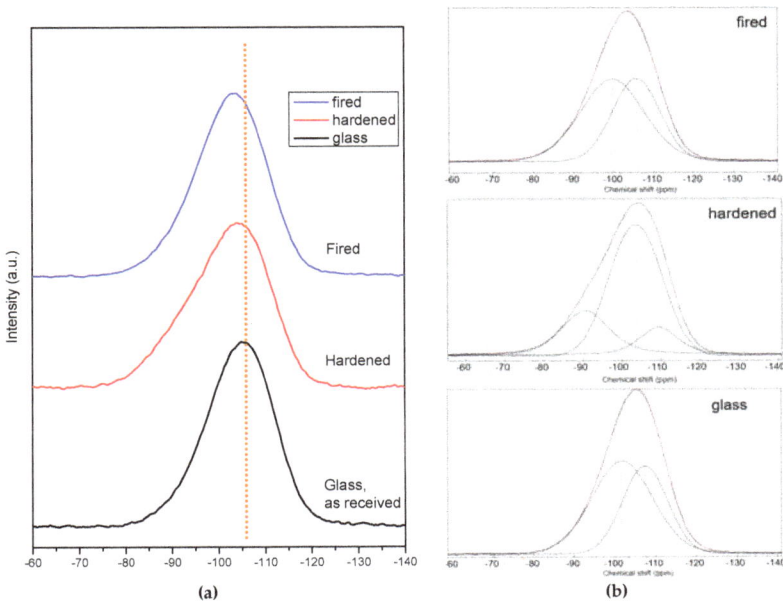

Figure 4. ^{29}Si NMR studies of BSG glass: (**a**) comparison of spectra in the as received state, after alkali activation (in 'green' foams, NaOH/KOH activation) and after firing (in final foams); (**b**) deconvolution analysis: blue lines, from experimental spectra, overlap with fitted curves, in red.

In Figure 5a, the ^{27}Al spectra reveal a slight change in the coordination of aluminium ions, passing from the as received state to the hardnened and fired states: the main peak became more symmetrical, at about 47 ppm, likely due to an increase of tetrahedral coordination (signals for 5- and 6-fold co-ordinations are known to prevail at lower chemical shifts [31]).

Finally, a more significant evidence of the structural transformation with alkali activation and firing was provided by ^{11}B NMR spectroscopy, as shown in Figure 5b. In BSG, boron oxide evidently acts as network former in a double configuration, forming BO_3 as well as BO_4 units, the latter made possible by charge compensation with surrounding alkali ions, like in the case of AlO_4 units. The two configurations, trigonal and tetrahedral, were maintained with alkali activation and firing but with a different balance. Consistent with the FTIR observations, the tetrahedral coordination was promoted [19].

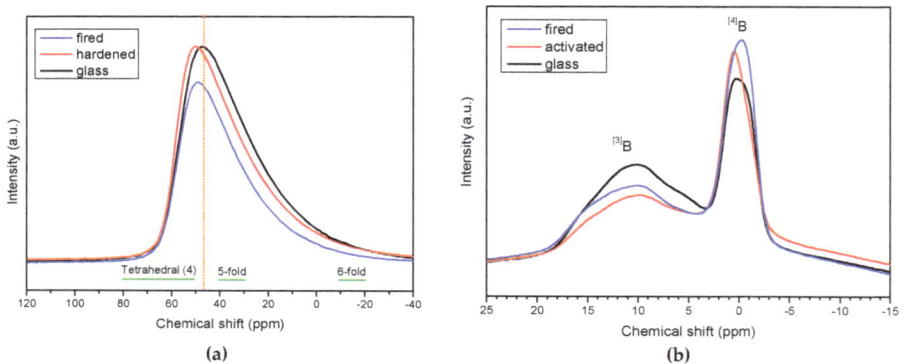

Figure 5. ^{27}Al (**a**) and ^{11}B (**b**) Nuclear magnetic resonance (NMR) studies BSG glass in the as received state, after alkali activation (in 'green' foams, NaOH/KOH activation) and after firing (in final foams).

We can posit, at the end, that alkali activation led to a complex amorphous hydrated gel with a boro-alumino-silicate structure (embedding alkali, for charge compensation), amorphous or slightly crystalline; the related spectroscopic signals were weak, due the limited dissolution of the starting glass. The firing treatment, besides causing dehydration, determined an 'absorption' of the gel in the molecular structure of BSG: extra Na^+ and K^+ ions from the activating solutions were incorporated in the glass structure in the formation of extra BO_4 and AlO_4 units.

The open-celled morphology obtained in the 'green' state was confirmed upon firing, for both activating solutions, as shown by Figure 6a,b. The adopted firing temperature was evidently too low for a substantial softening of BSG, 700 °C being the minimum temperature for the sintering of the specific glass (an optimum temperature for glass sintering can be estimated as 50 °C above the dilatometric softening temperature [32], which is 650 °C for BSG [17]). In addition, the limited amount of hydrated phase prevented the foams from a secondary foaming and reshaping of pores (observed for soda-lime glass [5]); the release of water vapour is the likely reason just for small micro-sized pores (darker spots) visible on the surface of struts (Figure 6c,d).

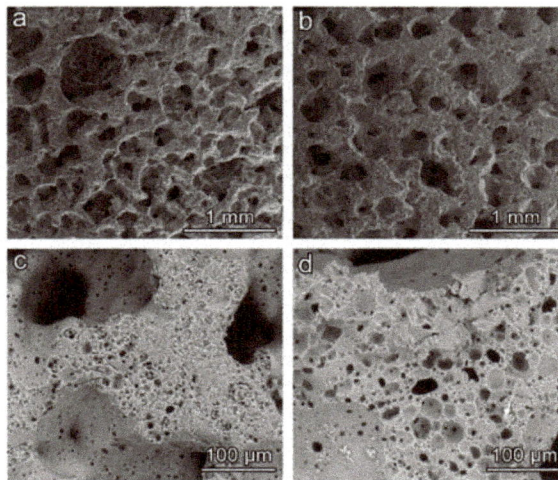

Figure 6. Morphology of BSG-derived glass foams (surfactant: Triton X-100): (**a**,**c**) NaOH activation; (**b**,**d**) NaOH/KOH activation.

Table 1 shows that the produced foams exhibited a remarkable compressive strength, considering the open-celled structure. According to the well-known Gibson-Ashby (GA) model [33], the compressive strength of cellular solid, σ_c, depends on the relative density ($\rho_{rel} = 1 - P$, where P is the total porosity), as follows:

$$\sigma_c \approx \sigma_{bend} \cdot f(\Phi, \rho_{rel}) = \sigma_{bend} \cdot [C \cdot (\Phi \cdot \rho_{rel})^{3/2} + (1 - \Phi) \cdot \rho_{rel}], \tag{1}$$

where σ_{bend} is the bending strength of the solid phase, C is a dimensionless calibration constant ($\sim$0.2) and f is a 'structural function'. The quantity $(1 - \Phi)$ expresses the mechanical contribution of solid positioned at cell faces, reasonably limited when open porosity is dominant. If we neglect this contribution ($\Phi \sim 1$), the observed compressive strength could be correlated to a bending strength well exceeding 100 MPa, that is, far above the measured values for pore-free sintered BSG [17].

Table 1. Physico-mechanical properties of BSG-derived glass foams produced under different conditions.

Activation	Surfactant	Density (g/cm³)	Total Porosity, P (%)	Open Porosity (%)	Compressive Strength (MPa)
NaOH		0.75 ± 0.01	66 ± 1	55 ± 3	4.3 ± 0.3
NaOH/KOH	TritonX-100	0.81 ± 0.02	68 ± 3	52 ± 4	7.7 ± 0.4
NaOH/KOH	Tween 80	0.79 ± 0.01	69 ± 2	55 ± 3	7.2 ± 0.3
NaOH/KOH	SLS	0.98 ± 0.01	61 ± 2	49 ± 2	8.2 ± 0.5

The favourable strength/density correlation is further testified by the bending strength data illustrated by the Ashby's plot shown in Figure 7. The plot was traced by means of the CES (Cambridge Engineering Selector, EduPack2017 [34]) and related database concerning ceramic foams. The experimental data from bending strength determinations are highlighted in yellow. Panels, designed to a given applied bending moment, are lighter and lighter with increasing index I, defined as [35]:

$$I = \sigma_f^{1/2} / \rho, \tag{2}$$

where σ_f is the real failure stress (bending strength, also known as 'modulus of rupture'); materials with the same I value stay on the same selection line, in the strength/density chart (see solid line in Figure 7) [35]. The position of newly developed foams in the chart (aligned above the line) makes them good candidate materials for strong lightweight panels, compared to most commercial ceramic foams, with the exception of commercial glass foams (which exploit the mechanical contribution of cell faces, being close-celled [8]). The open-celled morphology could be exploited for the infiltration of a secondary phase (e.g., elastomers, in ballistic protection composites [36]) or for filter manufacturing.

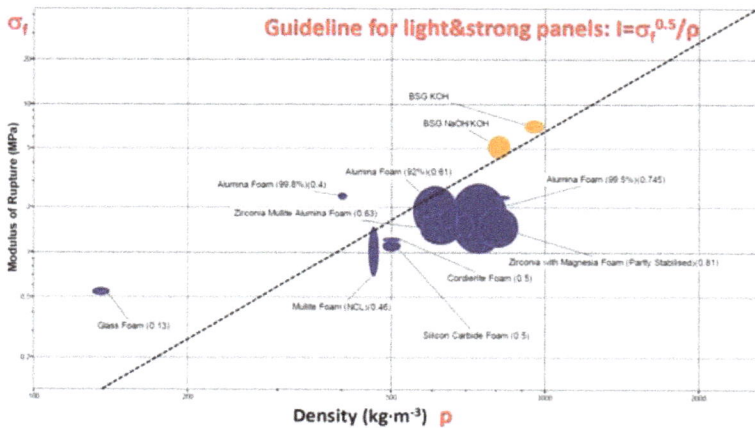

Figure 7. Bending strength/density chart for selected BSG-derived foams compared to commercial ceramic foams.

The proposed approach to cellular glasses is interesting for its inherent flexibility. As evidenced by the experiments with soda-lime glass and bioglass [5,15], the solid content and the duration of both activation and drying stages may affect the viscosity of glass slurries and thus modify the cellular structure in the 'green' state. These changes will be the reasonable focus of future investigations, especially at a semi-industrial scale. Figure 8 actually illustrates that the morphology could be tuned even by simply changing the chemistry of the adopted surfactant. The differences in cell size in the derived foams obtained with different surfactants are evident, showing a significant size difference in the macropores detected. Again, despite the open-celled morphology, the crushing strength remained remarkable (see Table 1).

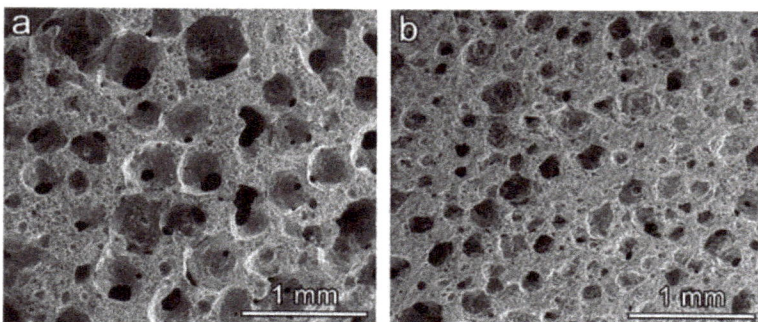

Figure 8. Examples of BSG-derived foams from different surfactants; (**a**) Tween 80; (**b**) SLS.

4. Conclusions

We may conclude that:

(1) Alkali activation of glass slurries, followed by low temperature sintering, was successfully applied to glass from discarded pharmaceutical vials;

(2) Owing to the specific glass chemistry, the hardening of slurries was not caused by the formation of a C-S-H gel but occurred by formation of a (mostly) amorphous hydrated boro-alumino-silicate gel, in limited quantities;

(3) Alkali activation and subsequent firing determined slight but measurable changes in the molecular structure of the adopted boro-alumino-silicate glass, with enhancement of BO_4 and AlO_4 units (by incorporation of 'activating' alkali ions);

(4) The cellular structure could be tuned depending on the chemistry of activating solution but also on the chemistry of surfactants used in the foaming of activated glass slurries;

(5) The relatively low firing temperature and the limited quantity of hydrated phases favored the retention of the open-celled morphology, developed upon foaming of activated slurries;

(6) The developed foams, in all processing conditions, exhibited a favorable strength/density correlation.

Author Contributions: For this paper, E.B. formulated research ideas, supervised the experiments on borosilicate glass foams and planned the structure of the article. A.R.R. performed the general experimentation except NMR measurements and relative data interpretation, which were done by S.T. and J.T. and bending strength test, which were done by G.T.. A.R.R., S.T., I.D. and E.B. have written and edited the article.

Funding: The authors acknowledge the support of the European Community's Horizon 2020 Programme through a Marie Skłodowska-Curie Innovative Training Network ('CoACH-ETN", g.a. No. 642557).

Acknowledgments: The author thank Nuova Ompi S.rl. (Dr. Fabiano Nicoletti), Piombino Dese (Padova, Italy) for supplying borosilicate glass. CIISB research infrastructure project LM2015043 funded by MEYS CR is gratefully acknowledged for the financial support of the measurements at the CF Josef Dadok National NMR Centre.

Conflicts of Interest: The authors declare no conflict of interest.

References

1. Beerkens, R.; Kers, G.; van Santen, E. Recycling of post-consumer glass: energy savings, CO_2 emission reduction, effects on glass quality and glass melting. In *71st Conference on Glass Problems*; John Wiley & Sons: Hoboken, NJ, USA, 2011; pp. 167–194.

2. Ashby, M.F. Chapter 6—Eco-data: Values, sources, precision. In *Materials and the Environment*, 2nd ed.; Ashby, M.F., Ed.; Butterworth-Heinemann: Amsterdam, the Netherlands, 2013; pp. 119–174.

3. Bonifazi, G.; Serranti, S. Imaging spectroscopy based strategies for ceramic glass contaminants removal in glass recycling. *Waste Manag.* **2006**, *26*, 627–639. [CrossRef] [PubMed]

4. Farcomeni, A.; Serranti, S.; Bonifazi, G. Non-parametric analysis of infrared spectra for recognition of glass and glass ceramic fragments in recycling plants. *Waste Manag.* **2008**, *28*, 557–564. [CrossRef] [PubMed]

5. Rincón, A.; Giacomello, G.; Pasetto, M.; Bernardo, E. Novel 'inorganic gel casting' process for the manufacturing of glass foams. *J. Eur. Ceram. Soc.* **2017**, *37*, 2227–2234. [CrossRef]

6. Rincón, A.; Marangoni, M.; Cetin, S.; Bernardo, E. Recycling of inorganic waste in monolithic and cellular glass-based materials for structural and functional applications. *J. Chem. Technol. Biotechnol.* **2016**, *91*, 1946–1961. [CrossRef] [PubMed]

7. Iacocca, R.G.; Allgeier, M. Corrosive attack of glass by a pharmaceutical compound. *J. Mater. Sci.* **2007**, *42*, 801–811. [CrossRef]

8. Scarinci, G.; Brusatin, G.; Bernardo, E. Glass foams. In *Cellular Ceramics: Structure, Manufacturing, Properties and Applications*; Scheffler, M., Colombo, P., Eds.; Wiley-VCH Verlag GmbH & Co. KGaA: Weinheim, Germany, 2005; pp. 158–176.

9. Meylan, G.; Ami, H.; Spoerri, A. Transitions of municipal solid waste management. Part II: Hybrid life cycle assessment of Swiss glass-packaging disposal. *Resour. Conserv. Recycl.* **2014**, *86*, 16–27. [CrossRef]

10. Sepulveda, P.; Jones, J.R.; Hench, L.L. In vitro dissolution of melt derived 45S5 and sol-gel derived 58S bioactive glasses. *J. Biomed. Mater. Res.* **2002**, *61*, 301–311. [CrossRef] [PubMed]

11. Jones, J.R.; Hench, L.L. Effect of surfactant concentration and composition on the structure and properties of sol-gel-derived bioactive glass scaffolds for tissue engineering. *J. Mater. Sci.* **2003**, *38*, 1–8. [CrossRef]

12. Wu, Z.Y.; Hill, R.G.; Yue, S.; Nightingale, D.; Lee, P.D.; Jones, J.R. Melt-derived bioactive glass scaffolds produced by a gel-cast foaming technique. *Acta Biomater.* **2011**, *7*, 1807–1816. [CrossRef]

13. Novajra, G.; Perdika, P.; Pisano, R.; Miola, M.; Bari, A.; Jones, J.R.; Detsch, R.; Boccaccini, A.R.; Vitale-Brovarone, C. Structure optimisation and biological evaluation of bone scaffolds prepared by co-sintering of silicate and phosphate glasses. *Adv. Appl. Ceram.* **2015**, *114*, 48–55. [CrossRef]

14. Elsayed, H.; Rincón Romero, A.; Ferroni, L.; Gardin, C.; Zavan, B.; Bernardo, E. Bioactive glass-ceramic scaffolds from novel 'inorganic gel casting' and sinter-crystallization. *Materials* **2017**, *10*, 171. [CrossRef] [PubMed]

15. Elsayed, H.; Rincón Romero, A.; Molino, G.; Vitale Brovarone, C.; Bernardo, E. bioactive glass-ceramic foam scaffolds from "inorganic gel casting" and sinter-crystallization. *Materials* **2018**, *11*, 349. [CrossRef] [PubMed]

16. Rabelo Monich, P.; Rincón Romero, A.; Höllen, D.; Bernardo, E. Porous glass-ceramics from alkali activation and sinter-crystallization of mixtures of waste glass and residues from plasma processing of municipal solid waste. *J. Clean. Prod.* **2018**, *188*, 871–878. [CrossRef]

17. Bernardo, E.; Scarinci, G. Sintering behaviour and mechanical properties of Al_2O_3 platelet-reinforced glass matrix composites obtained by powder technology. *Ceram. Int.* **2004**, *30*, 785–791. [CrossRef]

18. Provis, J.L. Geopolymers and other alkali activated materials: why, how, and what? *Mater. Struct.* **2014**, *47*, 11–25. [CrossRef]

19. Taveri, G.; Tousek, J.; Bernardo, E.; Toniolo, N.; Boccaccini, A.R.; Dlouhy, I. Proving the role of boron in the structure of fly-ash/borosilicate glass based geopolymers. *Mater. Lett.* **2017**, *200*, 105–108. [CrossRef]

20. Reben, M.; Wacławska, I.; Paluszkiewicz, C.; Środa, M. Thermal and structural studies of nanocrystallization of oxyfluoride glasses. *J. Therm. Anal. Calorim.* **2007**, *88*, 285–289. [CrossRef]

21. Elbatal, H.A.; Azooz, M.A.; Saad, E.A.; Ezzeldin, F.M.; Amin, M.S. Corrosion behavior mechanism of borosilicate glasses towards different leaching solutions evaluated by the grain method and FTIR spectral analysis before and after gamma irradiation. *Silicon* **2018**, *10*, 1139–1149. [CrossRef]

22. El-Egili, K. Infrared studies of Na_2O-B_2O_3-SiO_2 and Al_2O_3-Na_2O-B_2O_3-SiO_2 glasses. *Phys. B Condens. Matter* **2003**, *325*, 340–348. [CrossRef]

23. Fernández-Jiménez, A.; Palomo, A. Mid-infrared spectroscopic studies of alkali-activated fly ash structure. *Microporous Mesoporous Mater.* **2005**, *86*, 207–214. [CrossRef]

24. Ming, L.; Zhou, H.; Zhu, H.; Yue, Z.; Zhao, J. Microstructure and dielectric properties of glass/Al_2O_3 composites with various low softening point borosilicate glasses. *J. Mater. Sci. Mater. Electron.* **2012**, *23*, 2130–2139. [CrossRef]

25. Wan, J.; Cheng, J.; Lu, P. The coordination state of B and Al of borosilicate glass by IR spectra. *J. Wuhan Univ. Technol. Mater. Sci. Ed.* **2008**, *23*, 419–421. [CrossRef]

26. Hou, Z.X.; Wang, S.H.; Xue, Z.L.; Lu, H.R.; Niu, C.L.; Wang, H.; Sun, B.; Su, C. Crystallization and microstructural characterization of B_2O_3-Al_2O_3-SiO_2 glass. *J. Non Cryst. Solids* **2000**, *356*, 201–207. [CrossRef]

27. Larin, A.V.; Leherte, L.; Vercauteren, D.P. Approximation of the Mulliken-type charges for the oxygen atoms of all-siliceous zeolites. *Chem. Phys. Lett.* **1998**, *287*, 169–177. [CrossRef]

28. Malfait, W.J.; Halter, W.E.; Verel, R. ^{29}Si NMR spectroscopy of silica glass: T1 relaxation and constraints on the Si-O-Si bond angle distribution. *Chem. Geol.* **2008**, *256*, 269–277. [CrossRef]

29. Kinney, D.R.; Chuang, I.S.; Maciel, G.E. Water and the silica surface as studied by variable-temperature high-resolution ^1H NMR. *J. Am. Chem. Soc.* **1993**, *115*, 6786–6794. [CrossRef]

30. Duxson, P.; Provis, J.L.; Lukey, G.C.; Separovic, F.; van Deventer, J.S.J. ^{29}Si NMR study of structural ordering in aluminosilicate geopolymer gels. *Langmuir* **2005**, *21*, 3028–3036. [CrossRef]

31. Risbud, S.H.; Kirkpatrick, R.J.; Taglialavore, A.P.; Montez, B. Solid-state NMR Evidence of 4-, 5, and 6-Fold Aluminum Sites in Roller-Quenched SiO_2-Al_2O_3 Glasses. *J. Am. Ceram. Soc.* **1987**, *70*, 10–12. [CrossRef]

32. Ray, A.; Tiwari, A.N. Compaction and sintering behaviour of glass-alumina composites. *Mater. Chem. Phys.* **2001**, *67*, 220–225. [CrossRef]

33. Gibson, L.J.; Ashby, M.F. *Cellular Solids, Structure and Properties*; Cambridge University Press: Cambridge, UK, 1999.

Materials **2018**, *11*, 2545

34. CES EduPack 2017 User Manual and Getting_Started_Guide. Available online: https://www.grantadesign. com (accessed on 1 July 2018).

35. Ashby, M.F. *Materials Selection in Mechanical Design*; Butterworth-Heinemann: Amsterdam, the Netherlands, 2017.

36. Colombo, P.; Zordan, F.; Medvedovski, E. Ceramic-polymer composites for ballistic protection. *Adv. Appl. Ceram.* **2006**, *105*, 78–83. [CrossRef]

Article

Spectroscopic Properties of Er^{3+}-Doped Particles-Containing Phosphate Glasses Fabricated Using the Direct Doping Method

Pablo Lopez-Iscoa [1] , Nirajan Ojha [2] , Ujjwal Aryal [2], Diego Pugliese [1] , Nadia G. Boetti [3] , Daniel Milanese [1,4] and Laeticia Petit [2,5,*,†]

[1] Dipartimento di Scienza Applicata e Tecnologia (DISAT) and INSTM UdR Torino Politecnico, Politecnico di Torino, Corso Duca degli Abruzzi 24, 10129 Torino, Italy; pablo.lopeziscoa@polito.it (P.L.-I.); diego.pugliese@polito.it (D.P.); daniel.milanese@polito.it (D.M.)
[2] Laboratory of Photonics, Tampere University, Korkeakoulunkatu 3, 33720 Tampere, Finland; nirajan.ojha@tut.fi (N.O.); ujjwal.aryal@student.tut.fi (U.A.)
[3] Fondazione LINKS—Leading Innovation & Knowledge for Society, Via P. C. Boggio 61, 10138 Torino, Italy; boetti@ismb.it
[4] Istituto di Fotonica e Nanotecnologie-Consiglio Nazionale delle Ricerche (IFN-CNR), Caratterizzazione e Sviluppo di Materiali per la Fotonica e l'Optoelettronica (CSMFO) Lab., Via alla Cascata 56/C, 38123 Povo (TN), Italy
[5] nLIGHT Corporation, Sorronrinne 9, 08500 Lohja, Finland
*   Correspondence: laeticia.petit@tut.fi; Tel.: +358-50-447-8481
†   Current Address: Laboratory of Photonics, Tampere University of Technology, Korkeakoulunkatu 3, 33720 Tampere, Finland.

Received: 4 December 2018; Accepted: 21 December 2018; Published: 3 January 2019

Abstract: The effect of the incorporation of Er$_2$O$_3$-doped particles on the structural and luminescence properties of phosphate glasses was investigated. A series of different Er$_2$O$_3$-doped TiO$_2$, ZnO, and ZrO$_2$ microparticles was synthesized using soft chemistry and then added into various phosphate glasses after the melting at a lower temperature than the melting temperature. The compositional, morphological, and structural analyses of the particles-containing glasses were performed using elemental mapping by field emission-scanning electron microscopy (FE-SEM) with energy dispersive x-ray spectrometry (EDS) and x-ray diffraction (XRD). Additionally, the luminescence spectra and the lifetime values were measured to study the influence of the particles incorporation on the spectroscopic properties of the glasses. From the spectroscopic properties of the glasses with the composition 50P$_2$O$_5$-40SrO-10Na$_2$O, a large amount of the Er$_2$O$_3$-doped particles is thought to dissolve during the glass melting. Conversely, the particles were found to survive in glasses with a composition 90NaPO$_3$-(10 − x)Na$_2$O-xNaF (with x = 0 and 10 mol %) due to their lower processing temperature, thus clearly showing that the direct doping method is a promising technique for the development of new active glasses.

Keywords: phosphate glass; oxyfluoride phosphate glass; Er$_2$O$_3$-doped particles; direct particle doping; Er^{3+} luminescence property

---

1. Introduction

Among rare-earth (RE) ions, erbium (Er^{3+}) ions have been widely studied as dopants in different host matrices due to their emission at around 1.5 µm, which makes them suitable for applications such as fiber lasers and amplifiers for telecommunications [1]. Moreover, their up-conversion properties enable them to convert the infrared (IR) radiation into red and green emissions. Thus, Er^{3+}-doped materials show many other applications such as photovoltaics [2], display technologies [3], and medical diagnostics [4].

Phosphate glasses are very well known for their suitable mechanical and chemical properties, homogeneity, good thermal stability, and excellent optical properties [5–9]. As compared to silicate glasses, they possess low glass transition (T_g) (400–700 °C) and crystallization (T_p) (800–1400 °C) temperatures, which facilitate their processing and fabrication by melt-quenching technique [10,11]. Phosphate glasses also exhibit high RE ions solubility [12], leading to quenching phenomenon occurring at very high concentrations of RE [9,13]. Due to their outstanding optical properties, RE-doped phosphate glasses have recently become appealing for the engineering of photonic devices for optical communications [14], laser sources, and optical amplifiers [9,13,15–17]. Moreover, due to their low phonon energies and wide optical transmission window from Ultraviolet (UV) to mid-IR regions, fluoride phosphate glasses are considered to be good glass candidates for Er^{3+} doping [18].

It is well known that the luminescence properties as well as the solubility of RE ions in glassy hosts can be significantly impacted by parameters such as the mass, covalency or charge of the ligand atoms [19]. The luminescence properties of Er^{3+}-doped phosphate glasses with compositions in mol % of (0.25 $Er_2O_3$2013(0.5 $P_2O_5 - 0.4$ $SrO - 0.1$ Na_2O)$_{100-x}$ − ($TiO_2/Al_2O_3/ZnO)_x$), with x = 0 and 1.5 mol %, were previously investigated and discussed as a function of the glass composition [20]. It was found that the Er^{3+} ions spectroscopic properties depended on the glass structure connectivity, which changed the Er^{3+} ions solubility. The intensity of the emission at 1.5 µm could be increased if the phosphate network is depolymerized. Recently, the impact of the nucleation and growth on the Er^{3+} spectroscopic properties of these glasses was reported in Reference [21]. Indeed, due to the crystalline environment surrounding the RE ions, the RE-doped glass-ceramics have shown to combine glass properties (large flexibility of composition and geometry) with some advantages of the RE-doped single crystals (higher absorption, emission, and lifetimes) [22]. Upon heat treatment, the crystals precipitated from the surface of the glasses and the composition of the crystals depended on the glass composition. The crystals were found to be Er^{3+} free except in the glass with x = 0. With this study, it was shown that the heat treatment does not necessarily lead to the bulk precipitation of Er^{3+}-doped crystals, which should increase the spectroscopic properties of the glass.

Therefore, a new route was developed in order to control the local environment of the RE ions independently of the glass composition: the direct doping method [23]. With this technique, the particle matrix allows high RE content in a high dispersion state, thus avoiding the quenching effect independently of the glass composition [24]. Thus, innovative particles-containing glasses with specific particle compositions and nanostructures can be achieved by adding RE-doped particles in the glass batch after the melting process. Recently, up-conversion (UC) was obtained from phosphate glasses which contain only 0.01 at % of Er^{3+} and 0.06 at % of Yb^{3+} by adding $NaYF_4$:Er^{3+}, Yb^{3+} nanoparticles (NPs) in the glass after the melting [25]. However, as explained in Reference [26], the main challenge with this novel route of preparing glasses is to balance the survival and dispersion of the particles in the glasses. The particles should be thermally stable at the temperature they are added in the glass melt to ensure their survival within the glass during the glass preparation.

Here, new Er^{3+}-doped particles-containing phosphate glasses were prepared in order to fabricate phosphate glasses with enhanced emission at 1.5 µm as compared to standard Er^{3+}-doped phosphate glasses. Oxyfluoride phosphate glass family was also considered due to its low processing temperature. At first, the synthesis and characterization of different Er_2O_3-doped particles are presented. The concentration of Er_2O_3 in the particles was kept high to ensure that luminescence can be detected after embedding the particles into the glasses. The particles were prepared by the sol-gel method. TiO_2, ZnO, and ZrO_2 were selected as crystalline hosts for the Er^{3+} ions due to their high melting points and low phonon energies, which provide high thermal stability and high luminescence properties [27–32]. Then, the technique of incorporation of the particles into phosphate glasses with various compositions was addressed. Finally, a full characterization of the morphological and luminescence properties of the different particles-containing glasses was reported.

2. Materials and Methods

2.1. Particles Synthesis

The synthesis of the TiO_2 particles doped with 14.3 mol % of Er_2O_3 was reported in Reference [33]. Specifically, 21.28 g of titanium(IV) butoxide reagent grade (97% Sigma–Aldrich, Saint Louis, MO, USA) were dissolved in ethanol (100 mL) and then added dropwise into a mixture of deionized water (2 mL), ethanol (>99.8% Sigma–Aldrich, Saint Louis, MO, USA), and 8.37 g of erbium (III) acetate (>99.9% Sigma–Aldrich, Saint Louis, MO, USA). Once the addition was completed, the solution was heated at a reflux temperature of 90 °C and left under reflux for 1 day. Finally, the precipitates were collected by centrifugation, washed with ethanol for several times, and dried at 100 °C for 1 day. The as-prepared sample was further annealed in air at 800 °C for 2 h.

The ZnO particles were synthesized following the process described in Reference [34]. The concentration of Er_2O_3 was 14.3 mol % as in the TiO_2 particles. Then, 13.5 g of zinc acetate dihydrate ($Zn(CH_3COO)_2 \cdot 2H_2O$)) (99.999% Sigma–Aldrich, Saint Louis, MO, USA) and 7.8 g of erbium chloride hexahydrate ($ErCl_3$ $6H_2O$) (Sigma–Aldrich, Saint Louis, MO, USA, 99.9% purity) were used as starting materials. The precursors were dissolved in ethanol (>99.8% Sigma–Aldrich, Saint Louis, MO, USA)-deionized water (50–50% in volume) with polyvinylpyrrolidone (PVP K 30, average Mw 40,000, Sigma–Aldrich, Saint Louis, MO, USA) as a surfactant and stirred for 30 min until a clear solution was formed. Then, the sodium hydroxide 0.1 M (Fluka Massachusetts, MA, USA) was added until reaching a pH of 9, which is optimal for nucleation. The solution was heated up to 90 °C and stirred for 4 h to get fine precipitation. The obtained precipitation was washed 4 times with deionized water and centrifuged. The precipitation was collected and dried at 80 °C for 4 h. Finally, the sample was calcined at 1000 °C for 2 h.

The ZrO_2 particles were synthesized with 7 mol % of Er_2O_3 as in Reference [35]. More in detail, 28 mL of aqueous solution of 0.1 M erbium chloride hexahydrate ($ErCl_3 \cdot 6H_2O$, 99.9%, Sigma Aldrich, Saint Louis, MO, USA), and 1.83 mL of 0.1 M sodium bicarbonate ($NaHCO_3$, 99.7%, Fluka) were added to 50 mL of absolute ethanol (>99.8% Sigma Aldrich, Saint Louis, MO, USA) while stirring at 60 °C. Then, 9.17 mL of zirconium (IV) butoxide solution (($Zr(OBu)_4$) 80 wt % in 1-butanol) (Sigma-Aldrich, Saint Louis, MO, USA) was added and kept under stirring for 2 h. The obtained colloidal solution was centrifuged and washed three times with ethanol (>99.8% Sigma–Aldrich, Saint Louis, MO, USA) at 9000 rpm for 30 min. Lastly, the final product was calcined at 1000 °C for 2 h.

2.2. Particles-Containing Glasses Preparation

The particles-containing glasses were prepared by incorporating the aforementioned particles in the host glasses with compositions (in mol %): $50P_2O_5$-$40SrO$-$10Na_2O$, labeled as SrNaP glass, and $90NaPO_3$-$(10 - x)Na_2O$-$xNaF$ with x = 0 and 10, labeled as NaPF0 and NaPF10, respectively. The glasses were synthesized in a quartz crucible using $Na_6O_{18}P_6$ (Alfa-Aesar, technical grade), Na_2CO_3 (Sigma–Aldrich, >99.5%), $Sr(PO_3)_2$, and NaF (Sigma–Aldrich, 99.99%). $Sr(PO_3)_2$ precursor was independently prepared using $SrCO_3$ (Sigma-Aldrich, Saint Louis, MO, USA, ≥99.9%), and $(NH_4)_2HPO_4$ as raw materials heated up to 850 °C. Prior to the melting, the glass NaPF0 was heat treated at 400 °C for 30 min to decompose Na_2CO_3 and evaporate CO_2. Details on the direct doping process of the NaPF0 and NaPF10 glasses and of the SrNaP glass can be found in References [26] and [36], respectively. The particles (1.25 wt %) were incorporated in the SrNaP glass at 1000 °C after melting at 1050 °C for 20 min. After 5 min, the glasses were quenched. The NaPF0 and NaPF10 glasses were melted at 750 °C for 5 min, then the particles were added at 550 °C and finally the glass melts were quenched after a 3 min dwell time.

After quenching, all the melts were finally annealed at 40 °C below their respective T_g for 5 h in order to release the residual stress. All the glasses were cut and optically polished or ground, depending on the characterization technique.

2.3. Characterization Techniques

The thermal stability of the particles was measured by thermogravimetric analysis (TGA) using a Perkin Elmer TGS-2 instrument (PerkinElmer Inc., Waltham, MA, USA). The measurement was carried out in a Pt crucible at a heating rate of 10 °C/min in a controlled atmosphere (N_2 flow), featuring an error of ± 3 °C. The sample was approximately 10 mg of grounded particles.

The composition and morphology of the particles were determined using a field emission scanning electron microscope (FE-SEM, Zeiss Merlin 4248, Carl Zeiss, Oberkochen, Germany) equipped with an Oxford Instruments X-ACT detector and energy dispersive spectroscopy systems (EDS) (Oxford Instruments, Abingdon-on-Thames, UK). The composition and morphology of the particles-containing glasses were determined using FE-SEM/EDS (Carl Zeiss Crossbeam 540 equipped with Oxford Instruments X-Max^N 80 EDS detector) (Instruments, Abingdon-on-Thames, UK). The images were taken at the surface of the glasses, previously cut and optically polished. The samples were coated with a thin carbon layer before the EDS elemental mapping. The elemental mapping analysis of the composition of the samples was performed by using the EDS within the accuracy of the measurement (± 1.5 mol %).

The crystalline phases were identified using the X-ray diffraction (XRD) analyzer Philips X'pert (Philips, Amsterdam, Netherlands) with Cu K_α X-ray radiation (λ = 1.5418 Å). Data were collected from 2θ = 0 up to 60° with a step size of 0.003°.

The emission spectra in the 1400–1700 nm range were measured with a Jobin Yvon iHR320 spectrometer (Horiba Jobin Yvon SAS, Unterhaching, Germany) equipped with a Hamamatsu P4631-02 detector (Hamamatsu Photonics K.K., Hamamatsu, Japan) and a filter (Thorlabs FEL 1500, Thorlabs Inc., Newton, NJ, USA). Emission spectra were obtained at room temperature using an excitation monochromatic source at 976 nm, emitted by a single-mode fiber pigtailed laser diode (CM962UF76P-10R, Oclaro Inc., San Jose, CA, USA).

The fluorescence lifetime of Er^{3+}:$^4I_{13/2}$ energy level was obtained by exciting the samples with a fiber pigtailed laser diode operating at the wavelength of 976 nm, recording the signal using a digital oscilloscope (Tektronix TDS350, Tektronic Inc., Beaverton, OR, USA) and fitting the decay traces by single exponential. All lifetime measurements were collected by exciting the samples at their very edge to minimize reabsorption. Estimated error of the measurement was ±0.2 ms. The detector used for this measurement was a Thorlabs PDA10CS-EC (Thorlabs Inc., Newton, NJ, USA). The samples used for the spectroscopic measurements were previously cut and optically polished.

3. Results and Discussion

Er_2O_3-doped TiO_2, ZnO, and ZrO_2 particles were synthesized using the sol-gel technique. The morphology of the particles was assessed by FE-SEM analysis (see Figure 1).

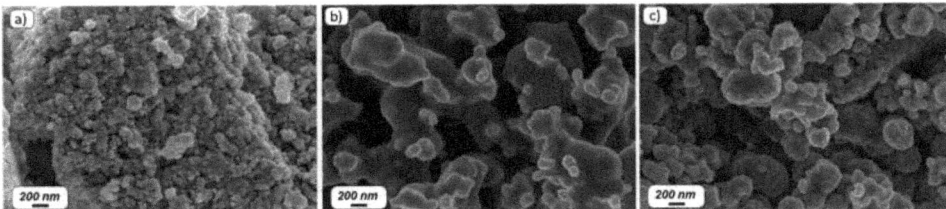

Figure 1. FE-SEM micrographs of Er_2O_3-doped TiO_2 (**a**), ZnO (**b**), and ZrO_2 (**c**) particles.

The Er^{3+}-doped particles show agglomerates of nanoparticles with irregular shape. As can be seen from Figure 1, the agglomerates are formed by small particles with a size of ~50–100 nm for the TiO_2, ~100–400 nm for the ZnO, and ~100–300 nm for the ZrO_2 particles. Additionally, based on the

EDS analysis, the composition of the particles was in agreement with the nominal one within the accuracy of the measurement.

The XRD patterns of the Er^{3+}-doped particles are presented in Figure 2.

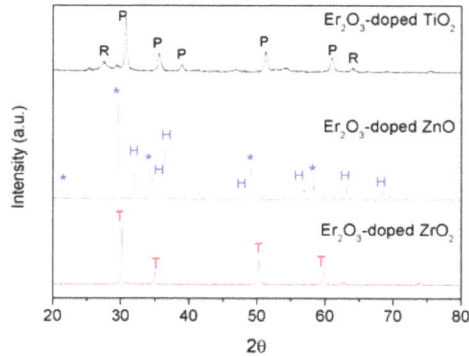

Figure 2. XRD patterns of the Er_2O_3-doped TiO_2, ZnO, and ZrO_2 particles, with the reference patterns of the pyrochlore (P), rutile (R), hexagonal (H), Zn extra phase (*), and tetragonal (T) crystalline phases.

The XRD pattern of the Er_2O_3-doped TiO_2 particles showed the presence of rutile phase (Inorganic Crystal Structure Database (ICSD) file No. 00-021-1276) and also the pyrochlore phase (ICSD file No. 01-073-1647), which seems to be the major phase as reported in Reference [37] when adding Er_2O_3. The XRD pattern of the Er_2O_3-doped ZnO particles revealed the presence of ZnO hexagonal wurtzite phase (ICSD file No. 89-1397), together with an extra unidentified phase. Similar results were reported by Rita John et al. [29], where the same unidentified phase was found in the XRD analysis of highly Er_2O_3-doped ZnO particles. The ZrO_2 tetragonal crystalline phase (ICSD file No. 66787) is present in the Er_2O_3-doped ZrO_2 sample, in agreement with References [38,39]. It should be pointed out that no peaks related to Er_2O_3 were observed in the XRD measurements of all the samples.

Additionally, the fluorescence lifetime values of the Er^{3+}:$^4I_{13/2}$ level upon 976 nm excitation of the Er_2O_3-doped TiO_2 and Zr_2O particles were (0.1 ± 0.2) and (1.0 ± 0.2) ms, respectively. However, in the case of the ZnO particles, the lifetime value could not be measured due to low IR emission from the particles. The high concentration of Er_2O_3 in the particles is expected to lead to concentration quenching resulting in a very short lifetime [9].

The thermal stability of the particles was assessed by TGA analysis (see Figure 3).

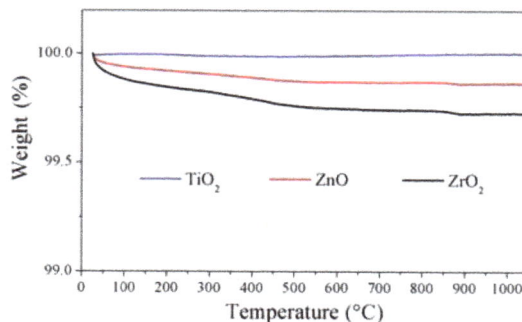

Figure 3. TGA analysis of the Er_2O_3-doped TiO_2, ZnO, and ZrO_2 particles.

Less than 1% of weight loss from all the investigated particles was detected when the temperature increased to 1050 °C, thus indicating that the particles should not suffer any degradation when added to the glass melt.

The SrNaP glasses were fabricated by incorporating the Er_2O_3-doped particles at 1000 °C after the melting as in References [36,40]. The ZnO and ZrO_2 particles-containing SrNaP glasses exhibited a slight pink coloration, while TiO_2 particles-containing glass displayed a purple color. Agglomerates of particles were found in the glasses by naked eyes. No diffraction peaks of any crystalline phase were found in the particles-containing glasses (data not shown), confirming that the number of particles in the glasses was too small to be detected.

The morphology and the composition of the particles found at the surface of the glasses were analyzed using FE-SEM/EDS analysis. The FE-SEM pictures and elemental mapping of Er_2O_3-doped TiO_2, ZnO, and ZrO_2 particles found in the SrNaP glasses are shown in Figure 4a–c, respectively.

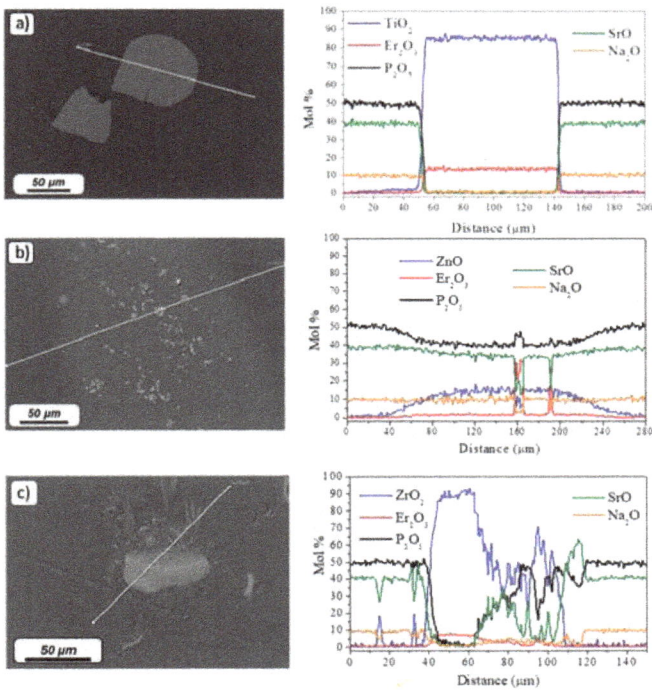

Figure 4. FE-SEM pictures with their corresponding elemental mapping of Er_2O_3-doped TiO_2 (**a**), ZnO (**b**), and ZrO_2 (**c**) particles-containing SrNaP glasses. The direction of the scan starts at the circle of the white line (corresponding to 0 μm).

Few microparticles with a size of ~100 μm were found in the Er_2O_3-doped TiO_2 particles-containing glass matrix. The composition of the glass matrix and of the particles are in accordance with the theoretical ones. The remaining TiO_2 particles maintained their compositional integrity in their center after the melting process, confirming the survival of some Er_2O_3-doped TiO_2 particles in the glass. However, most of the Er_2O_3-doped TiO_2 particles are suspected to degrade during the glass preparation. Regarding the Er_2O_3-doped ZnO particles-containing glasses, a bright area rich in Zn surrounds small particles with a size of ~15 μm. These particles contain mostly Er and P, thus indicating that Zn^{3+} ions diffused from the particles into the glass matrix during the glass preparation leading most probably to the precipitation of $ErPO_4$ crystals (see Figure 4b). Therefore, the Er$_2$O$_3$-doped ZnO particles are also expected to degrade in the glass during its preparation, leading to the diffusion of Zn and Er in the glass matrix which is in agreement with the pink coloration of the glass itself. Figure 4c shows the mapping of the Er$_2$O$_3$-doped ZrO$_2$ particles-containing glasses. Particles with a size of ~50 μm were found. These particles also maintained their compositional integrity in their center. However, contrary to the other particles, Sr-rich crystals were observed around the particles as seen in Reference [36], confirming that crystals with specific composition such as ZrO$_2$ and SrAl$_2$O$_4$ precipitated due to the decomposition of the particles.

The normalized emission spectra of the Er$_2$O$_3$-doped TiO$_2$, ZnO, and ZrO$_2$ particles and of their corresponding particles-containing SrNaP glasses are reported in Figure 5.

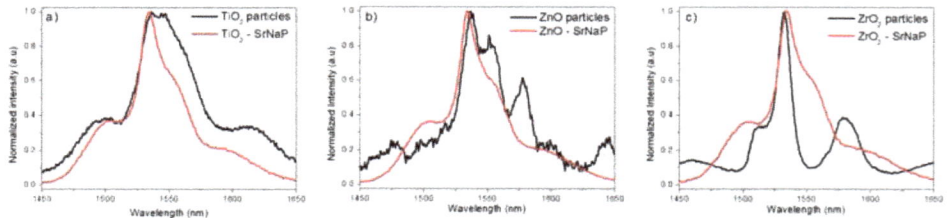

Figure 5. Emission spectra of the Er$_2$O$_3$-doped TiO$_2$ (**a**), ZnO (**b**), and ZrO$_2$ (**c**) particles and of their corresponding particles-containing SrNaP glasses.

The particles-containing glasses exhibit a different emission shape compared to that of the particles alone, indicating that the neighboring ions arrangement around the Er^{3+} ions is different after embedding the particles in the glasses. The emission spectra of the glasses are typical of the emission band assigned to the Er^{3+} transition from $^4I_{13/2}$ to $^4I_{15/2}$ in glass. Additionally, the investigated glasses show similar emission in shape, indicating that the environment of Er^{3+} ions is similar in the investigated glasses, as suspected from the Er^{3+}:$^4I_{13/2}$ lifetime values measured at 976 nm (~4.0 ± 0.2 ms) in the three investigated glasses. It should be pointed out that these lifetimes are longer than the lifetime of the particles alone. Therefore, the absence of sharp peaks in the IR emission spectra together with the long Er^{3+}:$^4I_{13/2}$ lifetime values and the absence of up-conversion emission from the glasses clearly suggest that a large number of particles suffered an important degradation during the melting process. This experimental evidence is in agreement with the color of the glasses, as reported in References [26,36,40]. Therefore, the Er^{3+} ions environment is expected to change from crystalline to glassy when the particles are incorporated into the glasses increasing the bandwidth of the emission band and the Er^{3+}:$^4I_{13/2}$ lifetime.

In order to limit the degradation of the particles during the glass preparation, new glasses with a lower melting temperature and with a composition of 90NaPO$_3$-(10 − x)Na$_2$O-xNaF, with x = 0 (NaPF0) and 10 (NaPF10) (in mol %), were investigated as an alternative host for the Er$_2$O$_3$-doped particles. The optimization of the direct doping method can be found in Reference [25]; the Er$_2$O$_3$-doped particles were added at 550 °C and the dwell time was 3 min. As opposed to the NaSrP glasses, the Er$_2$O$_3$-doped TiO$_2$- and ZnO-containing glasses exhibited a light pink coloration, whereas the Er$_2$O$_3$-doped ZrO$_2$-containing glasses were colorless. Some agglomerates could be observed in the glasses with naked eye.

The normalized emission spectra of the NaPF0 and NaPF10 glasses are reported in Figure 6. For the sake of comparison, the spectra of the particles alone are also shown.

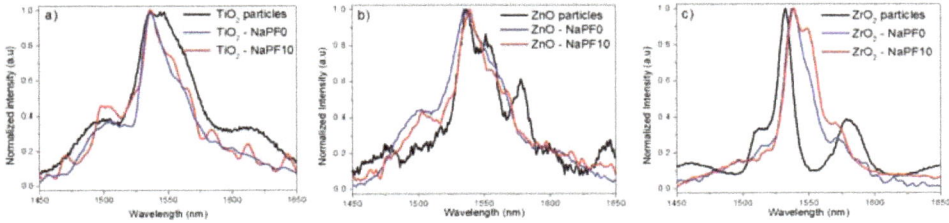

Figure 6. Normalized infrared emission spectra of the Er_2O_3-doped TiO_2 (**a**) ZnO (**b**), and ZrO_2 (**c**) particles and of their corresponding particles-containing NaPF0 and NaPF10 glasses with 1.25 wt % of particles.

Similarly to what previously reported for the NaSrP glasses, the spectra of the NaPF glasses were also different compared to the spectra of the particles alone. However, one can also notice that the shape of the emission depends on the glass composition: the emission band of the NaFP0 glasses was similar to the emission bands presented in Figure 5, while the NaFP10 glasses exhibit different emission bands depending on the particles. It should be pointed out the narrow bandwidth of the emission band of the Er_2O_3-doped ZrO_2 particles-containing NaPF glasses.

In contrast with the SrNaP glasses, the NaPF glasses, except for the TiO_2-containing NaPF0 glass, exhibit up-conversion emission confirming the survival of the particles. However, as observed for the emission at around 1.5 μm, the shape of the visible emission is modified after adding the particles into the glass (see Figure 7), thus confirming that the site of the Er^{3+} ions in the particles is changed after embedding the particles into the glass. According to Reference [26], the change in the Er^{3+} site can be related to the partial decomposition of the particles during the glass preparation.

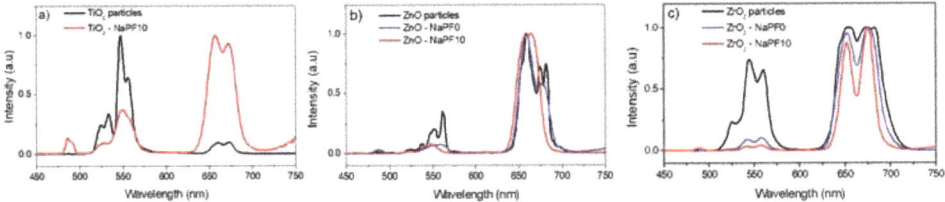

Figure 7. Normalized up-conversion emission spectra of the Er_2O_3-doped TiO_2 (**a**), ZnO (**b**), and ZrO_2 (**c**) particles and of the corresponding particles-containing NaPF0 and NaPF10 glasses with 1.25 wt % of particles.

The Er^{3+}:$^4I_{13/2}$ lifetime values of the NaPF0 and NaPF10 glasses are listed in Table 1.

Table 1. Er^{3+}:$^4I_{13/2}$ lifetime values of the Er_2O_3-doped TiO_2, ZnO and ZrO_2 particles and of the particles-containing NaPF0 and NaPF10 glasses.

Sample Code	Er^{3+}:$^4I_{13/2}$ Lifetime ± 0.2 ms		
	Particles Alone	Particles-Containing Glasses	
		NaPF0	NaPF10
Er_2O_3-doped TiO_2 particles	0.1	2.3	0.7
Er_2O_3-doped ZnO particles	n.a.	1.4	1.3
Er_2O_3-doped ZrO_2 particles	1.0	1.6	1.4

Compared to the NaSrP glasses, the NaPF glasses exhibit shorter Er^{3+}:$^4I_{13/2}$ lifetimes. These lifetimes are, however, longer than those of the particles alone. They are also longer than the lifetime of an Er^{3+}-doped NaPF10 glass prepared using a standard melting process with Er_2O_3

doping level similar to that considered for the particles-containing glasses (Er^{3+}:$^4I_{13/2}$ lifetime equal to (0.6 ± 0.2) ms). Therefore, the local environment of the Er^{3+} ions is expected to be modified after embedding the particles into the glasses. It does not correspond to the local environment of Er^{3+} ions in an amorphous matrix, confirming the survival of the Er^{3+}-doped particles into the glasses. Additionally, the Er^{3+}:$^4I_{13/2}$ lifetime depends on the glass composition: independently of the particles, the Er^{3+}:$^4I_{13/2}$ lifetimes of the NaPF10 glasses are the shortest, thus suggesting that the particles are less degraded in the NaPF10 glass than in the NaPF0 one. Similar results were reported in Reference [25].

As performed for the NaSrP glasses, FE-SEM/EDS analysis was used to check the morphology and the composition of the particles. In all the glasses, the composition of the glass matrices was found to be in agreement with the theoretical one within the accuracy of the measurement (± 1.5 mol %). No crystals around the particles were observed in the glasses, confirming that the crystals formation depends on the glass matrix host (see Figure 8).

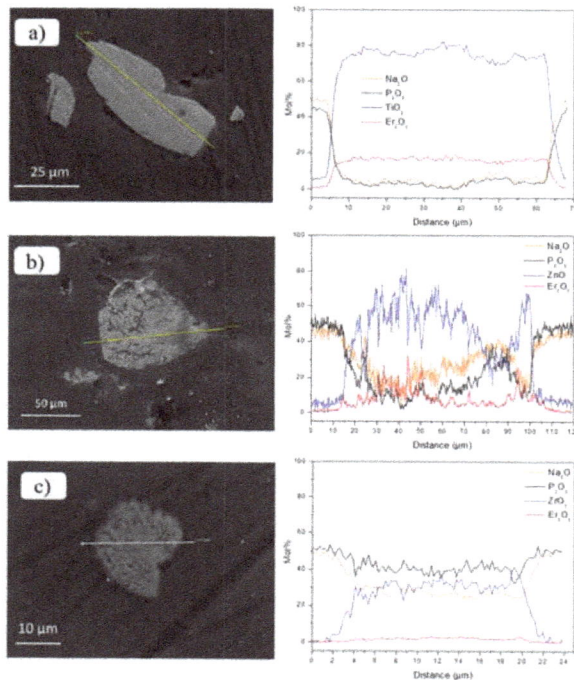

Figure 8. FE-SEM pictures with their corresponding elemental mapping of Er_2O_3-doped TiO_2 (**a**), ZnO (**b**), and ZrO_2 (**c**) particles-containing NaPF10 glasses. The direction of the scan starts at the circle of the white line (corresponding to 0 µm).

Very few TiO_2 particles were found in the NaPF0 glass, while a large amount of them could be observed in the NaPF10 glass. These particles, as well as the ZnO and ZrO_2 ones, maintained their compositional integrity in their center in both glasses (see Figure 8), thus confirming the survival of the Er_2O_3-doped particles in the glasses as suspected from their spectroscopic properties. However, ~5 mol % of TiO_2 and ZnO were detected in the glass matrix ~5 µm close to the particles, whereas a lower amount of ZrO_2 (<3 mol %) was detected around the particles (~1 µm) in the glass matrix. Therefore, as suspected from the color, the shape of the IR and Visible emissions and the Er^{3+}:$^4I_{13/2}$ lifetimes of the particles-containing glasses, the TiO_2 and ZrO_2 particles are suspected to degrade the most and the least in the NaPF glasses, respectively, as confirmed by FE-SEM/EDS analysis.

4. Conclusions

Particles-containing glasses were fabricated using the direct doping method, which consisted of the incorporation of Er_2O_3-doped TiO_2, ZnO, and ZrO_2 particles into the glasses after their melting. Although the particles were found to be thermally stable up to 1050 °C, when added at 1000 °C into the glass with composition $50P_2O_5$-$40SrO$-$10Na_2O$ (in mol %), very few ones are suspected to survive, as evidenced by the color (pink or purple depending on the composition of the particles), the FE-SEM/EDS analysis, the broad emission band centered at around 1.5 μm, the absence of the up-conversion and the long Er^{3+}:$^4I_{13/2}$ lifetime of the particles-containing glasses. By lowering the melting temperature and so the doping temperature to 550 °C, it was possible to limit the decomposition of the particles into the glass melt as evidenced by the color of the glasses (light pink or even colorless depending on the particles composition). Particles-containing glasses within the $90NaPO_3$-$(10 - x)Na_2O$-$xNaF$ system exhibiting up-conversion were successfully synthesized. From the narrow emission band centered at around 1.5 μm and the Er^{3+}:$^4I_{13/2}$ lifetime values of the particles-containing glasses, the particles, especially the ZrO_2 ones, are expected to survive in the NaPF10 glass, thus confirming that the direct doping technique can be profitably employed to fabricate novel varieties of glasses with controlled local environment of the Er^{3+} ions when located in crystals.

Author Contributions: L.P. and D.M. conceived and designed this work. N.G.B. carried out the spectroscopy measurements. D.P. supported the experimental activity related to the synthesis of the particles. P.L.-I. prepared the particles and performed their morphological and structural characterization. N.O. and U.A. melted the particles-containing glasses. All the authors discussed and analyzed the results, contributed to writing and approved the final manuscript.

Funding: This research was funded by the European Union's Horizon 2020 research and innovation program under the Marie Sklodowska–Curie grant agreement No. 642557.

Acknowledgments: The authors acknowledge the financial support of the Academy of Finland (Academy Projects-308558) and the support from Politecnico di Torino through the Interdepartmental Center PhotoNext. Turkka Salminen is also acknowledged for his help with FE-SEM analysis.

References

1. Miniscalco, W.J. Erbium-doped glasses for fiber amplifiers at 1500 nm. *J. Lightw. Technol.* **1991**, *9*, 234–250. [CrossRef]

2. De Wild, J.; Meijerink, A.; Rath, J.K.; van Sark, W.G.J.H.M.; Schropp, R.E.I. Upconverter solar cells: Materials and applications. *Energy Environ. Sci.* **2011**, *4*, 4835–4848. [CrossRef]

3. Rapaport, A.; Milliez, J.; Bass, M.; Cassanho, A.; Jenssen, H. Review of the properties of up-conversion phosphors for new emissive displays. *J. Disp. Technol.* **2006**, *2*, 68–78. [CrossRef]

4. Wang, M.; Mi, C.-C.; Wang, W.-X.; Liu, C.-H.; Wu, Y.-F.; Xu, Z.-R.; Mao, C.B.; Xu, S.K. Immunolabeling and NIR-excited fluorescent imaging of HeLa cells by using $NaYF_4$:Yb,Er upconversion nanoparticles. *ACS Nano* **2009**, *3*, 1580–1586. [CrossRef] [PubMed]

5. Campbell, J.H.; Suratwala, T.I. Nd-doped phosphate glasses for high-energy/high-peak-power lasers. *J. Non-Cryst. Solids* **2000**, *263–264*, 318–341. [CrossRef]

6. Jiang, S.; Myers, M.J.; Rhonehouse, D.L.; Hamlin, S.J.; Myers, J.D.; Griebner, U.; Koch, R.; Schonnagel, H. Ytterbium-doped phosphate laser glasses. In Proceedings of the 1997 SPIE International Conference on Photonics West, San Jose, CA, USA, 8–14 February 1997.

7. Gapontsev, V.P.; Matitsin, S.M.; Isineev, A.A.; Kravchenko, V.B. Erbium glass lasers and their applications. *Opt. Laser Technol.* **1982**, *14*, 189–196. [CrossRef]

8. Jiang, S.; Myers, M.; Peyghambarian, N. Er^{3+} doped phosphate glasses and lasers. *J. Non-Cryst. Solids* **1998**, *239*, 143–148. [CrossRef]

9. Pugliese, D.; Boetti, N.G.; Lousteau, J.; Ceci-Ginistrelli, E.; Bertone, E.; Geobaldo, F.; Milanese, D. Concentration quenching in an Er-doped phosphate glass for compact optical lasers and amplifiers. *J. Alloys Compd.* **2016**, *657*, 678–683. [CrossRef]

10. Seneschal, K.; Smektala, F.; Bureau, B.; Le Floch, M.; Jiang, S.; Luo, T.; Lucas, J.; Peyghambarian, N. Properties and structure of high erbium doped phosphate glass for short optical fibers amplifiers. *Mater. Res. Bull.* **2005**, *40*, 1433–1442. [CrossRef]

11. Campbell, J.H.; Hayden, J.S.; Marker, A. High-power solid-state lasers: A laser glass perspective. *Int. J. Appl. Glass Sci.* **2011**, *2*, 3–29. [CrossRef]

12. Weber, M.J. Science and technology of laser glass. *J. Non-Cryst. Solids* **1990**, *123*, 208–222. [CrossRef]

13. Boetti, N.G.; Scarpignato, G.C.; Lousteau, J.; Pugliese, D.; Bastard, L.; Broquin, J.-E.; Milanese, D. High concentration Yb-Er co-doped phosphate glass for optical fiber amplification. *J. Opt.* **2015**, *17*, 065705. [CrossRef]

14. Jiang, S.; Luo, T.; Hwang, B.-C.; Smekatala, F.; Seneschal, K.; Lucas, J.; Peyghambarian, N. Er^{3+}-doped phosphate glasses for fiber amplifiers with high gain per unit length. *J. Non-Cryst. Solids* **2000**, *263–264*, 364–368. [CrossRef]

15. Knowles, J.C. Phosphate based glasses for biomedical applications. *J. Mater. Chem.* **2003**, *13*, 2395–2401. [CrossRef]

16. Hofmann, P.; Voigtlander, C.; Nolte, S.; Peyghambarian, N.; Schulzgen, A. 550-mW output power from a narrow linewidth all-phosphate fiber laser. *J. Lightw. Technol.* **2013**, *31*, 756–760. [CrossRef]

17. Hwang, B.-C.; Jiang, S.; Luo, T.; Seneschal, K.; Sorbello, G.; Morrell, M.; Smektala, F.; Honkanen, S.; Lucas, J.; Peyghambarian, N. Performance of high-concentration Er^{3+}-doped phosphate fiber amplifiers. *IEEE Photonics Technol. Lett.* **2001**, *13*, 197–199. [CrossRef]

18. Adam, J.-L.; Matecki, M.; L'Helgoualch, H.; Jacquier, B. Non-radiative emissions in Er^{3+}-doped chloro-fluoride glasses. *Eur. J. Solid State Inorg. Chem.* **1994**, *31*, 337–349.

19. Tikhomirov, V.K.; Naftaly, M.; Jha, A. Effects of the site symmetry and host polarizability on the hypersensitive transition $^3P_0 \rightarrow {}^3F_2$ of Pr^{3+} in fluoride glasses. *J. Appl. Phys.* **1999**, *86*, 351–354. [CrossRef]

20. Lopez-Iscoa, P.; Petit, L.; Massera, J.; Janner, D.; Boetti, N.G.; Pugliese, D.; Fiorilli, S.; Novara, C.; Giorgis, F.; Milanese, D. Effect of the addition of Al$_2$O$_3$, TiO$_2$ and ZnO on the thermal, structural and luminescence properties of Er^{3+}-doped phosphate glasses. *J. Non-Cryst. Solids* **2017**, *460*, 161–168. [CrossRef]

21. Lopez-Iscoa, P.; Salminen, T.; Hakkarainen, T.; Petit, L.; Janner, D.; Boetti, N.G.; Lastusaari, M.; Pugliese, D.; Paturi, P.; Milanese, D. Effect of partial crystallization on the structural and luminescence properties of Er^{3+}-doped phosphate glasses. *Materials* **2017**, *10*, 473. [CrossRef] [PubMed]

22. Dantelle, G.; Mortier, M.; Vivien, D.; Patriarche, G. Influence of Ce^{3+} doping on the structure and luminescence of Er^{3+}-doped transparent glass-ceramics. *Opt. Mater.* **2006**, *28*, 638–642. [CrossRef]

23. Zhao, J.; Zheng, X.; Schartner, E.P.; Ionescu, P.; Zhang, R.; Nguyen, T.-L.; Jin, D.; Ebendorff-Heidepriem, H. Upconversion nanocrystal-doped glass: A new paradigm for photonic materials. *Adv. Opt. Mater.* **2016**, *4*, 1507–1517. [CrossRef]

24. Boivin, D.; Föhn, T.; Burov, E.; Pastouret, A.; Gonnet, C.; Cavani, O.; Collet, C.; Lempereur, S. Quenching investigation on new erbium doped fibers using MCVD nanoparticle doping process. In Proceedings of the SPIE 7580, Fiber Lasers VII: Technology, Systems, and Application (SPIE LASE 2010), San Francisco, CA, USA, 23–28 January 2010.

25. Ojha, N.; Tuomisto, M.; Lastusaari, M.; Petit, L. Upconversion from fluorophosphate glasses prepared with NaYF$_4$:Er^{3+},Yb^{3+} nanocrystals. *RSC Adv.* **2018**, *8*, 19226–19236. [CrossRef]

26. Nguyen, H.; Tuomisto, M.; Oksa, J.; Salminen, T.; Lastusaari, M.; Petit, L. Upconversion in low rare-earth concentrated phosphate glasses using direct NaYF$_4$:Er^{3+}, Yb^{3+} nanoparticles doping. *Scr. Mater.* **2017**, *139*, 130–133. [CrossRef]

27. Johannsen, S.R.; Roesgaard, S.; Julsgaard, B.; Ferreira, R.A.S.; Chevallier, J.; Balling, P.; Ram, S.K.; Nylandsted Larsen, A. Influence of TiO$_2$ host crystallinity on Er^{3+} light emission. *Opt. Mater. Express* **2016**, *6*, 1664–1678. [CrossRef]

28. Bahtat, A.; Bouazaoui, M.; Bahtat, M.; Mugnier, J. Fluorescence of Er^{3+} ions in TiO$_2$ planar waveguides prepared by a sol-gel process. *Opt. Commun.* **1994**, *111*, 55–60. [CrossRef]

29. John, R.; Rajakumari, R. Synthesis and characterization of rare earth ion doped nano ZnO. *Nano-Micro Lett.* **2012**, *4*, 65–72. [CrossRef]

30. Choi, S.R.; Bansal, N.P. Mechanical behavior of zirconia/alumina composites. *Ceram. Int.* **2005**, *31*, 39–46. [CrossRef]

31. Zhang, Y.; Chen, J.; Hu, L.; Liu, W. Pressureless-sintering behavior of nanocrystalline ZrO_2–Y_2O_3–Al_2O_3 system. *Mater. Lett.* **2006**, *60*, 2302–2305. [CrossRef]

32. Maeda, N.; Wada, N.; Onoda, H.; Maegawa, A.; Kojima, K. Spectroscopic properties of Er^{3+} in sol–gel derived ZrO_2 films. *Thin Solid Films* **2003**, *445*, 382–386. [CrossRef]

33. Lopez-Iscoa, P.; Pugliese, D.; Boetti, N.G.; Janner, D.; Baldi, G.; Petit, L.; Milanese, D. Design, synthesis, and structure-property relationships of Er^{3+}-doped TiO_2 luminescent particles synthesized by sol-gel. *Nanomaterials* **2018**, *8*, 20. [CrossRef] [PubMed]

34. Jayachandraiah, C.; Krishnaiah, G. Erbium induced raman studies and dielectric properties of Er-doped ZnO nanoparticles. *Adv. Mater. Lett.* **2015**, *6*, 743–748. [CrossRef]

35. Freris, I.; Riello, P.; Enrichi, F.; Cristofori, D.; Benedetti, A. Synthesis and optical properties of sub-micron sized rare earth-doped zirconia particles. *Opt. Mater.* **2011**, *33*, 1745–1752. [CrossRef]

36. Ojha, N.; Nguyen, H.; Laihinen, T.; Salminen, T.; Lastusaari, M.; Petit, L. Decomposition of persistent luminescent microparticles in corrosive phosphate glass melt. *Corros. Sci.* **2018**, *135*, 207–214. [CrossRef]

37. Li, J.-G.; Wang, X.-H.; Kamiyama, H.; Ishigaki, T.; Sekiguchi, T. RF plasma processing of Er-doped TiO_2 luminescent nanoparticles. *Thin Solid Films* **2006**, *506–507*, 292–296. [CrossRef]

38. Patra, A.; Friend, C.S.; Kapoor, R.; Prasad, P.N. Upconversion in Er^{3+}:ZrO_2 nanocrystals. *J. Phys. Chem. B* **2002**, *106*, 1909–1912. [CrossRef]

39. Maschio, S.; Bruckner, S.; Pezzotti, G. Synthesis and sintering of zirconia-erbia tetragonal solid solutions. *J. Ceram. Soc. Jpn.* **1999**, *107*, 1111–1114. [CrossRef]

40. Ojha, N.; Laihinen, T.; Salminen, T.; Lastusaari, M.; Petit, L. Influence of the phosphate glass melt on the corrosion of functional particles occurring during the preparation of glass-ceramics. *Ceram. Int.* **2018**, *44*, 11807–11811. [CrossRef]

Article

Shear Performance at Room and High Temperatures of Glass–Ceramic Sealants for Solid Oxide Electrolysis Cell Technology

Hassan Javed [1,*], **Antonio Gianfranco Sabato [1]**, **Ivo Dlouhy [2]**, **Martina Halasova [2]**,
Enrico Bernardo [3], **Milena Salvo [1]**, **Kai Herbrig [4]**, **Christian Walter [4]** and **Federico Smeacetto [5]**

[1] Department of Applied Science and Technology, Politecnico di Torino, Corso Duca Degli Abruzzi 24, 10129 Turin, Italy; antonio.sabato@polito.it (A.G.S.); milena.salvo@polito.it (M.S.)

[2] Institute of Physics of Materials, Zizkova 22, 61662 Brno, Czech Republic; idlouhy@ipm.cz (I.D.); halasova@ipm.cz (M.H.)

[3] Department of Industrial Engineering, Università degli Studi di Padova, via F. Marzolo 9, 35131 Padova, Italy; enrico.bernardo@unipd.it

[4] Sunfire GmbH, Gasanstaltstraße 2, 01237 Dresden, Germany; Kai.Herbrig@sunfire.de (K.H.); Christian.Walter@sunfire.de (C.W.)

[5] Department of Energy, Politecnico di Torino, Corso Duca Degli Abruzzi 24, 10129 Turin, Italy; federico.smeacetto@polito.it

* Correspondence: hassan.javed@polito.it

Received: 21 December 2018; Accepted: 15 January 2019; Published: 18 January 2019

Abstract: To provide a reliable integration of components within a solid oxide electrolysis cell stack, it is fundamental to evaluate the mechanical properties of the glass–ceramic sealing materials, as well as the stability of the metal–glass–ceramic interface. In this work, the mechanical behavior of two previously developed glass–ceramic sealants joined to Crofer22APU steel is investigated at room temperature, 650 °C, and 850 °C under shear load. The mechanical properties of both the glass–ceramics showed temperature dependence. The shear strength of Crofer22APU/glass–ceramic/Crofer22APU joints ranged from 14.1 MPa (20 °C) to 1.8 MPa (850 °C). The elastic modulus of both glass–ceramics also reduced with temperature. The volume fraction of the crystalline phases in the glass–ceramics was the key factor for controlling the mechanical properties and fracture, especially above the glass-transition temperature.

Keywords: glass–ceramic; shear strength; elastic modulus; SOFC; SOEC

---

1. Introduction

Solid oxide electrolysis cells (SOECs) are a promising technology to produce hydrogen through the electrolysis of water. Among the various components of an SOEC stack, the stability and performance of the glass–ceramic sealants are key factors to determine and control the overall efficiency of the system [1,2]. Because of the operational thermal regime of the whole stack, the joined components are subjected to the change of the acting stress, from almost compressive and/or tensile to almost shear. In laminate structures, this means that not only the SOEC materials, but also their interfaces often play a crucial role [3,4].

Due to cyclic temperature working conditions, there is a possibility of stress generation at the Crofer22APU/glass–ceramic interface and/or within the glass–ceramic joint. If the stresses increase to a critical level, either debonding at the Crofer22APU/glass–ceramic sealant interface or within the glass–ceramic can occur, thus causing gas leakage. Therefore, besides the thermal, chemical, thermomechanical and electrical stability of glass sealants, it is also important to analyze their response to mechanical loading under conditions corresponding to operational ones [5–10]. The high-temperature mechanical behavior of glass–ceramic sealants is crucial due to possible deterioration of these properties, especially if the working temperature is higher than the glass-transition temperature (T_g) [11–13]. Recently, a few researchers have studied the mechanical properties of glass–ceramic sealants at room and high temperatures [14–21]. Selçuk et al. [18] employed three different testing methods to investigate the shear strength of glass–ceramic sealants, without specifying the glass-based composition: single lap offset (SLO) under compression, single lap (SL) under compression and asymmetrical 4-point bending test (A4PB). They found a significant variation in the apparent shear strength obtained by the different methods; specifically, the shear strength measured in the SLO configuration was relatively low (around 7 MPa), likely due to significant normal tensile stresses perpendicular to the joint. Stephens et al. [22] and Lin et al. [13] investigated the tensile and shear properties between the Crofer22APU interconnect and glass–ceramic sealants from room temperature to 800 °C. Stephens et al. tested a barium–calcium–aluminosilicate-based glass-sealing material (G18); tensile and torsion tests were performed to characterize the interfacial shear strength between the G18 glass and the Crofer22APU. The mechanical strength of the joint decreased by almost 50% with an increase in temperature from 25 °C to 800 °C [22]. Lin et al. evaluated the joint strength of a $BaO–B_2O_3–Al_2O_3–SiO_2$ glass–ceramic joined to the Crofer22H specimen, at room temperature and at 800 °C under shear and tensile loading. They evaluated the effect of ageing and pre-oxidation of the Crofer22H and found that the tensile joint strength is lower if the fracture involves delamination at the interface between the steel substrate and $BaCrO_4$ layer, formed by the reaction between BaO from the glass and Cr from the steel. Anyhow, by increasing the testing temperature, the shear strength reduced from 7 MPa (25 °C) to 4 MPa (800 °C) [13].

In a paper by López et al. [23], the flexural strength of two glass–ceramic (one containing Ba and one containing Sr) bars using a three-point bending setup after different ageing times was measured. Although the glass–ceramics were not interfaced with a metallic interconnect, interesting results were obtained in terms of the comparison between the mechanical properties of the two different glass–ceramic compositions. The authors discussed the mechanical behavior regarding different thermal ageing times of the glass–ceramics and their microstructural evolution. The glass–ceramics containing SrO exhibited higher flexural strength than the glass–ceramics with BaO [23].

In our work, the shear strength of the Crofer22APU/glass–ceramic/Crofer22APU joined samples was studied by using two SrO-containing glass-based systems, designed for a working temperature of 850 °C. The Crofer22APU/glass–ceramic/Crofer22APU samples were investigated under shear load at room temperature, 650 °C and 850 °C. The elastic modulus of the glass–ceramics was also measured from room temperature to 650 °C by vibration method.

2. Materials and Methods

Two previously developed glass–ceramic systems, further labeled as HJ3 and HJ4, are employed to investigate their mechanical behavior in contact with the Crofer22APU interconnect. The T_g of the HJ3 and HJ4 glasses was 722 °C and 736 °C, respectively, as measured by differential thermal analysis (DTA). Previously performed XRD analysis showed the presence of $Sr_2Al_2SiO_7$, $Ca_{0.75}Sr_{0.2}Mg_{1.05}(Si_2O_6)$ and $Ca_2Mg(Si_2O_7)$ phases in the HJ3 glass–ceramic, while the HJ4 glass–ceramic had $SrSiO_3$ and SiO_2 phases after joining [24].

The coefficients of thermal expansion (CTE) and the softening behavior of both the glass–ceramics were measured by dilatometer (Netzsch, DIL 402 PC/4, Selb, Germany), at a heating rate of 5 °C/min. The dilatometer analyses were performed on the glass–ceramic pellets (diameter 1 cm), prepared by pressing the glass powder in a steel mold, followed by a heat treatment in static air. Quantitative XRD analyses based on the Rietveld method were not feasible for the as-joined HJ3 and HJ4 glass–ceramics due to the complex crystalline phases and the corresponding XRD patterns. Consequently, in order to determine the relative quantities of the crystalline phases in the HJ3 and HJ4 as-joined glass–ceramics, an estimation could be made on the relative weight balance between the crystals in the glass–ceramics and the internal standard (ZnO), introduced in a defined quantity (20 wt.%).

Therefore, the semi-quantitative analysis was performed by means of a Match software package (version 1.10, Crystal impact, Bonn, Germany), operating based on the reference intensity ratio method (RIR method) [25].

For mechanical characterization under quasistatic shear loading conditions, the Crofer22APU/glass–ceramic/Crofer22APU joined samples were prepared. Figure 1 illustrates the sample configuration (including dimensions) subsequently tested under shear load. Before joining, both plates of Crofer22APU were made plane parallel and polished to obtain the desired dimensions with a tolerance of $\pm$ 0.1 mm. Each plate was cleaned by using acetone and subsequently the glass was deposited in the form of a slurry containing glass particles in ethanol (70:30 wt.%). The joining of the HJ3 glass–ceramic with the Crofer22APU was carried out at 950 °C for 1 h at a heating rate of 5 °C/min, while for the HJ4 system the joining was processed at 950 °C for 5 h at a heating rate of 2 °C/min. The glass–ceramics in the joining region had a thickness of 600 µm $\pm$ 50. After joining, the samples were again gently polished for a few minutes to make sure that both steel plates were perfectly parallel to each other.

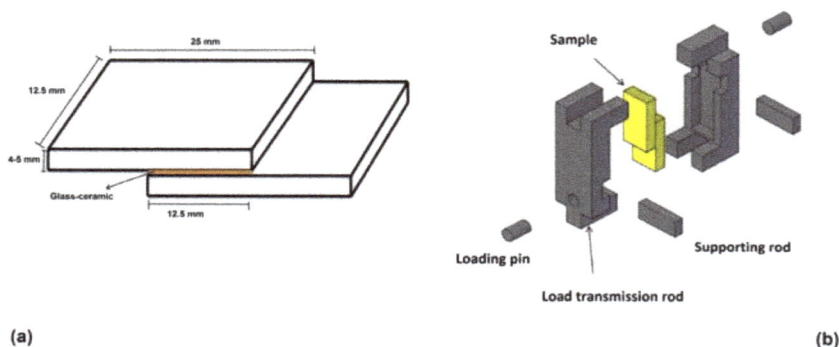

Figure 1. Illustration of Crofer22APU/glass–ceramic/Crofer22APU samples with glass–ceramic joint for shear testing (**a**) and setup for testing the sample under shear load (**b**).

Quasistatic shear testing was carried out at a constant machine cross-head rate of 50 µm/min. The loading fixture developed for the experiments is also shown in Figure 1. The red arrows in Figure 1 indicate the direction of the applied load. Tests were conducted at three different testing temperatures, namely room temperature, 650 °C and 850 °C. The displacement of the joined plates was quantified by using a high-temperature extensometer located outside the furnace. The Zwick/Roell–Messphysik Kappa 50kN test system with a Maytec inert gas high-temperature chamber was used for the experiments. The joint area of each sample was measured after the shear test by using a light microscope with CCD camera and image analysis. The shear stress was then calculated by dividing the applied load by the real joint area. All tests were conducted in an argon atmosphere. Before each test, the sample was heated to the desired temperature and kept at that temperature for 3 h, to make the temperature homogenous throughout the heating zone of the chamber. The temperature was measured by a thermocouple attached directly to the sample. To obtain statistically representative data, at least three samples of both compositions were tested at each temperature. The post mortem analysis of broken samples was carried out by scanning electron microscope (SEM, Merlin ZEISS, Munich, Germany). For this purpose, cross sections of the Crofer22APU/glass–ceramic interfaces were metallographically polished up to 1 µm by diamond paste and investigated by SEM after being coated with gold.

The elastic modulus of the pure glass–ceramics was measured by vibration method, at temperatures ranging from room temperature to 650 °C, during the heating and cooling cycles. For this purpose, thin rectangular samples of glass–ceramics with dimensions of 20 mm $\times$ 2 mm $\times$ 2 mm were prepared. The high-temperature impulse excitation technique, HT1600 system (IMCE, Belgium) was applied for these analyses. The elastic modulus was determined by measuring the resonant frequency of the sample at the given temperature and then calculated from the specimen dimensions and density.

3. Results and Discussion

The previously performed DTA on these glass systems showed a significant difference in crystallization behaviors. No crystallization peak was detected during the DTA analysis of the HJ4 glass, thus indicating that the crystallization was probably not sufficient to be detected. However, the DTA analysis of the HJ3 glass showed a crystallization peak due to the sufficient crystallization in that system [25].

The dilatometer curves of the HJ3 and HJ4 glass–ceramics are shown in Figure 2. The CTE of the as-joined HJ3 and HJ4 glass–ceramics were $10.2 \times 10^{-6} \text{ K}^{-1}$ and $9.3 \times 10^{-6} \text{ K}^{-1}$ respectively, in the temperature range of 200 °C–500 °C. The dilatometer curve of the HJ3 as-joined glass–ceramic is quite linear, thus indicating that the HJ3 glass–ceramic did not become soft up to 1000 °C. Nevertheless, the dilatometer curve of the HJ3 glass–ceramic shows a slight change in the slope around 680 °C, which is most likely due to the T_g of the residual glassy phase. On the other hand, the dilatometer analysis of the HJ4 glass–ceramic shows that the HJ4 glass–ceramic became soft at around 800 °C with a T_g around 700 °C. These results further support the hypothesis that the as-joined HJ4 glass–ceramic has more residual glass than the as-joined HJ3 glass–ceramic.

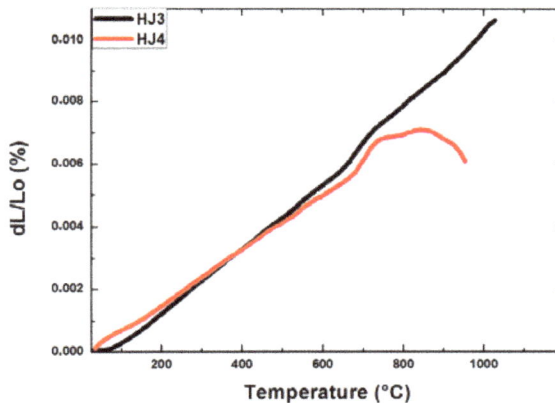

Figure 2. Dilatometer curves of as-joined HJ3 and HJ4 glass–ceramics. Measurements were carried out at a heating rate of 5 °C/min.

Figure 3a shows the XRD analyses performed on the as-joined HJ3 and HJ4 glass–ceramics. Due to the complex crystalline phases (especially in HJ3), some peaks are unidentified, as shown in Figure 3a,b shows the XRD patterns of both glass–ceramics after the inclusion of the standard (ZnO). Closer inspection of the XRD reported in Figure 3b shows the presence of a slightly more pronounced "amorphous halo" in the XRD pattern of the HJ4 glass–ceramic with respect to HJ3 system, thus undoubtedly indicating a relatively lower degree of crystallization in the HJ4 system. Even if the HJ4 system was processed for longer time in comparison with HJ3 (5 h vs. 1 h), it maintained a significant amount of amorphous phase, as further supported by the dilatometer curves of the as-joined HJ3 and HJ4 glass–ceramics shown in Figure 2, thus highlighting a significant difference between the two compositions. These results agree with the results obtained from the DTA, HSM and dilatometer analyses of these glass–ceramics i.e., the HJ4 glass–ceramic has a relatively higher quantity of residual glass than the HJ3 as-joined glass–ceramic.

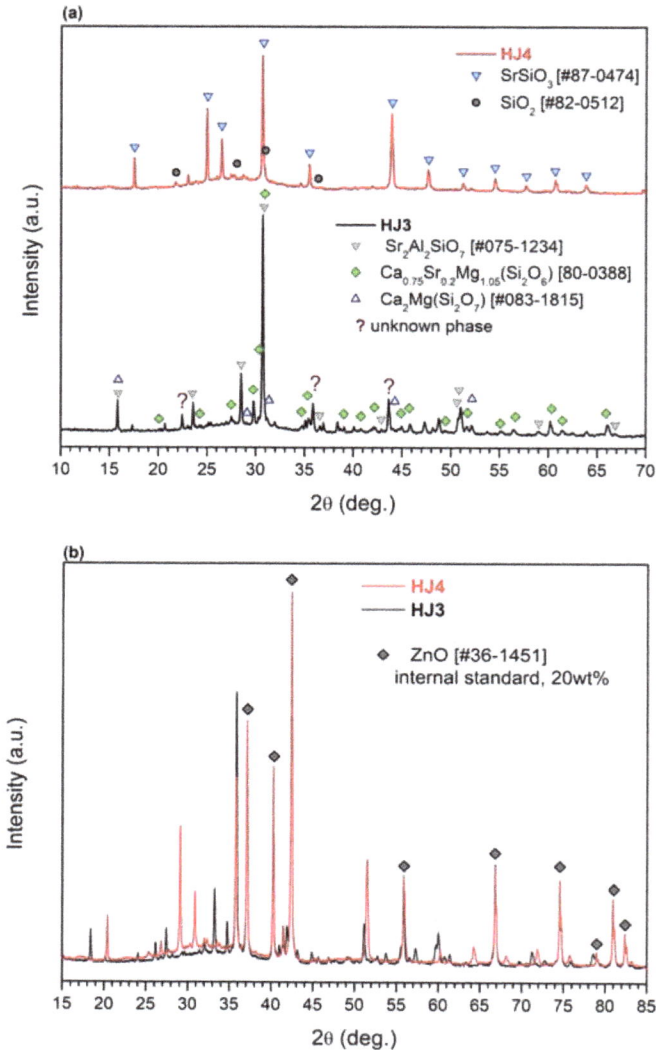

Figure 3. (**a**) Indexed XRD patterns of the HJ3 and HJ4 as-joined glass–ceramics (**b**) XRD pattern of as-joined glass–ceramics and ZnO standard.

Figure 4 shows the shear stress vs. load displacement curves for the HJ3 and HJ4 joint samples, tested at three different temperatures. For the sake of simplicity, only one curve has been shown for each temperature measurement. The traces reflect how the high temperatures significantly affect the deformation and fracture behavior of the joints made by the two different glass–ceramics. Please note that the slope of the linear part of the curves may be affected by an inexact determination of the joint area, as carried out by post mortem analysis of the fracture surfaces. When testing the Crofer22APU/glass–ceramic/Crofer22APU joints made of both glass–ceramics at room temperature and at 650 °C (lower than T_g), the fracture occurred in the linear mode once the applied stress reached maximum shear strength. This behavior reflects the almost elastic deformation and brittle response of the glass–ceramic joint. However, as expected, at 850 °C (T > T_g), the joints showed enhanced displacement under the applied load and extensive non-linear behavior. This effect is due to the stress relaxation and the softening and viscous flow of the residual glassy phase above T_g, as observed by Zhao et al. [26]. Although both systems were crystallized after the heat treatments during the joining process, the residual glassy phase seemed to be the main factor controlling the mechanical behavior of the joint. As both glass systems were designed for a working temperature of 850 °C, the stress relaxation phenomenon is favorable for reducing the thermal stresses that could be generated due to thermal mismatch at high temperature changes associated with the cell operation. Moreover, the residual glassy phase exhibiting viscous flow could also be beneficial for self-healing and consequently may enhance the long-term stability of the sealant. Chang et al. [12] observed the stress relaxation phenomenon for the as-joined and aged GC-9 glass–ceramic when tested in the temperature range of 650 °C–750 °C, and attributed it to the viscoelastic behavior of the residual glassy phase.

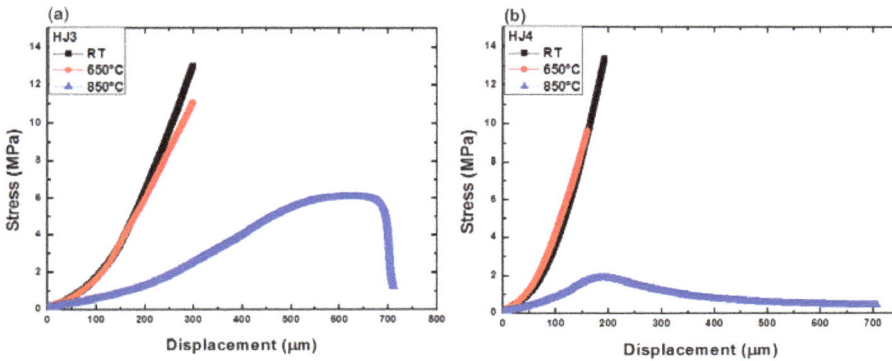

Figure 4. Shear stress vs. displacement curves for (**a**) HJ3 and (**b**) HJ4 joints, tested under shear loads at three different temperatures.

The shear strength values as determined and calculated from the stress displacement curves shown above (Figure 4) are given in Figure 5. Figure 5 shows the average shear strength obtained from three samples for each composition and at each test temperature. The shear strength of all the joints is seen to decrease with increasing testing temperatures. For the HJ3-based joints the strength reduced from 14.1 (at 25 °C) to 5.5 MPa (850 °C), while for the HJ4-based joints, the shear strength dropped from 13.9 MPa (25 °C) to 1.8 MPa (850 °C). The reduction observed in the strength with increasing temperature was according to expectations and is due to the softening of residual glassy phase at high temperature and its viscous flow (creep deformation). Because of the lower volume fraction of the crystalline phases within the HJ4, its shear strength reduced more drastically compared with the HJ3 glass–ceramic when increasing the testing temperature from 650 °C to 850 °C. Osipova et al. [17] also investigated the shear strength of glass–ceramics at room temperature, 600 °C and 800 °C and found similar behavior of enhanced reduction in shear strength at 800 °C, due to the softening of the remaining glassy phase.

It is worth highlighting that the HJ3-based joints fractured cohesively (fracture occurred within the glass–ceramic joint), while the HJ4-based joints fractured in an adhesive manner (fracture occurred at the Crofer22APU/glass–ceramic interface) for all the three testing temperatures.

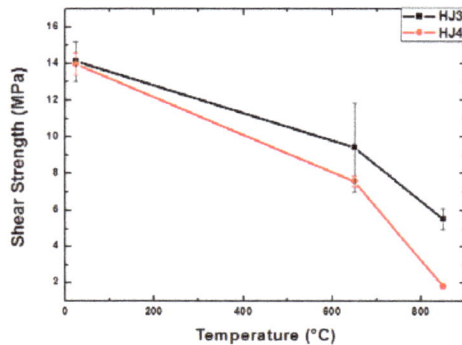

Figure 5. Comparison of shear strength of both glass systems as the function of test temperature.

Figure 6 shows the SEM images of the fractured surfaces of the joints based on both the glass systems. For the HJ4-based system, the adhesive fracture resulted in the detachment of the glass–ceramic from one of the joined Crofer22APU plates, therefore, the shown SEM images correspond to the Crofer22APU plate containing all the HJ4 glass–ceramic. The adhesive fracture occurred at one of the two Crofer22APU/HJ4 glass–ceramic interfaces, which resulted in the complete delamination of the HJ4 glass–ceramic at one interface. The corresponding SEM images (Figure 6a,b) of the Crofer22APU/HJ3/Crofer22APU samples tested at room temperature and 650 °C respectively, show that the glass–ceramics completely adhered throughout the joining area of both the Crofer22APU plates. However, for the HJ3 system-based joint tested at 850 °C, the glass–ceramic was also partially detached from one of the Crofer22APU plates, as shown in Figure 6c. The fracture was probably initiated within the glass–ceramic bulk as in the case of low temperature testing, and then propagated to the interface.

Furthermore, overall thermal expansion coefficient (CTE) of a glass–ceramic, the CTE of individual crystal phases in that glass–ceramic system can play a key role in controlling the fracture behavior as they may create the stress concentration regimes, either within the glass–ceramics and/or at interface.

The HJ3 glass–ceramic has CTE of 10.2×10^{-6} K^{-1} that is closely matching CTE of the Crofer22APU (12×10^{-6} K^{-1}) [27]; however, despite the presence of $Sr_2Al_2SiO_7$ as main crystalline phase with a CTE of 1.1×10^{-6} K^{-1} [28] that could generate localized stresses within the HJ3 glass–ceramic, the mechanical strength was found to be similar to HJ4 system (where $Sr_2Al_2SiO_7$ is not present).

The HJ4 system has $SrSiO_3$ as main crystalline phase having a CTE of 10.9×10^{-6} K^{-1} [29], thus reducing localized stresses generation within the glass–ceramic. Anyway, the as-joined HJ4 glass–ceramic has a CTE of 9.3×10^{-6} K^{-1} which is lower than that of Crofer22APU, therefore, it can lead to generate stresses at Crofer22APU/HJ4 glass–ceramic interface and make adhesive fracture more favorable under the externally applied load.

Furthermore, the presence of defects (pores or bubbles etc.) within the glass–ceramics can also lead to initiation of crack. The high degree of devitrification in the HJ3 glass–ceramic increases the possibility to generate residual micro porosity due to enhanced viscosity cause by crystallization. These micro pores could be another possible reason to initiate the crack within the HJ3 glass–ceramic. Hasanabadi et al. [9] also tested the Crofer22APU/glass–ceramic/Crofer22APU joints under torsion shear conditions and found that the presence of pores as major reason for crack initiation. On the other hand, even if the higher amount of the residual glassy phase in the HJ4 glass–ceramic minimizes the porosity due to its viscous behavior, the room temperature mechanical strength is not improved with respect to the HJ3 system, likely due to higher residual thermal stresses at the Crofer22APU/HJ4 glass–ceramic interface.

The lower mechanical strength for HJ4 with respect to HJ3 at 650 °C could be attributed to the presence of cristobalite (SiO_2) phase in the HJ4 glass–ceramic [24], that has different polymorphs and cause volume expansion around 250 °C, thus determining micro defects in the glass–ceramic at the interface with Crofer22APU.

The SEM images of fractured surfaces show the presence of micro pores in the HJ3 glass–ceramic (Figure 6a–c); however, the HJ4 glass–ceramic seems denser with slight residual porosity (Figure 6d–f).

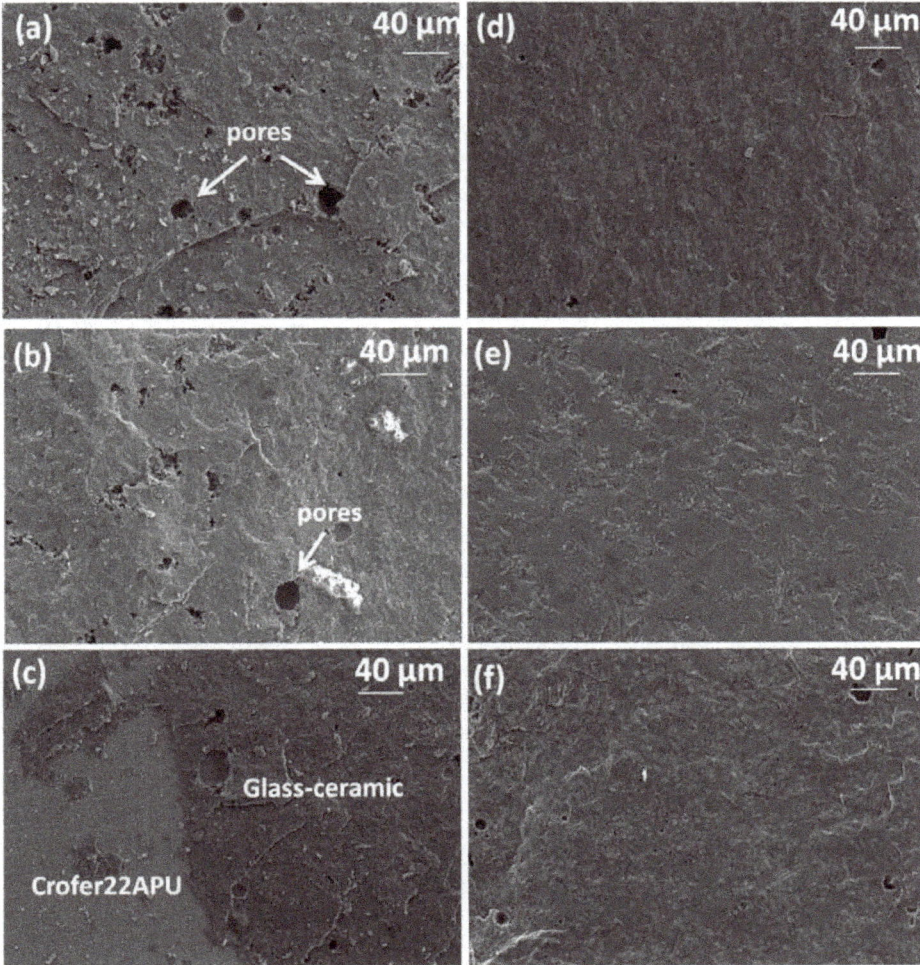

Figure 6. SEM images of top morphology of fracture surfaces of broken joint material; samples (**a**) HJ3 tested at RT, (**b**) HJ3 tested at 650 °C, (**c**) HJ3 tested at 850 °C, (**d**) HJ4 tested at RT, (**e**) HJ4 tested at 650 °C, (**f**) HJ4 tested at 850 °C.

Figure 7 shows the SEM images of the Crofer22APU/glass–ceramic interface for broken samples for both the glass systems tested at three different temperatures under shear load. For the HJ4 system, the images correspond to the Crofer22APU/HJ4 contained all the joined glass–ceramic well adhered to Crofer22APU surface after the fracture. The SEM images showed no cracks at glass–ceramics/Crofer22APU interfaces tested at room temperature as well as at high testing temperatures, thus demonstrating a good thermomechanical compatibility. A uniform microstructure of both the glass–ceramics was observed. For a particular glass system, the Crofer22APU/glass–ceramic interfaces showed a similar morphology after being tested at three different temperatures. Further details about the microstructure of both the glass–ceramics can be found elsewhere [24].

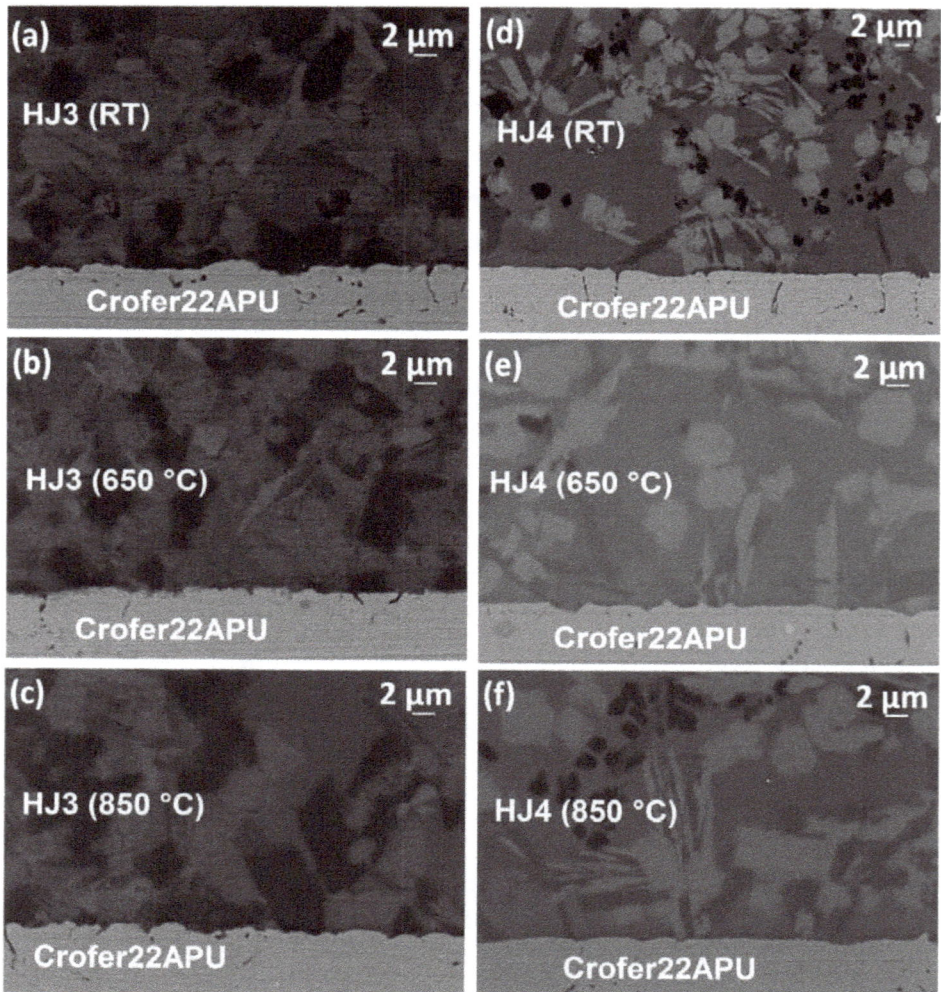

Figure 7. SEM images of interface of Crofer22APU with (**a**) HJ3 tested at RT, (**b**) HJ3 tested at 650 °C, (**c**) HJ3 tested at 850 °C, (**d**) HJ4 tested at RT, (**e**) HJ4 tested at 650 °C, (**f**) HJ4 tested at 850 °C. Details about the microstructure of both the glass–ceramics can be found elsewhere [24].

The elastic moduli of the as-joined glass–ceramics for both the glass systems are shown in Figure 8. Measurements were performed from room temperature to 650°C and elastic modulus was measured both during the heating and cooling. Above 650 °C, the obtained resonance frequencies were not sufficient to measure the reliable values due to the possible softening of the glass–ceramics. The direct comparison shows that the HJ3 glass–ceramic (Figure 8a) has higher values of elastic modulus as compared with the HJ4 system (Figure 8b). The higher volume fraction of crystals in the HJ3 glass–ceramic made this system stiffer than the HJ4 glass–ceramic. Similar behavior was observed by Milhans et al. [30], where reduced modulus was observed to increase with increasing volume content of the crystalline phases in the glass–ceramics, as measured by nanoindentation. Zhao et al. [31] also compared the elastic modulus of different glass–ceramic systems having different amounts of the residual glassy phases. The glass–ceramic with less quantity of residual glassy phase showed higher elastic modulus. For the HJ3 glass–ceramic, the elastic modulus reduced from 100 GPa (at room temperature) to 92 GPa (650 °C), whereas for the HJ4 glass–ceramic it decreased from 80 GPa (room temperature) to 75 GPa (650 °C). The elastic modulus curve for the HJ4 glass–ceramic (Figure 8) also shows discontinuities around 230°C during heating cycle and around 230 °C–270 °C during cooling. This is due to the presence of cristobalite (SiO_2) phase in the HJ4 glass–ceramic [24,32]. Nevertheless, the obtained elastic moduli for both the systems (HJ3 and HJ4) are comparable or slightly higher than the elastic modulus of the glass–ceramics available in the literature (50–80 GPa) [8,22,30,31,33,34].

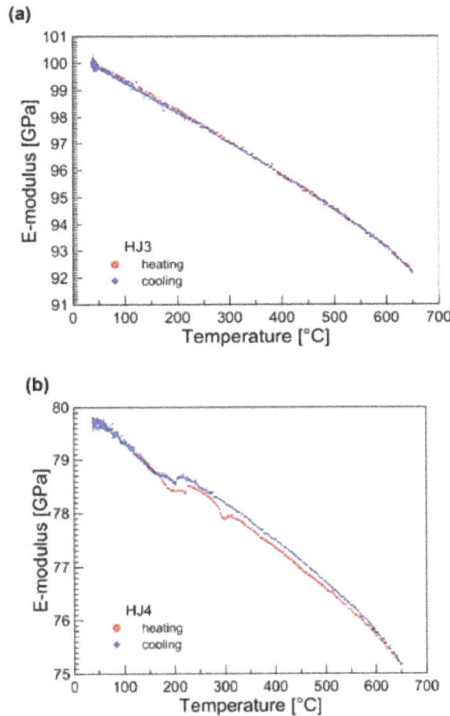

Figure 8. Elastic modulus of as-joined (**a**) HJ3 and (**b**) HJ4 glass–ceramics, measurement was done from room temperature to 650 °C.

4. Conclusions

Mechanical testing in shear conditions performed at room temperature, 650 °C and 850 °C enabled to quantify the temperature dependence of the investigated shear parameters. The mechanical strength of both glass–ceramic-joined samples tested at 650 °C was retained up to 60–70% of the RT values, while a substantial decrease was observed at 850 °C (T > T_g).

Similarly, as expected the elastic modulus of both the glass–ceramics also showed nearly linear reduction with temperature. The volume fraction of the crystalline phases has been found to be an important factor controlling the mechanical properties at high temperatures. The higher amount of residual glassy phase in the glass–ceramic reduces the high-temperature mechanical properties of the sealant but promotes the stress relaxation possibilities due to softening, which is beneficial to release the thermal stresses at high temperature.

This study provides insights for design and development of glass–ceramic sealants for SOEC working at 850 °C.

Author Contributions: H.J. prepared the samples for the mechanical testing and wrote the manuscript. A.G.S. conducted the SEM post mortem analysis. I.D. and M.H. performed the high-temperature mechanical testing. E.B. performed the XRD analysis and quantification of crystalline phases in the glass–ceramics. M.S. and K.H. helped in preparing the manuscript. C.W. and F.S. supervised the whole work and helped in interpretation of data. All the authors discussed the results and gave valuable inputs while preparing the final manuscript.

Funding: The project leading to this research has received funding from "The Fuel Cells and Hydrogen 2 Joint Undertaking under grant agreement No 700300 (GrInHy-Green Industrial Hydrogen via reversible high-temperature electrolysis)" and "European Union's Horizon 2020 research and innovation programme and Hydrogen Europe and N.ERGHY; The European Union's Horizon 2020 research and innovation programme under the Marie Skłodowska-Curie grant agreement No 642557 (CoACH-ETN)".

Conflicts of Interest: The authors declare no conflict of interest.

References

1. Kazempoor, P.; Braun, R.J. Hydrogen and synthetic fuel production using high temperature solid oxide electrolysis cells (soecs). *Int. J. Hydrog. Energy* **2015**, *40*, 3599–3612. [CrossRef]
2. Xiaoyu, Z.; James, E.O.B.; Robert, C.O.B.; Joseph, J.H.; Greg, T.; Gregory, K.H. Improved durability of soec stacks for high temperature electrolysis. *Int. J. Hydrogen Energy* **2013**, *38*, 20–28.
3. Boccaccini, D.N.; Sevecek, O.; Frandsen, H.L.; Dlouhy, I.; Molin, S.; Charlas, B.; Hjelm, J.; Cannio, M.; Hendriksen, P.V. Determination of the bonding strength in solid oxide fuel cells' interfaces by schwickerath crack initiation test. *J. Eur. Ceram. Soc.* **2017**, *37*, 3565–3578. [CrossRef]
4. Boccaccini, D.N.; Sevecek, O.; Frandsen, H.L.; Dlouhy, I.; Molin, S.; Cannio, M.; Hjelm, J.; Hendriksen, P.V. Investigation of the bonding strength and bonding mechanisms of sofcs interconnector–electrode interfaces. *Mater. Lett.* **2016**, *162*, 250–253. [CrossRef]
5. Cela Greven, B.; Gross-Barsnick, S.M.; Federmann, D.; Conradt, R. Strength evaluation of multilayer glass–ceramic sealants. *Fuel Cells* **2013**, *13*, 565–571. [CrossRef]
6. Hamid, A.; Parvin, A.; Dino, B.; Karsten, A. Fracture toughness of glass sealants for solid oxide fuel cell application. *Mater. Lett.* **2014**, *115*, 75–78.
7. Cela Greven, B.; Gross-Barsnick, S.; Koppitz, T.; Conradt, R.; Smeacetto, F.; Ventrella, A.; Ferraris, M. Torsional shear strength of novel glass-ceramic composite sealants for solid oxide fuel cell stacks. *Int. J. Appl. Ceram. Technol.* **2018**, *15*, 286–295. [CrossRef]
8. Bodhayan, D.; Mark, E.W.; Gene, B.A.; Scott, L.S. Mechanical and thermal characterization of a ceramic/glass composite seal for solid oxide fuel cells. *J. Power Sources* **2014**, *245*, 958–966.
9. Hasanabadi, M.F.; Faghihi-Sani, M.A.; Kokabi, A.H.; Malzbender, J. The analysis of torsional shear strength test of sealants for solid oxide fuel cells. *Ceram. Int.* **2017**, *43*, 12546–12550. [CrossRef]
10. Hamada, E.; Hassan, J.; Antonio, G.S.; Federico, S.; Enrico, B. Novel glass-ceramic sofc sealants from glass powders and a reactive silicone binder. *J. Eur. Ceram. Soc.* **2018**, *38*, 4245–4251.
11. Malzbender, J.; Mönch, J.; Steinbrech, R.W.; Koppitz, T.; Gross, S.M.; Remmel, J. Symmetric shear test of glass-ceramic sealants at sofc operation temperature. *J. Mater. Sci.* **2007**, *42*, 6297–6301. [CrossRef]
12. Chang, H.-T.; Lin, C.-K.; Liu, C.-K. Effects of crystallization on the high-temperature mechanical properties of a glass sealant for solid oxide fuel cell. *J. Power Sources* **2010**, *195*, 3159–3165. [CrossRef]
13. Liu, C.-K.; Liu, Y.-A.; Wu, S.-H.; Liu, C.-K.; Lee, R.-Y. Joint strength of a solid oxide fuel cell glass–ceramic sealant with metallic interconnect in a reducing environment. *J. Power Sources* **2015**, *280*, 272–288.
14. Hasanabadi, M.F.; Faghihi-Sani, M.A.; Kokabi, A.H.; Groß-Barsnick, S.M.; Malzbender, J. Room- and high-temperature flexural strength of a stable solid oxide fuel/electrolysis cell sealing material. *Ceram. Int.* **2019**, *45*, 733–739. [CrossRef]
15. Jeffrey, W.F. Sealants for solid oxide fuel cells. *J. Power Sources* **2005**, *147*, 46–57.

16. Hasanabadi, M.F.; Kokabi, A.H.; Faghihi-Sani, M.A.; Groß-Barsnick, S.M.; Malzbender, J. Room- and high-temperature torsional shear strength of solid oxide fuel/electrolysis cell sealing material. *Ceram. Int.* **2018**, *45*, 2219–2225. [CrossRef]

17. Osipova, T.; Wei, J.; Pećanac, G.; Malzbender, J. Room and elevated temperature shear strength of sealants for solid oxide fuel cells. *Ceram. Int.* **2016**, *42*, 12932–12936. [CrossRef]

18. Selçuk, A.; Atkinson, A. Measurement of mechanical strength of glass-to-metal joints. *Fuel Cells* **2015**, *15*, 595–603. [CrossRef]

19. Jacqueline, M.; Mohammed, K.; Xin, S.; Mehran, T.; Marwan, A.-H.; Garmestani, H. Creep properties of solid oxide fuel cell glass–ceramic seal g18. *J. Power Sources* **2010**, *195*, 3631–3635.

20. Selahattin, C. Influential parameters and performance of a glass-ceramic sealant for solid oxide fuel cells. *Ceram. Int.* **2015**, *41*, 2744–2751.

21. Smeacetto, F.; Miranda, A.D.; Ventrella, A.; Salvo, M.; Ferraris, M. Shear strength tests of glass ceramic sealant for solid oxide fuel cells applications. *Adv. Appl. Ceram.* **2010**, *114*, S70–S75. [CrossRef]

22. Stephens, E.V.; Vetrano, J.S.; Koeppel, B.J.; Chou, Y.; Sun, X.; Khaleel, M.A. Experimental characterization of glass–ceramic seal properties and their constitutive implementation in solid oxide fuel cell stack models. *J. Power Sources* **2009**, *193*, 625–631. [CrossRef]

23. Rodríguez-López, S.; Wei, J.; Laurenti, K.C.; Mathias, I.; Justo, V.M.; Serbena, F.C.; Baudín, C.; Malzbender, J.; Pascual, M.J. Mechanical properties of solid oxide fuel cell glass-ceramic sealants in the system bao/sro-mgo-b2o3-sio2. *J. Eur. Ceram. Soc.* **2017**, *37*, 3579–3594. [CrossRef]

24. Javed, H.; Sabato, A.G.; Herbrig, K.; Ferrero, D.; Walter, C.; Salvo, M.; Smeacetto, F. Design and characterization of novel glass-ceramic sealants for solid oxide electrolysis cell (soec) applications. *Int. J. Appl. Ceram. Technol.* **2018**, *15*, 999–1010. [CrossRef]

25. Hillier, S. Accurate quantitative analysis of clay and other minerals in sandstones by xrd: Comparison of a rietveld and a reference intensity ratio (rir) method and the importance of sample preparation. *Clay Miner.* **2000**, *35*, 291–302. [CrossRef]

26. Yilin, Z.; Jürgen, M. Elevated temperature effects on the mechanical properties of solid oxide fuel cell sealing materials. *J. Power Sources* **2013**, *239*, 500–504.

27. Ghosh, S.; Sharma, A.D.; Kundu, P.; Basuz, R.N. Glass-ceramic sealants for planar it-sofc: A bilayered approach for joining electrolyte and metallic interconnect. *J. Electrochem. Soc.* **2008**, *155*, B473–B478. [CrossRef]

28. Manu, K.M.; Ananthakumar, S.; Sebastian, M.T. Electrical and thermal properties of low permittivity sr2al2sio7 ceramic filled hdpe composites. *Ceram. Int.* **2013**, *39*, 4945–4951. [CrossRef]

29. Thieme, C.; Rüssel, C. Thermal expansion behavior of srsio3 and sr2sio4 determined by high-temperature x-ray diffraction and dilatometry. *J. Mater. Sci.* **2015**, *50*, 5533–5539. [CrossRef]

30. Milhans, J.; Ahzi, S.; Garmestani, H.; Khaleel, M.A.; Sun, X.; Koeppel, B.J. Modeling of the effective elastic and thermal properties of glass-ceramic solid oxide fuel cell seal materials. *Mater. Des.* **2009**, *30*, 1667–1673. [CrossRef]

31. Yilin, Z.; Jürgen, M.; Sonja, M.G. The effect of room temperature and high temperature exposure on the elastic modulus, hardness and fracture toughness of glass ceramic sealants for solid oxide fuel cells. *J. Eur. Ceram. Soc.* **2011**, *31*, 541–548.

32. Beals, M.D.; Zerfoss, S. Volume change attending low-to-high inversion of cristobalite. *J. Am. Ceram. Soc.* **1944**, *27*, 285–292. [CrossRef]

33. Liu, W.; Su, X.; Khaleel, M.A. Predicting young's modulus of glass/ceramic sealant for solid oxide fuel cell considering the combined effects of aging, micro-voids and self-healing. *J. Power Sources* **2008**, *185*, 1193–1200. [CrossRef]

34. Milhans, J.; Li, D.S.; Khaleel, M.; Sun, X.; Marwan, S.A.-H.; Adrian, H.; Garmestani, H. Mechanical properties of solid oxide fuel cell glass-ceramic seal at high temperatures. *J. Power Sources* **2011**, *196*, 5599–5603. [CrossRef]

materials

MDPI

Article

Electrolyte-Supported Fuel Cell: Co-Sintering Effects of Layer Deposition on Biaxial Strength

Alessia Masini [1,*], Thomas Strohbach [2], Filip Šiška [1], Zdeněk Chlup [1] and Ivo Dlouhý [1]

[1]   Institute of Physics of Materials, Academy of Science of the Czech Republic, 61662 Brno, Czech Republic; siska@ipm.cz (F.Š.); chlup@ipm.cz (Z.C.); idlouhy@ipm.cz (I.D.)

[2]   Sunfire GmbH, 01237 Dresden, Germany; Thomas.Strohbach@sunfire.de

*   Correspondence: masini@ipm.cz; Tel.: +420 532 290 336

Received: 13 December 2018; Accepted: 15 January 2019; Published: 18 January 2019

Abstract: The mechanical reliability of reversible solid oxide cell (SOC) components is critical for the development of highly efficient, durable, and commercially competitive devices. In particular, the mechanical integrity of the ceramic cell, also known as membrane electrolyte assembly (MEA), is fundamental as its failure would be detrimental to the performance of the whole SOC stack. In the present work, the mechanical robustness of an electrolyte-supported cell was determined via ball-on-3-balls flexural strength measurements. The main focus was to investigate the effect of the manufacturing process (i.e., layer by layer deposition and their co-sintering) on the final strength. To allow this investigation, the electrode layers were screen-printed one by one on the electrolyte support and thus sintered. Strength tests were performed after every layer deposition and the non-symmetrical layout was taken into account during mechanical testing. Obtained experimental data were evaluated with the help of Weibull statistical analysis. A loss of mechanical strength after every layer deposition was usually detected, with the final strength of the cell being significantly smaller than the initial strength of the uncoated electrolyte ($\sigma_0 \approx 800$ MPa and $\sigma_0 \approx 1800$ MPa, respectively). Fractographic analyses helped to reveal the fracture behavior changes when individual layers were deposited. It was found that the reasons behind the weakening effect can be ascribed to the presence and redistribution of residual stresses, changes in the crack initiation site, porosity of layers, and pre-crack formation in the electrode layers.

Keywords: SOC; mechanical strength; flexural biaxial test; ball-on-3-balls test; fractography; residual stresses

---

1. Introduction

Reversible solid oxide cells (SOCs) are devices able to produce synthetic fuels when operated in electrolysis mode (SOEC) and electricity when operated in fuel cell mode (SOFC). Recently, SOCs have been attracting a lot of interest because of their environmentally-friendly nature, their flexibility with respect to the fuel utilization and energy source integration, their capability and their surprisingly high overall efficiency [1–4]. They represent a promising tool towards a sustainable future.

A major threat hindering the successful commercialization of SOCs is their long-term reliability. Because of their high operating temperature (about 850 °C), the harsh oxidizing and reducing working atmospheres, while being subjected to external mechanical loads, their integrity is threatened [5,6].

A SOC device consists of ceramic, metallic, and glass components all stacked together. The mechanical failure of the ceramic cell, also known as MEA (membrane electrolyte assembly), would be detrimental for the proper functioning of the whole device. Being the component in which all the electrochemical reactions take place, its failure would inevitably lead to decreased performance of the entire stack. In order to ensure the efficiency of the SOC device, it is fundamental that the fuel and the oxidizing air are physically separated; their separation is ensured by the gastight electrolyte.

This way, the fuel is not directly burned off. Any kind of leakage would reduce the quantity of effectively utilized fuel and therefore lower the efficiency. If the electrolyte breaks, the necessary gas tightness is no longer maintained and the SOC performance is hindered [7]. This exemplifies the importance of ensuring the mechanical integrity of the cell and in particular of the electrolyte over the whole expected lifetime.

This work deals with the effect of the manufacturing process on the final strength of the ceramic cell. The design of this multi-layered ceramic system is mainly focused on the electrochemical properties necessary to make it an effective means for the production of electricity and synthetic fuels. However, it also requires certain robustness to be able to bear the severe mechanical and thermal stresses to which it is exposed during service [8]. The cell is a ceramic component mainly consisting of a dense electrolyte, which in this study is the supportive layer, embedded between two porous electrodes. During the production process, these functional layers are sintered together [9]. Because of the mismatch in the coefficient of thermal expansion, residual stresses will arise between the layers. Such stresses might be responsible for the formation of cracks in the porous electrodes, as well as for layer delamination, potentially compromising the overall strength of the ceramic cell [8,10,11]. The overall robustness of the cell under investigation, being of the electrolyte-support kind, depends on the properties of the electrolyte in the first place. Yet, its strength is also influenced by the features of the electrodes embedding the electrolyte and by the interfaces generated between them.

Despite most of the research activity for the development of fuel cells being devoted to the electrochemical aspects, there are some studies dealing with the fracture mechanics of materials involved in SOC technology. However, nearly all of them are focused on the characterization of individual materials [12–16] or they investigate either symmetrical systems or half-cell systems [7,10,17–20]. As already mentioned, interfacial bonding between layers can play a significant role in the mechanical response of the whole cell. Thus, it is of high importance to understand the fracture mechanism of such a fundamental SOC component in its totality, treating it as a whole system, thus taking into account co-sintering effects and interaction between layers. This challenging approach is the main novelty of the presented work.

It has already been reported that the strength of the electrolyte on the cell level is sensibly reduced [7,10]. The goal of this work is to understand the reasons behind this strength loss, aiming toward the improvement of the cell mechanical stability.

2. Materials and Methods

2.1. Specimen Preparation

The planar electrolyte-supported SOC cell investigated in this study was provided by Sunfire (Sunfire GmbH, Dresden, Germany) and consisted of four layers; its layout is schematically represented in Figure 1. The detailed composition of each layer, together with the nominal thickness, is reported in Table 1. The planar electrolyte was produced via tape casting [21] and provided by the company Kerafol (KERAFOL Keramische Folien GmbH, Eschenbach in der Oberpfalz, Germany); it consisted of dense 3 mol% Y_2O_3-stabilized ZrO_2 with a nominal thickness of 90 μm and it provided the mechanical support for the electrodes. Both the electrodes were manufactured at Sunfire (Sunfire GmbH, Dresden, Germany) via a screen printing process. The fuel electrode was a 27 μm thick porous $NiO/Gd_{0.1}Ce_{0.9}O_2$ cermet, while the air electrode consisted of two layers: a 10-μm thick $Gd_{0.2}Ce_{0.8}O_2$ barrier layer [22] and a $La_{0.6}Sr_{0.4}Co_{0.2}Fe_{0.8}O_{3-\delta}$ functional layer with a nominal thickness of 45 μm.

Figure 1. Schematic representation of the Sunfire MEA with its functional layers.

Table 1. List of the cell layers with their composition and nominal thickness.

Layer	Material	Composition	Thickness (μm)
Electrolyte	3YSZ	$(Y_2O_3)_{0.03}(ZrO_2)_{0.97}$	90
Barrier layer	20GDC	$Gd_{0.2}Ce_{0.8}O_2$	10
Fuel Electrode	NiO/10GDC	$(NiO)/(Gd_{0.1}Ce_{0.9}O_2)$	27
Air Electrode	LSCF	$La_{0.6}Sr_{0.4}Co_{0.2}Fe_{0.8}O_{3-\delta}$	45

All the layers added to the electrolyte are sintered together. Thus, the cell is a complex system made up of co-sintered individual layers and its overall properties are inevitably affected by the constraints arising between them. To investigate the effect of these constraints on the mechanical response of the cell, samples were taken out from each production step (i.e., after each layer deposition in the green state) prior to sintering. Then laminates having two, three, and four layers were sintered via the same profile used for the whole cell. This approach led to the production of three non-symmetric layered structures, each of them with a different number of layers, and enabled the detection of the interactions between them. The layered structures were named from SOC1, corresponding to the electrolyte with barrier layer, to SOC3, corresponding to the whole cell, as illustrated in Table 2. SOC0 refers to the monolithic bare 3YSZ (3% mol Ytttria Stabilized Zirconia) electrolyte support. Mechanical characterizations were performed after each layer deposition, (i.e., on each layered structure from SOC0 to SOC3); in order to take into account the non-symmetric and non-periodic layer placement, both laminate sides were subject to testing (i.e., eight configurations were evaluated in total).

Table 2. List of the layered structures characterized with their given names, brief description, and nominal thicknesses.

Sample	Name	Description	t (μm)
	SOC0	Electrolyte	90
	SOC1	Electrolyte + Barrier	100
	SOC2	Electrolyte + Barrier + Fuel Electrode	127
	SOC3	Electrolyte + Barrier + Fuel Electrode + Air Electrode	172

For the biaxial flexural testing, samples were extracted directly from SOC0–SOC3 as-sintered plates of dimensions 100×150 mm^2 according to Table 2. The plates, being extremely thin and fragile, were glued onto a rigid support and cut into 4 x 3 mm^2 rectangular specimens using a precision diamond saw Isomet 5000 (Buehler, Lake Bluff, IL, USA). The cutting speed was set to 7.8 mm/min in order to prevent the cracking of edges. The edges of the samples did not need any further polishing as, during the biaxial flexural test used, samples were subjected to tensile stresses concentrated in the central area located in between the loading balls; thus, any micro-cracks in correspondence of edges had no influence on the fracture load.

2.2. Biaxial Flexural Strength Test

The biaxial flexural strength was determined through the ball-on-3-balls bending (B3B) configuration. Details of the testing procedure and setup can be found elsewhere [7,23,24]. The measurements were performed at room temperature in air atmosphere on rectangular specimens, which were symmetrically supported by three balls on one side and loaded by a fourth ball placed in the center of the opposite side (see Figure 2); all the balls were made of hardened steel (E_b = 210 GPa; ν_b = 0.3) and had a diameter R_b = 2.38 mm, giving a support radius R_a = 1.3747 mm. All B3B tests were performed under displacement control in a universal testing machine INSTRON 8862 (Instron, Norwood, MA, USA) with the aid of a jig especially produced at IPM (Institute of Physics of Materials, Brno, Czech republic), following the design from ISFK (Institut of Structural and Functional Ceramics, Leoben, Austria) [11]. During the test, the required alignment between the specimen, loading ball, and supporting balls was ensured by a guide, which was carefully removed after pre-load. The load was then further increased until failure [25]. The crosshead speed of the test was set to 500 μm/min in order to achieve the fracture of the samples in less than 5 s. The test setup utilized is shown in Figure 3. The minimum of 45 valid tests was conducted. Each configuration (SOC0–SOC3) was tested on both sides to take into account two aspects:

1. The electrolyte had a different surface refinement at the top and bottom side due to the manufacturing process; one side was smoother and the other is rougher, depending on whether it was on the support or the doctor blade side. This aspect may have led to a difference in the strength between two sides even when the electrolyte was a dense monolithic ceramic;
2. The layered structures SOC1 to SOC3 had a non-symmetrical non-periodic layout.

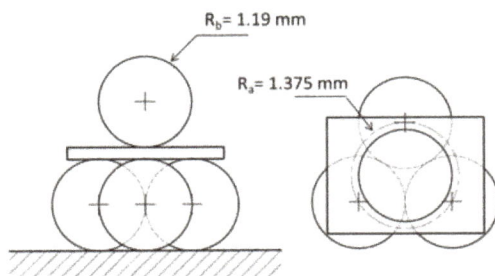

Figure 2. Scheme of the ball-on-3-balls test setup for biaxial testing.

Figure 3. The test setup used for the measurements of the bi-axial flexural strength via B3B.

Experimental data were evaluated according to the Weibull statistical analysis. This was to take into account the inherently scattered nature of the strength of brittle materials, which cannot be described by a single strength value, but by a strength distribution. The characteristic strength σ_0 and the Weibull modulus m of the Weibull distribution, together with their 95%-confidence intervals, were determined using the maximum-likelihood method, following the standard EN 843-5 [26]. Calculations were performed with the help of the statistical software Statgraphics Centurion 18 (Statgraphics Technologies, Inc., The Plains, VA, USA).

After the tests, fractographic analyses were performed for every data set in order to characterize the fracture mechanisms acting and to investigate the effect of the layered layout on the crack propagation. The fracture surfaces of specimens exhibiting the highest and lowest values within the dataset were observed. For the fractographic analyses, fractured specimens were mounted in the specially prepared holder via silver paste and coated with a thin carbon film in order to give them the required conductivity for enabling scanning electron microscopy (SEM) observations. A scanning electron microscope Tescan LYRA 3 XMU (Tescan Brno, s.r.o., Brno, Czech Republic) was used. All the observations were performed at a working distance of 9 mm with an acceleration voltage of 20 kV.

2.3. Determination of the Flexural Strength

The flexural strength (in N/mm^2) was determined from the experimental fracture force measured for each sample, via the equation:

$$\sigma_{max} = f \cdot \frac{F}{t^2}$$ (1)

where F (N) is the maximum load at fracture; t (mm) the thickness of the specimen; and f is a dimensionless factor depending on the geometry of the specimen, its Poisson's ratio, and the geometry of the test jig. Considering that the thickness is one of the most influential parameters for the estimation of the maximum stress, it was carefully measured in the center of all specimens (i.e., area where the maximum stress is located) before testing. To determine the f factor for each tested material configuration loaded using B3B, an FEM (Finite Elements Method) analysis was performed using the commercial software Abaqus/CAE6.13 (Dassault Systemes Simulia Corp., Providence, RI, USA). For the simulation, the rectangular samples and the balls were modelled using 3D deformable elements of the C3D8R type. Given the symmetry of the system, only half of the testing setup was modelled in order to save computational time. The chosen geometry and boundary conditions are illustrated in Figure 4. The mesh in the model was created in order to combine sufficient precision and reasonable computational demands. Therefore, the areas of contact between the balls and the cell were meshed more densely with the in-plane element size from 2 μm to 10 μm. The rest of the cell was meshed with increasing element size (up to 100 μm). The average through thickness element size was 4 μm; however, there were at least two elements through the thickness of the layer. The number of DOF (Degree of Freedom) for the cell ranged between 252 000 (SOC0) and 468 000 (SOC3). Siska et al. [27] showed that for elastic calculations of heterogeneous material, the mesh convergence is achieved at around 100 000 DOF. Therefore, the performed simulations were well conditioned in the sense of mesh convergence.

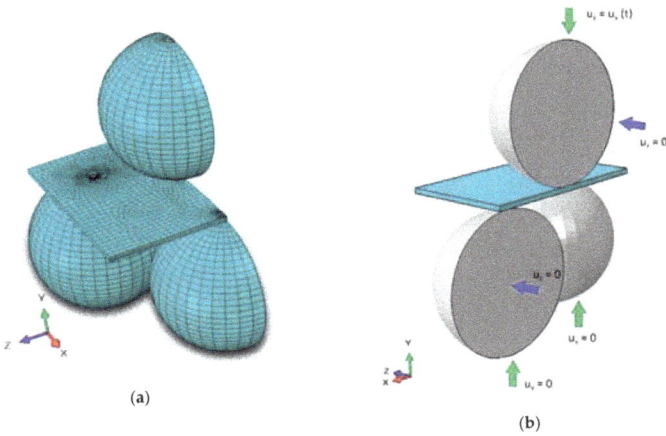

(a)

(b)

Figure 4. Finite Element (FE)-model example of the ball on three balls test assembly, half model: (a) view of the meshed model, and (b) outlined boundary conditions.

Material data used for the simulations are reported in Table 3. Elastic modulus E, Poisson's ratios ν, and densities ϱ were taken from Reference [4], while coefficients of thermal expansion α were measured via dilatometry or taken from literature [28,29].

Table 3. List of the cell layers with their composition and nominal thickness.

Layer	Material	E (GPa)	ν (-)	ϱ (g/cm^3)	α (K^{-1})
Electrolyte	3YSZ	202.5	0.27	6.05	10.8×10^{-6}
Barrier Layer	20GDC	120	0.26	4.02	12.5×10^{-6}
Fuel Electrode	NiO/10GDC	120	0.25	5.97	13.4×10^{-6}
Air Electrode	LSCF	80	0.30	2.36	16.6×10^{-6}

In Figure 5, an example of the first maximum principal stress distribution in the specimen during biaxial loading is represented. It can be observed that the maximum stress arose in the center of the tensile surface of the specimen (the red area), corresponding to the center of the three balls, and its intensity decreased sharply in the radial direction. Therefore, as the area loaded with the maximum tensile stress was a small portion of the volume of the sample, localized strength measurements could be carried out.

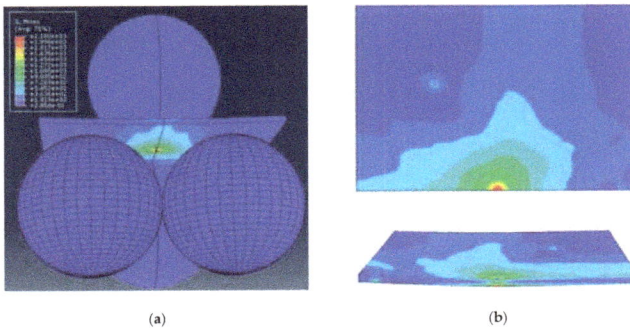

(a)

(b)

Figure 5. Example of the principal stress field in a rectangular plate specimen for a typical loading condition in a ball-on-3-balls test: (a) perspective view, and (b) view on the top plane and section view of the plate.

3. Results

3.1. Flexural Strength Measurement

The failure stresses of the samples in the flexural B3B configuration are illustrated in Figure 6 in the form of Weibull plots and summarized in Table 3, where characteristic strengths and Weibull modules are reported together with their calculated confidence intervals. For the electrolyte (SOC0), the characteristic strength ranged between $\sigma_0 = 1819$ MPa (rough side under tension) and $\sigma_0 = 1854$ MPa (smooth side under tension), with both Weibull moduli close to m ≈ 20, as shown in Figure 6a. The Weibull parameters determined for both SOC0 orientations were very similar, revealing that the surface quality had a statistically negligible influence on the mechanical response of the electrolyte but still was detectable by the method used.

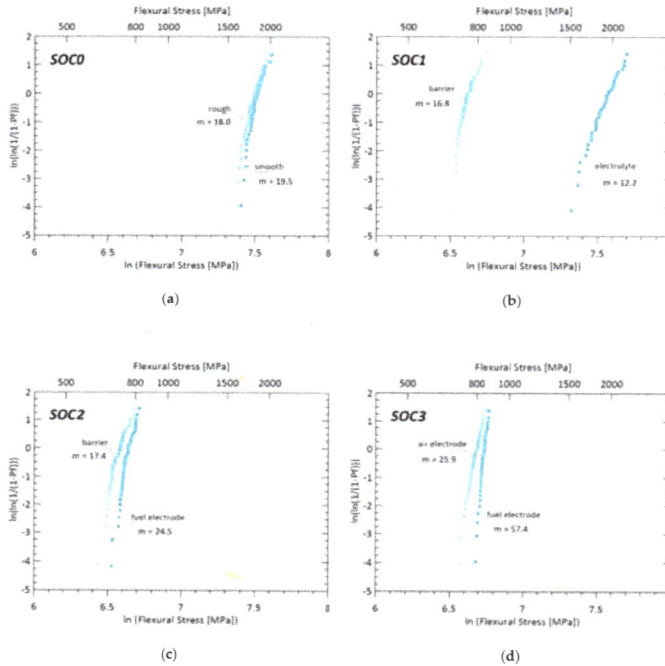

Figure 6. Weibull plots of the fracture stress distribution of the SOC0-SOC3 samples, obtained via ball-on-3-balls test. In each graph are represented two plots, one for each side of the sample under tension, with their respective Weibull modules. (**a**): Weibull plot for SOC0 (**b**): Weibull plot for SOC1 (**c**): Weibull plot for SOC2 (**d**): Weibull plot for SOC3.

For the SOC1 samples, corresponding to the electrolyte with addition of the barrier layer on the rough electrolyte side, the characteristic strength calculated was $\sigma_0 = 1956$ MPa when the electrolyte smooth side was on the tensile side and $\sigma_0 = 763$ MPa when the barrier was in tension. While the strength of the electrolyte side was comparable to the one obtained for the uncoated 3YSZ electrolyte, a pronounced strength decrease was obtained when the barrier layer was exposed to the tensile load. Even if the thickness of the GDC (Gadolinium Doped Ceria) barrier layer (≈10 μm) was small compared to the electrolyte thickness (≈90 μm), its effect on the resulting strength was remarkable.

The addition of the fuel electrode resulted in the approach of the strength distribution plots of the two orientations, while the GDC barrier side nearly maintained its previous strength ($\sigma_0 = 730$ MPa), and the electrolyte side, now coated with the NiO/GDC cermet layer, underwent a drastic weakening (from σ_0 = 1956 MPa to σ_0 = 775 MPa).

Finally, the presence of the LSCF air electrode had a minor influence on the mechanical response of the layered structure. The Weibull strengths determined were σ_0 = 843 MPa and σ_0 = 802 MPa when the fuel electrode and the air electrode were in tension, respectively.

Table 4 reports the values of the measured flexural strength and Weibull modulus of each sample, for both orientations.

Table 4. Compilation of Weibull parameters obtained from the biaxial bending test, including the 95% confidence intervals.

Sample	Tested Surface	σ_0 (MPa)	m
SOC0	Smooth	1854.4 (1818.8 | 1889.7)	19.5 (14.7 | 25.5)
	Rough	1818.9 (1782.2 | 1855.3)	18.0 (13.7 | 23.4)
SOC1	Electrolyte	1955.9 (1901.2 | 2010.7)	12.2 (9.4 | 15.7)
	Barrier	762.6 (747.2 | 777.9)	16.8 (13.0 | 21.5)
SOC2	Fuel Electrode	775.0 (762.4 | 783.3)	24.5 (19.0 | 31.1)
	Barrier	729.5 (715.4 | 743.5)	17.4 (13.5 | 22.1)
SOC3	Fuel Electrode	844.4 (838.9 | 849.7)	57.4 (43.4 | 74.8)
	Air Electrode	801.7 (790.9 | 812.4)	25.9 (19.9 | 33.3)

Comparing the stress levels in the bare electrolyte (SOC0) with the one of the whole cell (SOC3), it is clear that the strength of the electrolyte on the cell level was significantly reduced. The stress in the electrolyte at the failure of the cell was less than half that of the uncoated electrolyte. In order to understand the reasons behind the weakening phenomenon, fractographic analyses using scanning electron microscopy on selected specimens were performed.

3.2. Fractographic Analysis

The selected most significant (characteristic) fracture surfaces are reported below. The fracture surface in Figure 7 belongs to a 3YSZ specimen tested with the rough side in tension. The fracture initiation site is highlighted by the dashed line. From the high-magnification micrograph in Figure 7b, it is possible to state that the crack propagation in the vicinity of the free surface was mainly inter-granular, revealing a weak bonding strength between grains, probably caused by the manufacturing process; it then became rather trans-granular in the interior of the material. The same fracture mechanism was observed for electrolyte specimens tested with the smooth side on the tensile side of loading; this result is in agreement with the similar fracture strengths measured via the flexural test for both orientations. As expected, for all the tested single-layer 3YSZ electrolyte samples, the fracture initiated on the tensile surface of the specimen, within the area of maximum stress determined via FEM (see Figure 5).

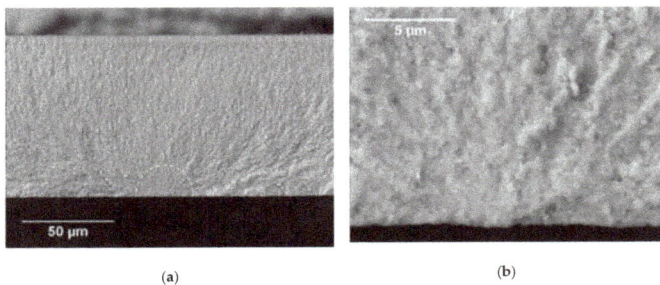

(a) (b)

Figure 7. SEM images of the fracture surface of a 3YSZ specimen tested in B3B with the rough surface in tension. The dashed line in (**a**) highlights the fracture initiation site. (**b**) Magnified image of the initiation site.

Figure 8 shows the fracture surface of two SOC1 specimens, consisting of the electrolyte with the addition of the GDC barrier layer on the rough surface.

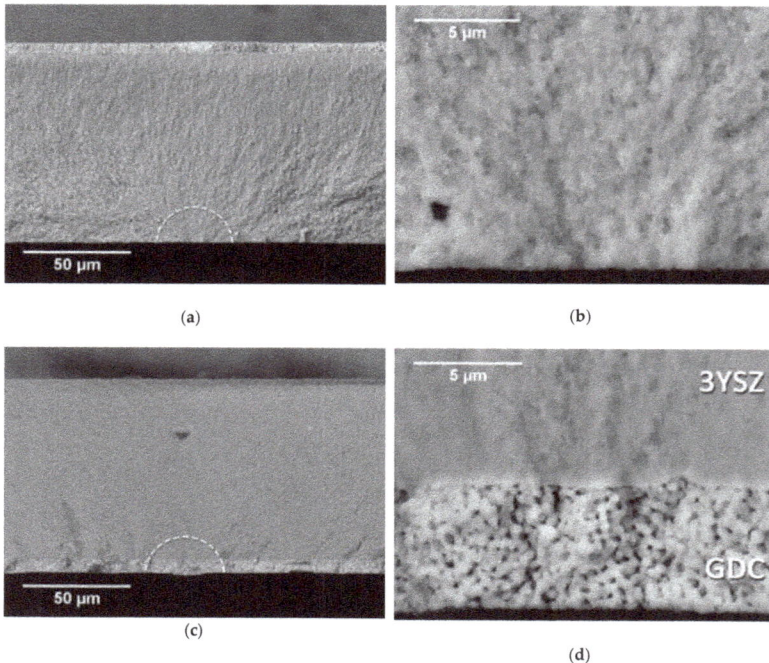

(a)   (b)

(c)

(d)

Figure 8. SEM images of the fracture surface of a SOC1 specimen tested in B3B with the electrolyte side in tension in (**a**) and (**b**), and with the GDC barrier layer in tension in (**c**) and (**d**). The fracture initiation areas are marked by the white dashed lines.

The micrographs in Figure 8a,b illustrate the fracture mechanism of this bi-layered specimen when tested with the electrolyte in tension. The fracture pattern was the same as the one observed for the single-layer electrolyte, meaning that the presence of the barrier layer on the top surface had no influence in the cracking mode. Hence, the similar characteristic strength measured during the flexural test. Figure 8c,d report the fracture surfaces belonging to SOC1 specimens, this time being tested with the GDC layer on the tensile side of the loading. The fracture seemed to initiate in correspondence with the outer surface of the barrier layer. The magnified image in Figure 8d clearly shows how the crack propagates with continuity through both the electrolyte and the barrier layer, revealing a high bonding strength between the two layers.

Figure 9 illustrates the fracture surface of a SOC2 specimen, tested with the GDC barrier layer on the tensile side of loading. Observing Figure 9a, it is possible to notice partial delamination of the fuel electrode in the compressive area (i.e., upper part of the specimen): a continuous crack runs along the interface between the YSZ electrolyte and the NiO/GDC fuel electrode. Figure 9b shows the fracture starting site located at the outer surface of the specimen and propagating from the GDC layer into the electrolyte; this behavior reveals a good bonding between the two layers like that already observed for the SOC1 samples. It should be noted that the presence of the fuel electrode on the compressive side of loading had no visible influence on the fracture behavior; indeed, the fracture surface looks very similar to the one shown in Figure 8. This is in agreement with the experimental results, which revealed that SOC1 and SOC2 samples had, to a good approximation, the same flexural strength when tested with the GDC layer on the tensile side of loading was. The specimen under investigation broke in two pieces, but a clear "third-branch" of the crack propagating through the electrolyte is present and highlighted by the arrows in Figure 9c.

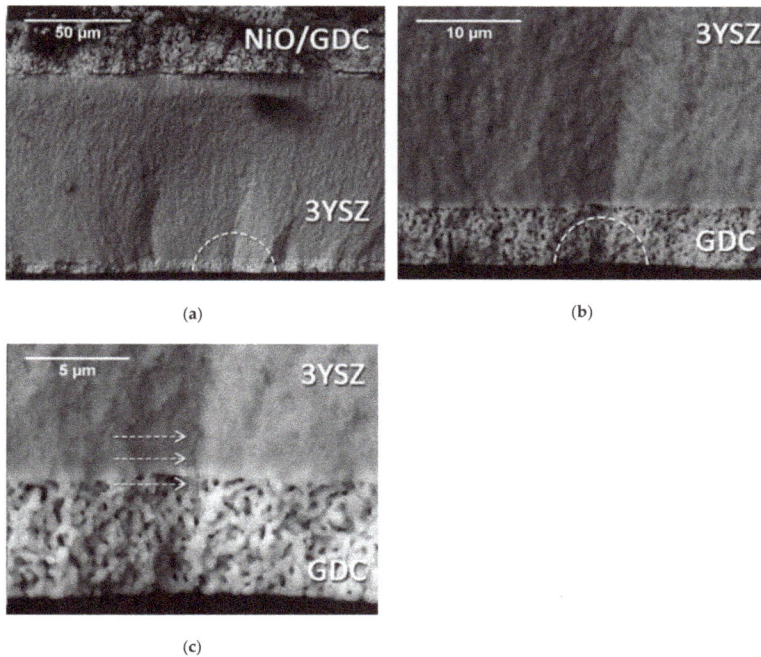

(a)

(b)

(c)

Figure 9. SEM micrographs of the fracture surface of a SOC2 specimen tested with the GDC barrier layer on the tensile side of loading. (**a**): Overview of the fracture surface. (**b**) Fracture initiation site located on the outer surface of the GDC layer. (**c**) The arrows indicate the "third-branch" of the crack.

An example of the fracture surface of a SOC2 specimen tested with the fuel electrode in tension is illustrated in Figure 10. This time, no delamination was observed. The fracture initiation site seemed to be located somewhere between the fuel electrode and the electrolyte; however, the porous nature of the electrode layer makes the exact identification of the initiation point impossible. As shown in Figure 10b, the crack propagated with continuity from the fuel electrode into the electrolyte, meaning that the bonding between these layers was strong enough and the crack did not deflect along the interface. This could explain the strength decrease measured for the SOC2 samples when tested with the fuel electrode in tension. Considering that the fuel electrode was weaker than the electrolyte, it would start cracking at lower applied stress; since these cracks were not able to deflect along the interface with the electrolyte because of the strong bonding, they would penetrate into the electrolyte. This mechanism seemed to have a detrimental effect on the strength of the layered structure analyzed. Similar fracture mechanism for SOC3 specimens tested on the same orientation (i.e. with the fuel electrode in tension) was observed.

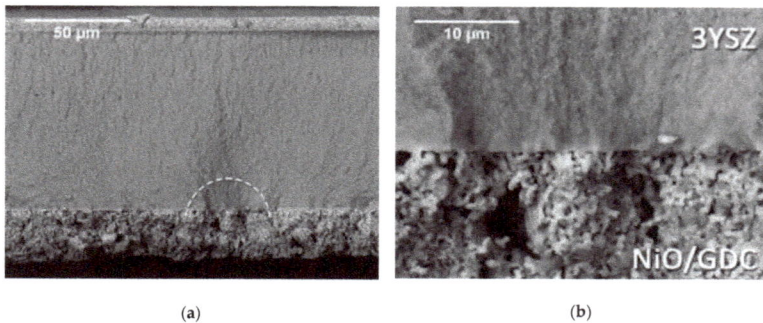

(a)

(b)

Figure 10. (a) Example of the fracture surface appearance of a SOC2 specimen tested with the fuel electrode on the tensile side of loading; fracture. (b) Detail of the fracture initiation area.

In Figure 10a, the effect of compressive stresses acting on the upper part of the specimen can be observed at the interface between the electrolyte and the barrier layer.

The micrographs in Figure 11 illustrate a typical fracture surface of the whole MEA (SOC3), tested with the air electrode on the tensile side of loading. The fracture initiation site was most likely located at the interface between the LSCF (Lanthanum Strontium Cobalt Ferrite) air electrode and the GDC barrier layer, in correspondence with the area of maximum stress calculated via FEM. A detail of this area is shown in Figure 11b. Before the failure, some cracks formed in the air electrode layer and they expanded up to the substrate interface. The pre-cracking of the air electrode was a consequence of the much lower strength of this layer in comparison to the electrolyte. Because of residual stresses derived from the mismatch of the thermal expansion coefficients, the air electrode encountered itself already in tension before the mechanical test; given that this layer was really porous and mechanically weak, it was not expected to support a much higher tensile stress during the flexural test. Therefore, it was likely to pre-crack at low applied stress levels. The dashed circle ellipse in Figure 11a highlights an example of a pre-crack starting at the surface and propagating to the interface with the barrier layer being out of the failure initiation site. Such cracks only formed locally, therefore their influence on the overall stiffness was negligible and the force–displacement curve would not show any evident deflection from linearity. However, they would act as stress concentrators during external loading and might lead to an early failure of the laminate, when in correspondence with defects in the substrate. As a result, the failure stress and the characteristic strength of the electrolyte on the cell level were much lower (more than twice) than those of the uncoated electrolyte samples, which is in good agreement with the literature [7,10]. However, the whole cell showed a strength increase in comparison to the strength values measured for SOC2 specimens.

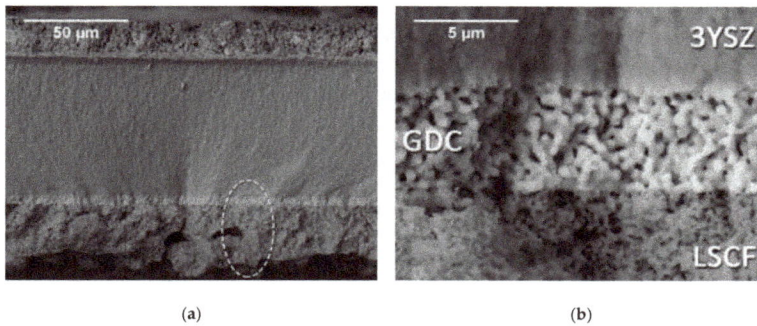

Figure 11. SEM images showing a typical fracture surface of a whole MEA (SO3) tested with the air electrode side in tension. The circle in (**a**) highlights a pre-crack in the air electrode. (**b**) Detail of the fracture initiation site at the interface between the LSCF air electrode and the GDC barrier layer.

4. Discussion

4.1. Effect of Residual Stresses

The cell was made up of co-sintered functional self-supported layers. As emerged from the fractographic analyses, the interfaces generated between adjacent layers might influence the fracture mechanism of the cell. However, layer interfaces were not the only factor responsible for the changes in the mechanical resistance; a significant role was ascribed to residual stresses developed during processing. Because of the CTE (Coefficient of Thermal Expansion) mismatch, residual stresses arose in the layers during cooling from the sintering temperature as the layers were bonded to each other and were not allowed to shrink freely. The main difference with layered systems studied in the literature where co-sintering of green bodies were usually investigated [30–32], contrary to the cell case where already sintered electrolyte is subjected to co-sintering with green bodies of added layers. Therefore, the multi-layered samples were not in a stress-free state at room temperature before mechanical loading during B3B tests. Some of the layers were already in tension while others were in compression. These residual stresses were responsible for a more fragile cell when handling it [33]. For the proper evaluation of the room-temperature strength of the materials under investigation, such residual stresses should be taken into account. The effect of residual stresses on the strength of bi-layer SOC materials has already been investigated; it has been reported that residual stresses could either strengthen or weaken the layered structure, depending on which of the layers is exposed to tensile loading. In fact, the residual stresses present in the layers redistribute the stress field created by an external load applied. However, literature data deals only with bi-layer or symmetrical structures; the influence of residual stresses on the mechanical integrity of the whole cell has not been reported yet.

Looking at the results reported in Table 4 for SOC1 samples, it can be observed that the strength of the bi-layered structure slightly increased when the electrolyte was on the tensile side of loading, but it drastically decreased when the electrolyte was on the compressive side. The reason for this behavior can be found in the residual stresses that developed in the sample after cooling down from the sintering temperature. A finite element model allowed for the estimation of the stresses inside the layers: at room temperature, the electrolyte was in a compressive state (about 50 MPa), while the GDC barrier layer was solicited by high tensile stresses of about 600 MPa. Therefore, the barrier layer was already in tension before the flexural test; hence, the applied stress necessary to reach failure was lower. On the contrary, the compressive stresses developed in the electrolyte acted against the tensile stresses applied during mechanical testing, resulting in a strength increase. The high residual stresses derived from the CTE mismatch in the non-symmetrical laminate would be partially released by elastic deformation of the whole laminate, resulting in non-planar (curved) samples. Moreover, the different strength of individual materials should be taken in to account depending on the fracture initiation place. Indeed, fracture might not occur in the vicinity of the tensile outer surface as is usual in case of monolithic materials tested (see SOC0 as an example).

The addition of the fuel electrode resulted again in nominal residual compressive stresses generated in the electrolyte (about 95 MPa) and tensile stresses in both the outer layers (465 MPa in the barrier layer and 90 MPa in the fuel electrode). Despite the addition of the new layer, a negligible difference between the mechanical response of SOC1 and SOC2 samples tested with the barrier layer on the tensile side of loading was observed. This might be a consequence of three aspects: first, the delamination of the fuel electrode when on the tensile side of loading (see Figure 9); second, the residual stresses acting in the electrolyte and barrier layer, which were of the same magnitude of those acting in SOC1 samples; finally, the significantly higher compliance of the electrode layer.

In contrast, when SOC2 samples were tested in the opposite orientation, the presence of the fuel electrode in a pre-tensed state played an important role in the strength of the tri-layered structure: the Weibull characteristic strength calculated for SOC2 samples became significantly smaller than the one determined for SOC1 samples. This strength loss was caused by the electrode porosity, which allowed cracking at lower applied stresses due to its weak nature (lower strength and stress concentration effect on pores) and the residual tensile stresses, which were independent of the porosity. Such cracks propagated across the thickness until reaching the electrolyte without fracture of the whole layered system; there they encountered a strong and brittle interface generated during the sintering process (see Figure 10). Thus, pre-cracks were not able to deflect along the interface and act as stress concentrators finally propagating into the electrolyte, with a consequent loss of mechanical strength.

With the addition of the air electrode, the stress distribution within the layers was analogous to the one described for SOC2 samples: the electrolyte was in compression, while all the other layers around it were in tension. In particular, the air electrode was pre-stressed with a tensile stress of about 180 MPa; given that its room-temperature strength was approximately 160 MPa [34,35], this would explain the cracks observed in Figure 11.

With respect to the final mechanical strength, the presence of the LSCF air electrode played a minor role. Indeed, the characteristic strength measured for SOC3 samples was slightly higher, but still of the same order of magnitude of the one obtained for the SOC2 samples. This is in accordance with the fuel electrode having low residual stresses, low elastic modulus, and low strength [10]. The increase was a consequence of the stress redistribution within the functional layers. This means that the pre-cracks forming in the air electrode layer for relatively low applied stresses were not able to penetrate into the electrolyte and they had no significant influence on the failure mechanism.

4.2. Stresses across the Thickness

All the analyzed samples fractured within the area of maximum applied stress indicated in Figure 5. For a better identification and explanation of the fracture initiation site across the section of the samples, an FEM was used. In particular, the FEM of the biaxial flexural test allowed the estimation of the stress distribution across the section corresponding to a certain applied load. Figure 12 shows how the stresses distribute across the section of every sample in a fully elastic regime with low surface curvature. In these results, an estimation of residual stresses developed during processing was taken into account. They were incorporated as a separate step in the model before calculating the biaxial flexural stresses. The cross-sectional distribution of the total stresses revealed that the maximum tensile stress was not always located at the surface under a tensile load. Indeed, in some cases, the maximum tensile stress arose at the interface between layers. This means that the fracture would not always initiate at the external surface, but it could start at the most stressed interface between layers. This was particularly evident for the SOC3 sample oriented with the air electrode on the tensile side of loading. In this case, the maximum tensile stress was located at the interface between the LSCF electrode and the GDC barrier layer and the tensile stresses in the air electrode were pretty small compared to those developed in the barrier layer. Therefore, if the tensile stress present in the air electrode did not exceed the material strength, the fracture would initiate at the interface between electrode and barrier layer. The FEM results were used to confirm and to explain the observed fracture behavior using fractographic analysis discussed previously in this work.

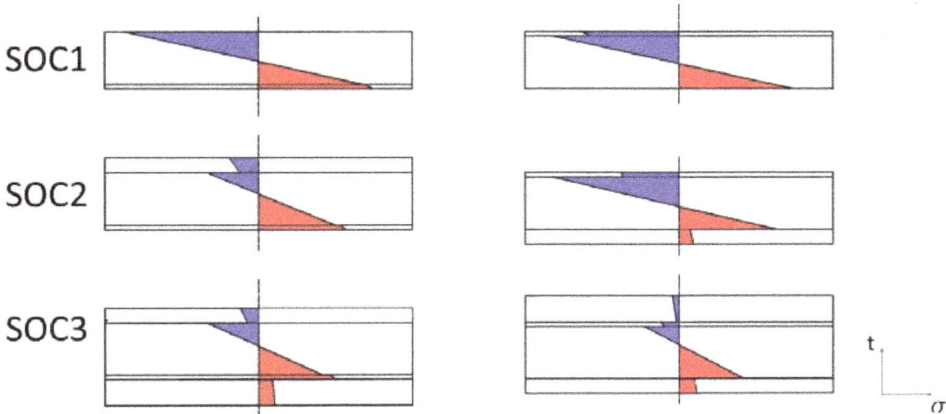

Figure 12. Stress distribution across the section of the layered samples tested in biaxial flexure. Compressive stresses are indicated in blue, tensile stresses in red.

5. Conclusions

In this work, electrolyte-supported SOC cells that are currently used in commercial stacks have been investigated with respect to their biaxial flexural strength at room temperature. It can be concluded that the strength of the supportive electrolyte on the cell level is sensibly reduced. The reason for the mechanical loss is a consequence of two main phenomena:

- The formation of strong interfaces and constraints between adjacent functional layers during manufacturing, and especially during the sintering process. Such interfaces, due to their high fracture energy, will impede the deflection of the crack formed in the porous electrodes to deflect at the interface with the electrolyte. The tip of such cracks may act as stress concentrators at the electrolyte interface and they might easily penetrate into the electrolyte, thus lowering the final strength.
- Residual stresses arising in the different layers of the cell as a consequence of the thermal expansion mismatch. Such stresses will redistribute with the addition of layers to the electrolyte and will act against or in favor of the externally applied load affecting the resulting strength.

Author Contributions: Conceptualization, A.M. and Z.C.; Methodology, A.M. and Z.C.; Software, F.Š.; Investigation, A.M. and Z.C.; Resources, Z.C. and T.S.; Data Curation, A.M. and Z.C.; Writing-Original Draft Preparation, A.M.; Writing-Review and Editing, Z.C., T.S., and I.D.; Supervision, Z.C.; Project Administration, I.D.; Funding Acquisition, I.D.

Funding: This research was funded by the European Union's Horizon 2020 research an innovation program under the Marie Skłodowska-Curie project "CoACH – Advanced glasses, Composites And Ceramics for High growth Industries. European Training Network (ETN)", Grant Number: 642557 and "GrInHy – Green Industrial Hydrogen", Grant Number: 700300.

Acknowledgments: The authors express their gratitude to Sunfire GmbH and to its R&D department for providing the experimental material and for the precious support. Special thanks to Prof. Raul Bermejo for his fundamental help in the B3B test arrangement and data interpretation.

Conflicts of Interest: The authors declare no conflict of interest.

References

1. Mitili, N.Q.; Mogensen, M.B. Reversible Solid Oxide Fuel Cell Technology for Green Fuel and Power Production. *Interface* **2013**, *22*, 55–62. [CrossRef]
2. Boudghene, A.; Traversa, E. Solid Oxide Fuel Cells (SOFCs): A Review of an Environmentally Clean a Deficient Source of Energy. *Renew. Sustain. Energy Rev.* **2002**, *6*, 433–455.
3. Gómez, S.Y.; Hotza, D. Current Developments in Reversible Solid Oxide Fuel Cells. *Renew. Sustain. Energy Rev.* **2016**, *61*, 155–174. [CrossRef]
4. Masini, A.; Šiška, F.; Ševeček, O.; Chlup, Z.; Dlouhý, I. Elastic Properties of Multi-Layered Ceramic Systems for SOCs. *Int. J. Appl. Ceram. Technol.* **2018**, *15*, 370–379. [CrossRef]
5. Reisert, M.; Aphale, A.; Singh, P. Solid Oxide Electrochemical Systems: Material Degradation Processes and Novel Mitigation Approaches. *Materials* **2018**, *11*, 2169. [CrossRef] [PubMed]
6. Kusnezoff, M.; Trofimenko, N.; Müller, M.; Michaelis, A. Influence of Electrode Design and Contacting Layers on Performance of Electrolyte Supported SOFC/SOEC Single Cells. *Materials* **2016**, *9*, 906. [CrossRef] [PubMed]
7. Fleischhauer, F.; Bermejo, R.; Danzer, R.; Mai, A.; Graule, T.; Kuebler, J. Strength of an Electrolyte Supported Solid Oxide Fuel Cell. *J. Power Sources* **2015**, *297*, 158–167. [CrossRef]
8. Sørensen, B.F.; Primdahl, S. Relationship between Strength and Failure Mode of Ceramic Multilayers. *J. Mater. Sci.* **1998**, *33*, 5291–5300. [CrossRef]
9. Dai, H.; Ning, Y.; He, S.; Zhang, X.; Guo, L.; Zhao, G. Surface Modification Allows High Performance for Solid Oxide Fuel Cells Fabricated by a Single-Step Co-Firing Process. *Fuel Cells* **2017**, *17*, 905–908. [CrossRef]
10. Selçuk, A.; Merere, G.; Atkinson, A. The Influence of Electrodes on the Strength of Planar Zirconia Solid Oxide Fuel Cells. *J. Mater. Sci.* **2001**, *36*, 1173–1182. [CrossRef]
11. Özkol, E.; Wätjen, A.M.; Bermejo, R.; Deluca, M.; Ebert, J.; Danzer, R.; Telle, R. Mechanical Characterisation of Miniaturised Direct Inkjet Printed 3Y-TZP Specimens for Microelectronic Applications. *J. Eur. Ceram. Soc.* **2010**, *30*, 3145–3152. [CrossRef]
12. Fleischhauer, F.; Terner, M.; Bermejo, R.; Danzer, R.; Mai, A.; Graule, T.; Kuebler, J. Fracture Toughness and Strength Distribution at Room Temperature of Zirconia Tapes Used for Electrolyte Supported Solid Oxide Fuel Cells. *J. Power Sources* **2015**, *275*, 217–226. [CrossRef]
13. Fleischhauer, F.; Bermejo, R.; Danzer, R.; Mai, A.; Graule, T.; Kuebler, J. High Temperature Mechanical Properties of Zirconia Tapes Used for Electrolyte Supported Solid Oxide Fuel Cells. *J. Power Sources* **2015**, *273*, 237–243. [CrossRef]
14. Selçuk, A.; Atkinson, A. Elastic Properties of Ceramic Oxides Used in Solid Oxide Fuel Cells (SOFC). *J. Eur. Ceram. Soc.* **1997**, *17*, 1523–1532. [CrossRef]
15. Atkinson, A.; Selçuk, A. Mechanical Behaviour of Ceramic Oxygen Ion-Conducting Membranes. *Solid State Ionics* **2000**, *134*, 59–66. [CrossRef]
16. Giraud, S.; Canel, J. Young's Modulus of Some SOFCs Materials as a Function of Temperature. *J. Eur. Ceram. Soc.* **2008**, *28*, 77–83. [CrossRef]
17. Liu, X.; Martin, C.L.; Bouvard, D.; Di Iorio, S.; Laurencin, J.; Delette, G. Strength of Highly Porous Ceramic Electrodes. *J. Am. Ceram. Soc.* **2011**, *94*, 3500–3508. [CrossRef]
18. Frandsen, H.L.; Ramos, T.; Faes, A.; Pihlatie, M.; Brodersen, K. Optimization of the Strength of SOFC Anode Supports. *J. Eur. Ceram. Soc.* **2012**, *32*, 1041–1052. [CrossRef]
19. Faes, A.; Frandsen, H.L.; Kaiser, A.; Pihlatie, M. Strength of Anode-Supported Solid Oxide Fuel Cells. *Fuel Cells* **2011**, *11*, 682–689. [CrossRef]
20. Wei, J.; Osipova, T.; Malzbender, J.; Krüger, M. Mechanical Characterization of SOFC/SOEC Cells. *Ceram. Int.* **2018**, *44*, 11094–11100. [CrossRef]
21. Moon, H.; Kim, S.D.; Hyun, S.H.; Kim, H.S. Development of IT-SOFC Unit Cells with Anode-Supported Thin Electrolytes via Tape Casting and Co-Firing. *Int. J. Hydrog. Energy* **2008**, *33*, 1758–1768. [CrossRef]
22. Exner, J.; Pöpke, H.; Fuchs, F.-M.; Kita, J.; Moos, R. Annealing of Gadolinium-Doped Ceria (GDC) Films Produced by the Aerosol Deposition Method. *Materials* **2018**, *11*, 2072. [CrossRef]
23. Börger, A.; Supancic, P.; Danzer, R. The Ball on Three Balls Test for Strength Testing of Brittle Discs: Part II: Analysis of Possible Errors in the Strength Determination. *J. Eur. Ceram. Soc.* **2004**, *24*, 2917–2928. [CrossRef]
24. Bermejo, R.; Supancic, P.; Krautgasser, C.; Morrell, R.; Danzer, R. Subcritical Crack Growth in Low Temperature Co-Fired Ceramics under Biaxial Loading. *Eng. Fract. Mech.* **2013**, *100*, 108–121. [CrossRef]

25. Danzer, R.; Supancic, P.; Harrer, W. Biaxial Tensile Strength Test for Brittle Rectangular Plates. *J. Ceram. Soc. Jpn.* **2006**, *114*, 1054–1060. [CrossRef]

26. EN 843-5. *Advanced Technical Ceramics—Monolithic Ceramics—Mechanical Tests at Room Temperature—Part 5: Statistical Analysis*; European Committee for Standardization: Brussels, Belgium, 1996.

27. Šiška, F.; Forest, S.; Gumbsch, P. Simulations of Stress-Strain Heterogeneities in Copper Thin Films: Texture and Substrate Effects. *Comput. Mater. Sci.* **2007**, *39*, 137–141. [CrossRef]

28. Hayashi, H.; Saitou, T.; Maruyama, N.; Inaba, H.; Kawamura, K.; Mori, M. Thermal Expansion Coefficient of Yttria Stabilized Zirconia for Various Yttria Contents. *Solid State Ionics* **2005**, *176*, 613–619. [CrossRef]

29. Tietz, F. Thermal Expansion of SOFC Materials. *Ionics* **1999**, *5*, 129–139. [CrossRef]

30. Chlup, Z.; Hadraba, H.; Drdlik, D.; Maca, K.; Dlouhy, I.; Bermejo, R. On the Determination of the Stress-Free Temperature for Alumina-Zirconia Multilayer Structures. *Ceram. Int.* **2014**, *40*, 5787–5793. [CrossRef]

31. Bermejo, R.; Pascual, J.; Lube, T.; Danzer, R. Optimal Strength and Toughness of Al2O3-ZrO2laminates Designed with External or Internal Compressive Layers. *J. Eur. Ceram. Soc.* **2008**, *28*, 1575–1583. [CrossRef]

32. Sglavo, V.M.; Paternoster, M.; Bertoldi, M. Tailored Residual Stresses in High Reliability Alumina-Mullite Ceramic Laminates. *J. Am. Ceram. Soc.* **2005**, *88*, 2826–2832. [CrossRef]

33. Charlas, B.; Frandsen, H.L.; Brodersen, K.; Henriksen, P.V.; Chen, M. Residual Stresses and Strength of Multilayer Tape Cast Solid Oxide Fuel and Electrolysis Half-Cells. *J. Power Sources* **2015**, *288*, 243–252. [CrossRef]

34. Huang, B.X.; Malzbender, J.; Steinbrech, R.W.; Singheiser, L. Mechanical Properties of La0.58Sr0.4Co0.2Fe0.8O3 -Δmembranes. *Solid State Ionics* **2009**, *180*, 241–245. [CrossRef]

35. Chou, Y.-S.; Stevenson, J.W.; Armstrong, T.R.; Pederson, L.R. Mechanical Properties of La1-XSrxCo0.2Fe0.8O3 Mixed-Conducting Perovskites Made by the Combustion Synthesis Technique. *J. Am. Ceram. Soc.* **2004**, *83*, 1457–1464. [CrossRef]

materials

MDPI

Article

Optical Fiber Sensors for the Detection of Hydrochloric Acid and Sea Water in Epoxy and Glass Fiber-Reinforced Polymer Composites

Cristian Marro Bellot [1], **Marco Sangermano** [1,]*[iD], **Massimo Olivero** [2][iD] **and Milena Salvo** [1][iD]

[1] Department of Applied Science and Technology, Politecnico di Torino, c.so duca degli Abruzzi 24, 10129 Torino, Italy; cristian.marrobellot@polito.it (C.M.B.); milena.salvo@polito.it (M.S.)
[2] Department of Electronics and Telecommunication, Politecnico di Torino, c.so duca degli Abruzzi 24, 10129 Torino, Italy; massimo.olivero@polito.it
* Correspondence: marco.sangermano@polito.it; Tel.: +39-0110904651

Received: 13 December 2018; Accepted: 23 January 2019; Published: 25 January 2019

Abstract: Optical fiber sensors (OFSs), which rely on evanescent wave sensing for the early detection of the diffusion of water and hydrochloric acid through glass fiber-reinforced polymers (GFRPs), have been developed and tested. Epoxy and GFRP specimens, in which these sensors were embedded, were subjected to tests in artificial sea water and hydrochloric acid. The sensors were able to detect the diffusion of chemicals through the epoxy and GFRP samples on the basis of a drop in the reflected signal from the tip of the optical sensor probe. Water and hydrochloric acid diffusion coefficients were calculated from gravimetric measurements and compared with the experimental response of the OFSs. Furthermore, mechanical tests were carried out to assess the influence of the sensors on the structural integrity of the GFRP specimens.

Keywords: evanescent wave optical fiber sensors; diffusion; glass fiber-reinforced polymers; testing and aging

1. Introduction

The perfect combination of low cost, high corrosion resistance, high strength, easy manufacturing, and easy scalability has designated glass fiber-reinforced polymers (GFRPs) as promising materials for applications in hostile environments, such as underwater applications and the oil and gas industry [1]. GFRPs have been an important part of the fast-increasing expansion of petrochemical companies. However, there is still a lack of research on their failures and their aging is difficult to predict. For this reason, a great deal of effort has been devoted to embedding sensors into GFRPs for structural health monitoring purposes [2,3]. The aim of producing "sensitive" GFRP composites is to promote efficient maintenance programs with reduced costs, because any degradation (e.g., due to the diffusion of salty water or other chemicals in the polymer matrix) may be detected before the structural integrity is compromised. Optical fiber sensors (OFSs) can represent an excellent aid in such a framework, since they are minimally invasive and can operate remotely (for up to kilometers) without the need of any electrical supply.

In this paper, we focused on the design and experimental testing of low-cost OFSs for use in the monitoring of the diffusion of corrosive media through the thickness of epoxy resin and GFRP samples. OFSs can cope with the requirements of GFRPs for the oil and gas industry, because their small form factor and intrinsic fire safety make them very attractive, not just as temperature sensors (i.e., in wells during crude extraction) or as structural health monitoring probes [4], but also as chemical detection sensors [5].

The OFSs here described rely on the evanescent wave sensor concept developed by the authors and described in References [6,7]. In this work, the sensors were developed to selectively detect the diffusion of either salty water or hydrochloric acid (HCl), two of the several corrosion agents in the oil and gas industry [8]. Epoxy and GFRP samples were prepared by embedding OFSs, and diffusion was measured both by means of gravimetric measurements and by recording the response of the OFSs with a custom-developed multichannel spectroscopic system. The samples equipped with optical fiber sensors were also mechanically characterized and their properties were compared with pristine samples in order to evaluate how the embedded sensors could affect the mechanical properties of GFRP composites.

2. Materials and Methods

2.1. Fabrication of Optical Fiber Sensors, Their Embedding into Epoxy and GFRP and the Interrogation Setup

The OFSs were fabricated from a commercial glass optical fiber (GIF 625 silica multimode fiber, Thorlabs). A 1.5-m-long optical fiber was cut from the spool for each sensor, and then one of the two tips was mechanically uncoated (i.e., the polymeric coating was removed by means of a stripping tool) and etched in hydrofluoric acid (HF) to remove the cladding, thus reducing the initial 125 μm fiber diameter to ~60 μm in order to expose the core—guiding the optical signal—to the surrounding environment. A reflective silver coating was then deposited onto the fiber tip by means of a quick chemical procedure based on Tollen's reagent [9]. The thickness of the silver coating was ~5 μm. In this way, the optical signal traveling into the fiber core and containing the sensing information could propagate back to the interrogation unit, thus realizing a single-ended sensor [7]. The sensitive section of some of the sensors was specifically coated to make these sensors selective to HCl. Since HCl dissolves aluminum, these sensors (here referred to as Al-OFSs) were prepared by depositing a ~100-nm-thick aluminum coating on the bare core of the etched fiber by means of magnetron sputtering (Kenosistec, Plasma RF type).

Sensitive epoxy specimens (with dimensions of 100 × 15 × 5 mm) were prepared by embedding the single-ended OFSs in the epoxy resin (Ampreg, Gurit, Switzerland) during the curing process (Figure 1). Details of the curing process are reported in Reference [7]. The sensors were embedded at different depths to detect chemical diffusion through the thickness of the polymer matrix and to evaluate the dynamics of the diffusion.

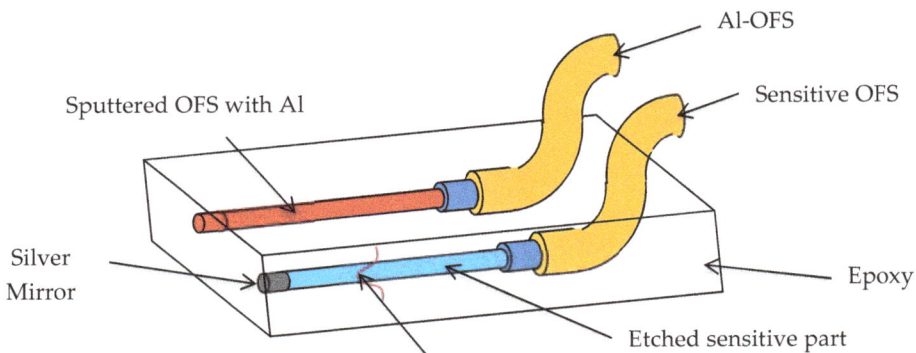

The evanescent field extends into the epoxy polymer matrix

Figure 1. Scheme of an embedded optical fiber sensor (OFS) and sputtered OFS with aluminum (Al-OFS) in an epoxy polymer matrix (adapted from Reference [10]).

GFRP samples were prepared, following a similar workflow, by means of vacuum bag molding infusion, using the same kind of epoxy resin that was used for the epoxy samples, and then introducing an Advantex E-CR glass fiber fabric (Owens Corning, USA). The glass fiber plies were laid at orientations of 0° and 90°; three optical fiber sensors were deposited on the ply with an inter-distance of 0.5 cm. A peel ply and a diffusion mesh were put on top of the glass fiber plies. A plastic foil seal was used to cover all the plies and butyl tape was used to seal the infusion. The epoxy resin was mixed at a ratio of 100:33 and degassed, as described in Reference [7]. Finally, it was infused and left for 24 h at room temperature (r.t.) before proceeding with the thermal curing at 80 °C for 5 h.

The sensors were monitored by means of the setup shown in Figure 2. Each sensor is optically supplied by a broadband source (Photonetics 3626BT), which emits a total maximum power of 25 mW over the 1500–1600 nm range. A fiber optics switch (JDS Uniphase 2 × 16 SB series) enables up to 16 sensors to be swapped; these sensors are connected to the switch by movable connectors (Thorlabs BFT-1). A fiber optic coupler separates the signal back-reflected from the sensor (which contains the sensing information) from the transmitted signal. The signal is processed and displayed on a portable spectrometer (Avantes AvaSpec-NIR256-1.7), which is able to measure spectra over the 1100–1700 nm range. The measured signal is compensated for by the background noise of the spectrometer and the spurious coupling between the source and the spectrometer. The system was designed for automatic switching among the sensors and logging of the spectra through a LabVIEW custom-developed program. The spectra may be recorded with a selectable sampling rate, and most of the experiments were carried out by recording spectra every 5 min, lasting from a number of hours to days. The inset in Figure 2 shows the arrangement of the experiments at high pressure (50 bars). In this case, the vessel is placed in a separate room for safety reasons. The fibers, which cross the wall between the two rooms, run through a Conax adaptor sealant. This adaptor has two holes in which up to four optical fibers per hole can be inserted. The adaptor was screwed directly onto the vessel, into which N$_2$ air gas was pumped. Moreover, the vessel was connected to a network system in order to monitor the pressure and temperature during the experiments.

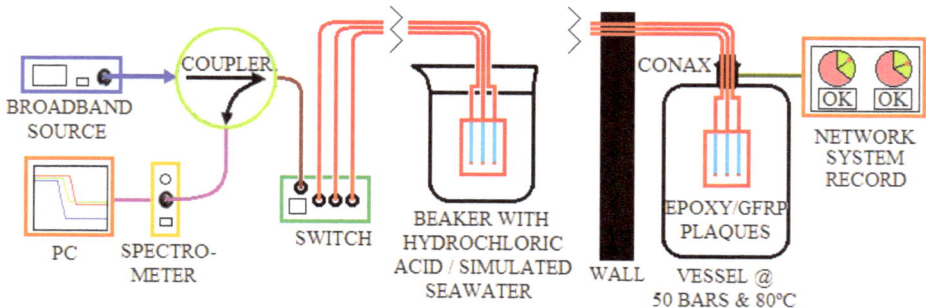

Figure 2. Interrogation setup. The inset shows the arrangement for high-temperature/high-pressure experiments (adapted from Reference [7]).

The diffusion coefficients were evaluated through gravimetric measurements (10 samples per test) for artificial sea water (according to ASTM D1141-98(2013) [11]) at 80 °C and at 80 °C/50 bars, as well as for HCl (37% Carlo Erba Reagents) at room temperature. The diffusion coefficient was calculated from the absorption curves, i.e., weight gain versus time, assuming, as an approximation, an ideal Fickian behavior using Equation (1) [12]:

$$D = \pi \left(\frac{h}{4 M_\infty} \right)^2 \left(\frac{M_2 - M_1}{\sqrt{t_2} - \sqrt{t_1}} \right)^2 \tag{1}$$

where h is the sample thickness, M_∞ is the weight gain at saturation, M_1 and M_2 are two experimental values on the linear section of the weight gain curve, and t_1 and t_2 are the corresponding time spots in s.

The expected diffusion length of the chemicals in the epoxy and GFRP was evaluated assuming Fick's model and considering constant diffusion coefficients (Equation (2)):

$$d = 2\sqrt{Dt} \tag{2}$$

where d is the diffusion length, t is the time at which the chemical reaches length d, and D is the diffusion constant.

Furthermore, the real part of the refractive index was measured with a refractometer (Metricon 2010/M, New Jersey, USA) on pristine and water-saturated epoxy samples.

2.2. Mechanical Characterization of GFRP

A tensile test and a three-point bending tests were carried out on the GFRPs, with and without embedded sensors, to assess their mechanical properties and investigate whether and how embedded OFSs could affect the composite. Two sets of samples, that is, 10 without OFSs and 10 with three embedded OFSs, were prepared. The distance between the embedded optical sensors was ~1 cm.

The tensile test was carried out according to the ASTM D3039/D3039M-17 standard [13] using a mechanical testing machine (Zwick 750, ZwickRoell GmbH and Co., Ulm, Germany). Samples (250 mm × 25 mm × 2 mm) were arranged in the middle of holding grips, which had a gauge length of 50 mm. The load was applied by moving the cross-head at a speed of 2 mm/min.

Samples (80 mm × 15 mm × 2 mm) subjected to three-point bending were tested according to the ISO 14125 standard [14] using a mechanical testing machine (Zwick 100, ZwickRoell GmbH and Co., Ulm, Germany). The load was applied by moving the cross-head at a speed of 1.5 mm/min.

3. Results and Discussion

3.1. Mechanical Properties

In order to demonstrate that the embedding of the OFSs does not affect the mechanical properties of the composites, the GFRPs were mechanically tested with and without embedded OFSs. The results of the tensile and three-point bending tests are summarized in Table 1. The results confirm that the optical fibers have no relevant effects on the mechanical properties of the composites and are in agreement with early studies [15], though attention should be paid if the composites are subjected to a fatigue load. Furthermore, more specific mechanical characterizations should be carried out to assess the possible effect of embedded OFSs on the final component properties.

Table 1. Mechanical test results of the tensile test and three-point bending test of GFRPs with and without embedded OFSs.

	Tensile Test		Three-Point Bending Test	
	σ (MPa)	E (GPa)	σ (MPa)	E (GPa)
GFRP	367 ± 15	22 ± 5	364 ± 24	17 ± 1
GFRP + OFSs	356 ± 23	20 ± 3	368 ± 26	17 ± 1

3.2. Characterization of Epoxy and GFRPs with Embedded OFSs in Artificial Sea Water at 80 °C

3.2.1. Gravimetric Measurements and Monitoring of Diffusion by Means of the Optical Fiber Sensors

OFSs were prepared and embedded in the epoxy matrix and in GFRP, as reported in Section 2.1. Figure 3a,b show an epoxy and a GFRP sample, respectively.

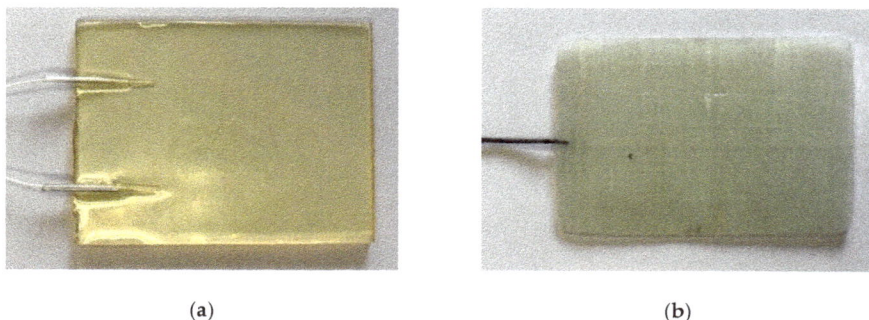

(a)   (b)

Figure 3. (a) Epoxy and (b) GFRP sample with embedded OFSs on the left-hand side.

The diffusion rates for the epoxy resin and GFRP were evaluated gravimetrically by immersing the samples into artificial sea water (ASW) under two different conditions: At 80 °C and at 80 °C under 50 bars of pressure. The diffusion values were also evaluated for the same samples, but this time immersed in HCl at room temperature. Figure 4 reports the weight gain versus time; the calculated diffusion coefficients are reported in Table 2. The diffusion coefficients were used to predict the diffusion time at a given depth using Equation (2). It is apparent from this table that the hydrostatic pressure affects the diffusion coefficient of water at 80 °C. Several studies have attempted to clarify the effect of hydrostatic pressure on water diffusion in composite materials [16–18], but the published results are contradictory. However, it is difficult to foresee, in general terms, the role of hydrostatic pressure, since the overall effect is due to a balance between the free-volume reduction (due to the increasing compression forces on the composites) and the increase of the absorption driving force (due to the increasing chemical potential in the liquid phase). In accordance with our study, Choqueuse at al. [17] showed that increasing the pressure results in faster water diffusion rates for epoxy-based materials.

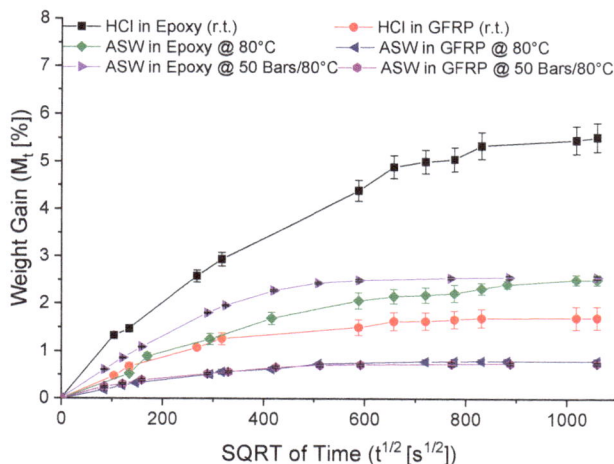

Figure 4. Weight increment of the epoxy and GFRP samples immersed in artificial sea water (ASW) at 80 °C, 80 °C under 50 bars of pressure and in HCl at room temperature.

Table 2. Diffusion coefficients calculated from Figure 4. ASW = artificial sea water.

	Epoxy	GFRP
ASW 80 °C	$8.8 \pm 0.4 \times 10^{-12}\ m^2/s$ [7]	$4.9 \pm 0.1 \times 10^{-12}\ m^2/s$
ASW 50 bars/80 °C	$2.9 \pm 0.1 \times 10^{-11}\ m^2/s$	$8.9 \pm 0.1 \times 10^{-12}\ m^2/s$
HCl r.t.	$1.4 \pm 0.4 \times 10^{-11}\ m^2/s$	$4.5 \pm 0.2 \times 10^{-12}\ m^2/s$

The epoxy and GFRP samples containing the embedded sensors were immersed in artificial sea water heated to 80 °C for 96 h. The reflected signal from the sensors was recorded every 5 min over a spectral range of between 1500 and 1600 nm. The recorded spectra did not show any change in shape, whereas a drop in intensity was observed and ascribed to water diffusion. As discussed in Reference [7], the spectral attenuation is probably due to a change in the optical properties, particularly in the absorption coefficient of the epoxy surrounding the exposed core. The absorption coefficient represents the imaginary part of the refractive index and it affects the attenuation of the sensing region of the embedded optical fiber. In previous works (e.g., References [6,19]), it was demonstrated that the increment of absorption in the infrared region of the spectrum, observed after water diffusion, could be ascribed to OH stretching. Furthermore, the real part of the refractive index measured on pristine and water-saturated samples was about 1.551 without a remarkable variation caused by water exposure, hence concluding that the attenuation of the optical signal is caused by an increment of the optical absorption of the epoxy in the recorded spectral range. The maximum intensity of the signal provided by the source occurred at 1532 nm; hence, only this wavelength was analyzed in long-term experiments, considering it as a satisfactory representation of the entire spectrum.

Figure 5a shows the signal intensity at 1532 nm recorded from three different OFSs embedded in the epoxy at different distances from the surface. The signals from the optical sensors dropped at different times according to the depths of the sensors. The first sensor, located at a depth of 2.5 ± 0.1 mm, showed a drop in the signal after approximately 43 h. The second sensor, located at a depth of 2.8 ± 0.1 mm, showed a drop in the signal after approximately 64 h. Finally, the third sensor, located at a depth of 3.0 ± 0.1 mm, showed a drop in the signal after approximately 73 h. The calculated diffusion times, related to the depth of the three sensors, were 44, 63 and 72 ± 1 h, respectively. Hence, the OFSs were able to detect the water diffusion through the epoxy thickness with a very good time accuracy, in accordance with the calculated diffusion time.

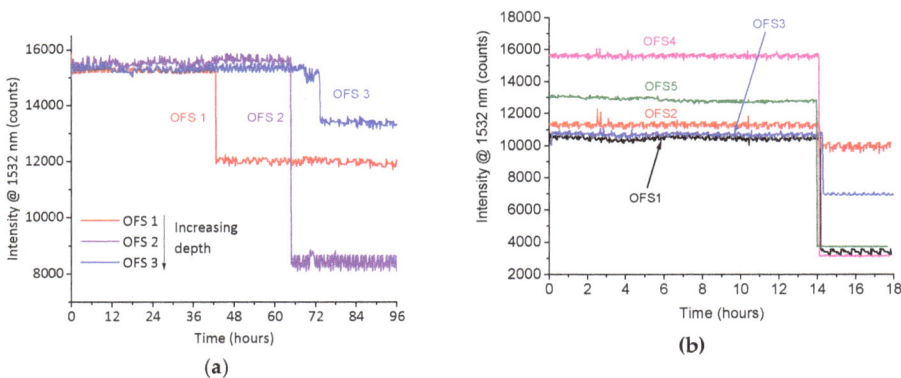

Figure 5. Signal intensity, at a wavelength of 1532 nm, from optical glass sensors embedded in: (**a**) Epoxy at different depths and (**b**) GFRP samples. The samples were immersed in artificial sea water at 80 °C.

Figure 5b shows the signal intensity drop at 1532 nm from five different OFSs embedded in GFRP, all of which were located at a distance of 0.9 ± 0.1 mm from the surface. As they were positioned at the same depth, the drop in the signals occurred at nearly the same time, i.e., after approximately 14 h. Assuming the approximation of a Fickian behavior, and using the diffusion coefficient reported in Table 2, the diffusion time for a thickness of 0.9 mm was 13 ± 1 h. This result shows that the OFSs also detected the water diffusion with a good accuracy when embedded in the composite.

3.2.2. High-Pressure Experiments

In order to investigate the use of the OFSs under harsh conditions, bare OFSs were subjected to high pressures. The reflected signal was continuously monitored by exposing the OFSs to pressures ranging from 1 up to 75 bars, with a ramp pressure of 5 bars per minute. No degradation of the response of the sensors was observed, thus showing that the OFSs could withstand high-pressure conditions.

On the basis of this preliminary test, epoxy and GFRP samples containing OFSs were tested in artificial sea water at 80 °C under a pressure of 50 bars. The response of an OFS embedded in epoxy is reported in Figure 6 as an example. In this case, the depth of the OFS, as measured with a microscope, was 0.9 ± 0.1 mm. The reflected light intensity at 1532 nm showed a sharp decrease after around 3 h. This time was in accordance with the theoretical diffusion time calculated to reach a depth of 1.1 mm. Figure 6 also reports the signal recorded from an OFS embedded in GFRP and a drop was observed after about 4 h. For this diffusion time, the expected diffusion depth was about 0.7 mm. The response of the OFSs in these tests are in rough agreement with the expected diffusion time, yet they prove the capability of non-destructive, real-time detection.

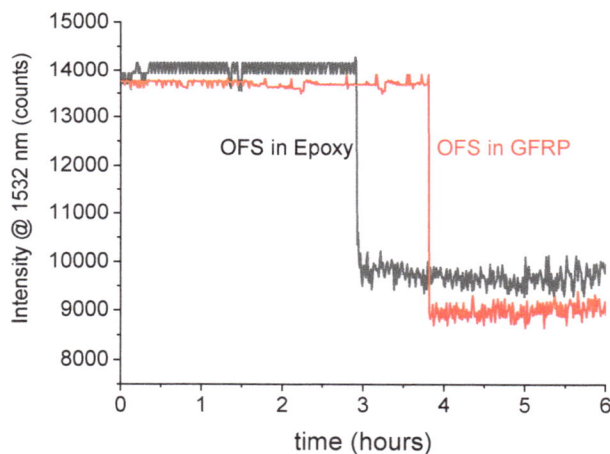

Figure 6. Signal intensity at a wavelength of 1532 nm from an OFS embedded in an epoxy and in a GFRP at depths of 0.9 ± 0.1 mm and 0.8 ± 0.1 mm, respectively. The samples were immersed in artificial sea water at 80 °C and 50 bars.

3.3. *Hydrocloric Acid Detection*

OFSs, conceived to be selectively sensitive to HCl, were fabricated as described in Section 2.1, but the exposed core was then coated with a ~100-nm layer of aluminum. These sensors are here referred to as Al-OFSs.

The calculated diffusion coefficients of HCl at room temperature in the pristine epoxy were measured by means of a gravimetric test, as reported in Figure 4 and Table 2. Furthermore, epoxy samples containing both standard OFSs and Al-OFSs were immersed in artificial sea water and HCl, respectively, at room temperature, for up to 24 h. The results of the diffusion experiment are shown in Figure 7. Figure 7a reports the signal intensity recorded at 1532 nm, as a function of time, for the sensitive epoxy sample immersed in artificial sea water. On the other hand, Figure 7b depicts the outcome for the sensitive sample subjected to HCl. It can be observed that the standard OFSs detected both water and HCl diffusion, whereas the Al-OFSs were only responsive to HCL, since HCl can solubilize the thin Al-layer, thus enabling the core to be sensitive to changes in the refractive index and absorption of the epoxy. This is a proof-of-concept that the integration of OFSs and Al-OFSs can be exploited to selectively detect the diffusion of water and acid.

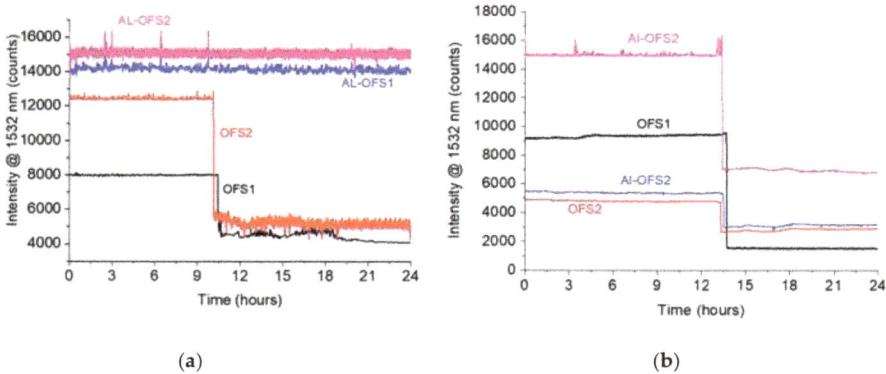

Figure 7. Signal intensity at a wavelength of 1532 nm from sensors embedded in the epoxy exposed to (**a**) artificial sea water and (**b**) HCl. The aluminum optical fiber sensors (Al-OFS) are sensitive to HCl, whereas they do not exhibit any response to artificial sea water.

Analogous experiments were then performed with GFRP samples. Figure 8 reports the response of two sensitive samples exposed to artificial sea water and HCl at room temperature. While the signal drop was not the same for all the sensors, they exhibited the same behavior as those in epoxy. The standard OFSs detected the diffusion of both water and HCl, whereas the Al-OFS were only effective in sensing the diffusion of HCl. The time response was once more in agreement with that expected from Fick's diffusion model.

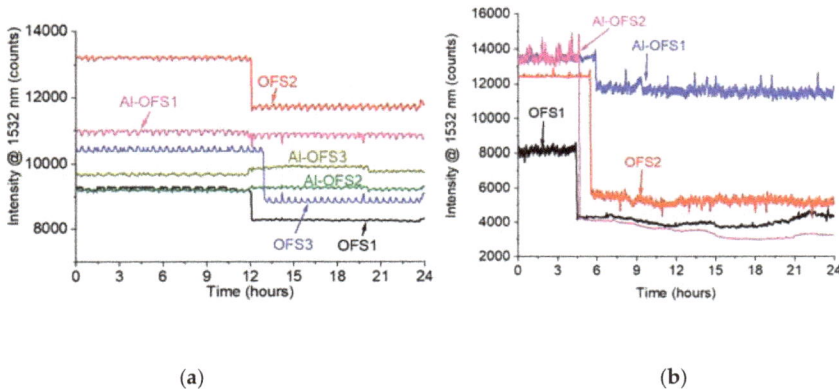

Figure 8. Signal intensity at a wavelength of 1532 nm from sensors embedded in GFRP exposed to (**a**) artificial sea water and (**b**) HCl. The Al-OFS are sensitive to HCl, whereas they do not exhibit any response to artificial sea water.

The fabrication process reproducibility of these sensors still demands some improvements, as can be observed by comparing the initial and final intensity levels of the signals collected from the different experiments (Figures 5–8). The initial reflectivity (i.e., the initial recorded signal) is strongly dependent on the quality of the mirror realized on the fiber's tip, as well as on the connection to the interrogation unit. Furthermore, the variation observed on the final levels could be ascribed to the effective length of the sensing region, which is not perfectly controlled during fabrication.

4. Conclusions

Novel, single-ended, optical fiber sensors for the detection of chemicals have been fabricated, embedded, and tested in an epoxy resin and glass fiber-reinforced polymers. These sensors, which rely on evanescent field sensing, have the additional advantages of being single-ended and of being able to work as probes. A procedure to embed the optical fiber sensors has been developed and experimentally optimized. A setup for the continuous monitoring of the sensors has been designed to remotely track the embedded sensors. The recorded optical data of samples immersed in artificial sea water at 80 °C showed that it is possible to detect water diffusion by observing decreases in the spectral reflection from the sensors. The results were then corroborated by means of gravimetric measurements and a Fickian diffusion model. Tests conducted for up to 100 hours showed that the response of the sensors embedded at increasing depths in the samples was in full agreement with the expected diffusion time. Experiments at high pressure, carried out in a special vessel, also exhibited acceptable agreement between the experimental data and the calculated diffusion time. Furthermore, tests at high pressure yielded some knowledge about the installation of optical fiber sensors in harsh environments and at hard-to-reach locations.

By tailoring the sensors, it was possible to selectively detect the diffusion of hydrochloric acid. To implement this feature, the standard fabrication process was upgraded by coating the etched fiber with a 100-nm aluminum layer, which reacted to the hydrochloric acid.

Embedding optical glass fibers in the GFRP composites did not influence their mechanical properties, as shown by means of a tensile test and a three-point bending test.

In short, this research has provided a proof-of-concept of a sensitive GFRP composite that could be used, for example, in a pipeline to monitor the diffusion of chemicals in real time and provide information in order to reduce maintenance services.

5. Patents

M. Salvo, M. Sangermano, C. Marro Bellot, M. Olivero, Optical Sensor, Process for Making Such Sensors and Evaluation System Comprising at Least One of Such Sensors, 2018WO–IT00059, 2017.

Author Contributions: Conceptualization, writing review and editing, M.S. (Milena Salvo), M.S. (Marco Sangermano) and M.O.; experimental investigation: C.M.B. (main contribution in tests) and M.O. (optical setup); preparation of the original draft, C.M.B. and M.O.; funding acquisition, M.S. (Milena Salvo).

Funding: The research leading to these results received funding from the European Union's Horizon 2020 research and innovation programme under the Marie Sklodowska-Curie grant agreement No. 642557.

Acknowledgments: The authors would like to thank M. Cavasin, S. Ivan and Element Material Technology Ltd. for their support with high-pressure tests and mechanical tests. The authors also thank the support from the PhotoNext initiative of Politecnico di Torino (www.photonext.polito.it).

Conflicts of Interest: The authors declare no conflict of interest.

References

1. Martin, R. Composite Materials: An Enabling Material for Offshore Piping Systems. In Proceedings of the Offshore Technology Conference, Huston, TX, USA, 6–9 May 2013; Offshore Technology Conference: Huston, TX, USA, 2013; pp. 1–26.
2. Majumder, M.; Gangopadhyay, T.K.; Chakraborty, A.K.; Dasgupta, K.; Bhattacharya, D.K. Fibre Bragg gratings in structural health monitoring—Present status and applications. *Sens. Actuators A Phys.* **2008**, *147*, 150–164. [CrossRef]

3. Kuang, K.S.C.; Kenny, R.; Whelan, M.P.; Cantwell, W.J.; Chalker, P.R. Embedded fibre Bragg grating sensors in advanced composite materials. *Compos. Sci. Technol.* **2001**, *61*, 1379–1387. [CrossRef]

4. Olivero, M.; Perrone, G.; Vallan, A.; Tosi, D. Comparative Study of Fiber Bragg Gratings and Fiber Polarimetric Sensors for Structural Health Monitoring of Carbon Composites. *Adv. Opt. Technol.* **2014**, *2014*, 1–8. [CrossRef]

5. Yin, S.; Ruffin, P.B.; Yu, F.T.S. *Fiber Optic Sensors*; CRC Press: Boca Raton, FL, USA, 2008; ISBN 9781420053654. Available online: https://www.crcpress.com/Fiber-Optic-Sensors/Yin-Ruffin-Yu/p/book/9781420053654 (accessed on 22 November 2018).

6. Milsom, B.; Olivero, M.; Milanese, D.; Giannis, S.; Martin, R.H.; Terenzi, A.; Kenny, J.; Ferraris, M.; Perrone, G.; Salvo, M. Glass optical fibre sensors for detection of through thickness moisture diffusion in glass reinforced composites under hostile environments. *Adv. Appl. Ceram.* **2015**, *114*, S76–S83. [CrossRef]

7. Marro Bellot, C.; Olivero, M.; Sangermano, M.; Salvo, M. Towards smart polymer composites: Detection of moisture diffusion through epoxy by evanescent wave optical fibre sensors. *Polym. Test.* **2018**, *71*, 248–254. [CrossRef]

8. Prabha, S.S.; Rathish, R.J.; Dorothy, R.; Brindha, G.; Pandiarajan, M.; Al-Hashem, A.; Rajendran, S. Corrosion Problems in Petroleum Industry and their solution. *Eur. Chem. Bull.* **2014**, *3*, 300–307. [CrossRef]

9. Montazer, M.; Alimohammadi, F.; Shamei, A.; Rahimi, M.K. In situ synthesis of nano silver on cotton using Tollens reagent. *Carbohydr. Polym.* **2011**, *87*, 1706–1712. [CrossRef]

10. Salvo, M.; Sangermano, M.; Marro Bellot, C.; Olivero, M. Optical Sensor, Process for Making Such Sensor and Evaluation System Comprising at Least One of Such Sensors. Patent Application 2018WO–IT00059, 24 April 2018.

11. ASTM D1141-98(2013). *Standard Practice for the Preparation of Substitute Ocean Water*; ASTM International: West Conshohocken, PA, USA, 2013.

12. Grammatikos, S.A.; Zafari, B.; Evernden, M.C.; Mottram, J.T.; Mitchels, J.M. Moisture uptake characteristics of a pultruded fibre reinforced polymer flat sheet subjected to hot/wet aging. *Polym. Degrad. Stab.* **2015**, *121*, 407–419. [CrossRef]

13. ASTM D3039/D3039M-17. *Standard Test Method for Tensile Properties of Polymer Matrix Composite Materials*; ASTM International: West Conshohocken, PA, USA, 2017.

14. *ISO 14125: Fibre-reinforced Plastic Composites—Determination of Flexural Properties*; ISO: Geneva, Switzerland, 1998.

15. Lee, D.C.; Lee, J.J.; Yun, S.J. The mechanical characteristics of smart composite structures with embedded optical fiber sensors. *Compos. Struct.* **1995**, *32*, 39–50. [CrossRef]

16. Grogan, D.; Flanagan, M.; Walls, M.; Leen, S.; Doyle, A.; Harrison, N.; Mamalis, D.; Goggins, J. Influence of microstructural defects and hydrostatic pressure on water absorption in composite materials for tidal energy. *J. Compos. Mater.* **2018**, *52*, 2899–2917. [CrossRef]

17. Choqueuse, D.; Davies, P. Ageing of composites in underwater applications. *Ageing Compos.* **2008**, *467*–498.

18. Le Gac, P.-Y.; Davies, P.; Choqueuse, D. Evaluation of Long Term Behaviour of Polymers for Offshore Oil and Gas Applications. *Oil Gas Sci. Technol. Rev. d'IFP Energies Nouv.* **2015**, *70*, 279–289. [CrossRef]

19. Milsom, B.; Olivero, M.; Milanese, D.; Roseman, M.; Giannis, S.; Martin, R.; Terenzi, A.; Ferraris, M.; Salvo, M. Development and integration of innovative glass fibre sensors into advanced composites for applications in hostile environments. In Proceedings of the European conference on Composites Materials, Selville, Spain, 22–26 June 2014.

materials

MDPI

Article

Development and Characterization of Bioactive Glass Containing Composite Coatings with Ion Releasing Function for Antibiotic-Free Antibacterial Surgical Sutures

Francesca E. Ciraldo [1], Kristin Schnepf [1], Wolfgang H. Goldmann [2] and Aldo R. Boccaccini [1,*]

[1] Department of Materials Science and Engineering, Institute of Biomaterials, University of Erlangen-Nuremberg, Cauerstraße 6, 91058 Erlangen, Germany; francesca.elisa.ciraldo@fau.de (F.E.C.); kristin_schnepf@yahoo.de (K.S.)

[2] Institute of Biophysics, Department of Physics, University of Erlangen-Nuremberg, Henkestraße 91, 91052 Erlangen, Germany; wgoldmannh@aol.com

* Correspondence: aldo.boccaccini@ww.uni-erlangen.de

Received: 11 January 2019; Accepted: 26 January 2019; Published: 30 January 2019

Abstract: Resorbable (Vicryl® Plus) sutures were coated with zinc-doped glass (Zn-BG) and silver-doped ordered mesoporous bioactive glass (Ag-MBG) particles by a dip coating technique. A multilayer approach was used to achieve robust coatings. The first coating was a polymeric layer (e.g., PCL or chitosan) and the second one was a composite made of BG particles in a polymer matrix. The coatings were characterized in terms of morphology by scanning electron microscopy (SEM), in vitro bioactivity, and antibacterial properties. Chitosan/Ag-MBG coatings showed the ability to form hydroxyl-carbonate-apatite on their surfaces after immersion in SBF. An antibacterial effect against Gram (+) and Gram (-) bacteria was confirmed, highlighting the potential application of the coated sutures for antibiotic-free approaches.

Keywords: Zinc; silver-doped mesoporous glass; chitosan; PCL; Vicryl Plus suture; dip coating

1. Introduction

Due to their widespread application in medicine, e.g., routine skin laceration or organ transplantation [1,2], sutures are one of the largest groups of medical devices implanted in humans (250 million per year only in the USA) [3]. Sutures can be produced using natural or synthetic materials and can be resorbable or non-resorbable [4]. Prominent synthetic bio-resorbable materials used for sutures are polyglycolic acid (PGA), polylactic acid (PLA), or polyglycolide lactide copolymers (PLGA). One example for PLGA sutures is Vicryl® (Ethicon Inc., Edinburgh, Scotland), a widespread braided suture made of polyglactin 910, copolymer of glycolide, and lactide moieties at a ratio of 90:10. Sutures like Vicryl® are often used because of their good mechanical properties, high reproducibility, and minimal tissue reaction. To further strengthen the bond between suture and tissue, several researchers have investigated the coating of sutures with bioactive substances such as bioactive glass [1,5–8]. Bioactive glasses (BGs) have been shown to bond strongly to bone as well as to soft tissues [9–11]. Stamboulis et al. [7] coated biodegradable sutures with bioactive glass (45S5 BG) to tailor biodegradability, to avoid a heterogeneous degradation and to improve the mechanical properties of the suture when exposed to body fluids. Pratten et al. [1] coated Mersilk® (Ethicon Inc., Edinburgh, Scotland) sutures with Ag-doped BG by a conventional slurry-dipping technique. Ag-doped BG was used to make the suture bioactive and antibacterial. The coated sutures showed antibacterial properties against *Staphylococcus epidermidis*, limiting the bacteria adhesion to the surface of the samples.

The use of antibacterial bioactive glasses is being considered to replace the controversially discussed triclosan [7,12–14] used in Vicryl® Plus (Ethicon Inc., Edinburgh, Scotland) sutures.

The introduction of a foreign device in the human body can induce inflammatory reactions or introduce pathogens on site, which in turn could cause infections. Studies have shown that in 2%–5% of all surgeries, surgical site infections (SSIs) can occur [6] with consequent delays in healing, inconvenience for the patient, and in some extreme cases, even leading to death. In addition, SSIs are estimated to cause extra healthcare costs of 1.5 billion dollars just in the USA alone [1,6]. Recent studies have proposed the use of local drug delivery systems to limit bacterial infection after surgeries [15]. The introduction of a local drug delivery system offers a reduction of toxicity and side effects and the possibility to have an effective, highly controlled release. Mesoporous bioactive glasses (MBGs), first reported in 2004 [16], have been proposed as optimal candidates as local drug delivery vehicles. Thanks to their textural properties (e.g., large surface area, ordered mesoporosity with pores in a range of 2–50 nm) MBGs can be used for the release of biologically active molecules with specific effects on cells [17–21].

In the present work, composite sutures were produced combining commercially available sutures (Vicryl® Plus) with silver containing ordered mesoporous bioactive glass (Ag-MBG) and with melt derived BG-doped with zinc (Zn-BG) particles applying, a dip coating technique. To improve the adhesion of the BG particles on the surface, two different polymers were used: chitosan and poly(ε-caprolactone) (PCL) and layered coatings were produced. Sutures were characterized in terms of acellular in vitro bioactivity by immersion in simulated body fluid (SBF) and antibacterial properties against Gram-positive *Staphylococcus carnosus* and Gram-negative *Escherichia coli*. The knot test was also performed to verify the stability and attachment of the coatings to the substrate. The BG/polymer-coated sutures developed in this study represent an innovative advance from the first BG-coated sutures reported earlier [5,6,8]. The novelty of this work is indeed the use of ordered mesoporous glass particles, which can be further loaded with growth factors or drugs (e.g., antibiotics or anti-inflammatories), and by the use of a layered polymeric matrix able to improve the adhesion of the BG particles on the surface and to modulate the release of antibacterial ions.

2. Results

2.1. Microstructural Observations

The slurry dipping technique was optimized for the coating of Vicryl® sutures with Ag-MBG or Zn-BG particles. Qualitative analyses of the morphology and uniformity of the coatings were performed by visual inspection and SEM observation of the glass particles attached on the surface of the samples. Figure 1 shows SEM micrographs of "as-received" and coated sutures. SEM observation confirmed that all coatings homogeneously covered the surface of the sutures.

The adhesion strength of the BG particles to the suture surface was not quantitatively evaluated. However, the coating stability was assessed by performing a knot test. SEM observation revealed that the majority of the coating was detached after the knot test, suggesting that the glass particles weakly adhered to the suture surface. Parts of the coating were peeled off, more at the chitosan/BG composite coating. On the other hand, the PCL/BG composite coating was more rigid and it was harder to perform the test. SEM micrographs documenting this behavior are shown in Figure 2.

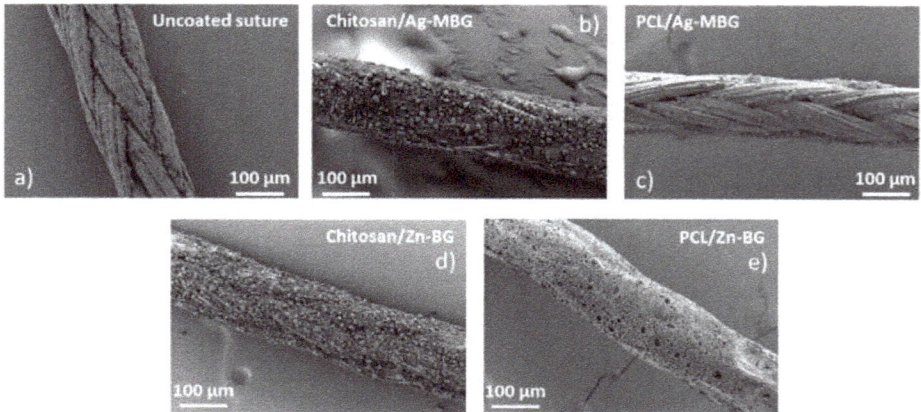

Figure 1. SEM micrographs showing the morphology of: (**a**) Uncoated Vicryl® Plus suture, (**b**) chitosan/Ag-MBG-coated suture, (**c**) PCL/Ag-MBG-coated suture, (**d**) Chitosan/Zn-BG-coated suture, and (**e**) PCL/Zn-BG-coated suture.

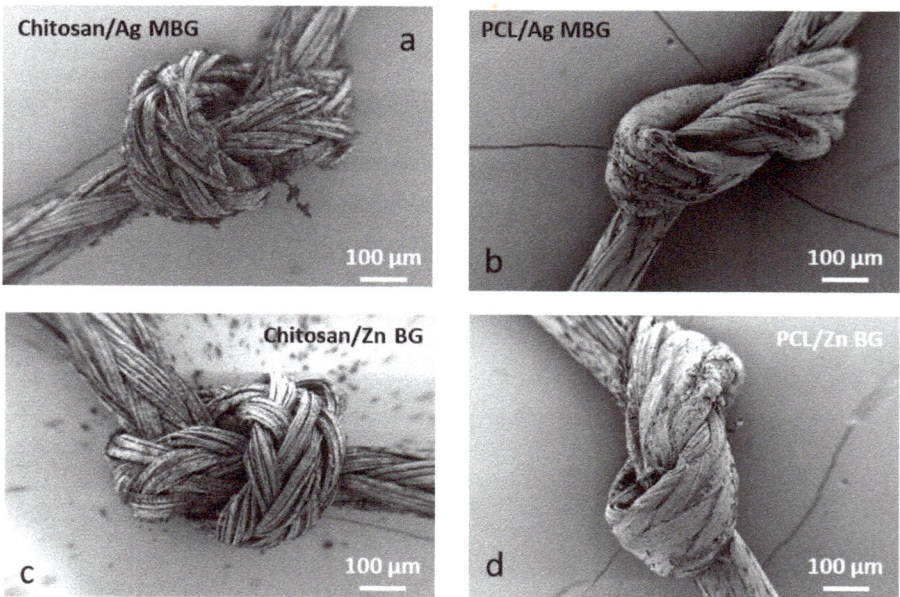

Figure 2. SEM micrographs of coated sutures after the knot test. (**a**) Chitosan/Ag-MBG, (**b**) PCL/Ag-MBG, (**c**) Chitosan/Zn-BG, and (**d**) PCL/Zn-BG.

2.2. Bioactivity Tests

In vitro bioactivity tests were performed by soaking the sutures in SBF for up to seven days. Results after three days are shown in Figure 3. After three days of immersion in SBF, the chitosan/Ag-MBG-coated suture was homogeneously covered by a layer, which appears to be hydroxyl-carbonate-apatite (HCA). On the other hand, PCL/Ag-MBG and PCL/Zn-BG-coated sutures did not show the presence of HCA even after 28 days of immersion in SBF. This result could be explained by the low amount of particles observed on the surface of the PCL samples. Moreover, similar results were shown by Miola et al. [22], who tested the ability of powdered Zn-MBG to form HCA, soaking it in SBF for up to 28 days, without finding any evidence of HCA formation.

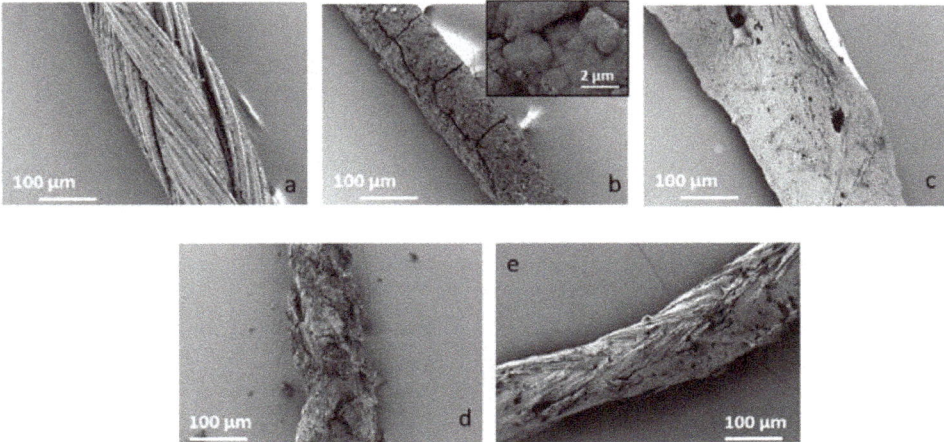

Figure 3. SEM micrographs of uncoated Vicryl® Plus sutures. (**a**) Chitosan/Ag-MBG coated suture, (**b**) PCL/Ag-MBG-coated suture, (**c**) chitosan/Zn-BG suture, (**d**) PCL/Zn-BG suture, and (**e**) suture after 3 days of immersion in SBF: The chitosan/Ag-MBG-coated suture developed a layer, which can be ascribed to hydroxyl-carbonate-apatite based on its morphology.

2.3. Antibacterial Tests

Qualitative antibacterial tests with coated sutures were performed. Agar diffusion tests were carried out to assess the antibacterial properties of the coated sutures against both Gram-positive and Gram-negative bacteria (Figure 4). Uncoated sutures were taken as reference. After 24 h of incubation at 37 °C and high humidity (~80%), an inhibition halo with no bacteria is clearly visible for chitosan/Zn-BG (Figure 4i,ii, sample d)) and chitosan/Ag-MBG (Figure 4i,ii, sample c)) coated sutures for both the selected strains. However, an inhibition zone is also present around the uncoated suture (Figure 4i,ii, sample a)) due to the presence of a triclosan layer in the as-received sutures. In addition, no inhibition zone was observed on samples coated with PCL and containing Ag-MBG and Zn-BG (Figure 4i,ii, samples b and e respectively). This result suggests that the polymeric coating reduces the antibacterial effect of triclosan and that the halos of PCL/BG and chitosan/BG samples are actually induced by the progressive release of Ag^+ and Zn^{2+} from the MBG and BG particles present in the coatings, respectively. Further studies should be performed to clarify the mechanism of antibacterial activity of Vicryl® Plus sutures combined with Ag^+ and Zn^{2+} ion release.

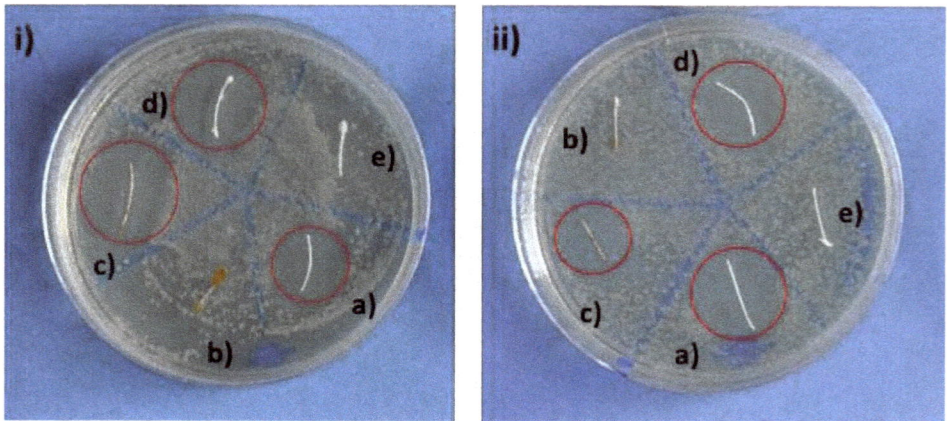

Figure 4. Agar-disk diffusion test with (**i**) Gram-positive (S. carnosus) and (**ii**) Gram-negative (E. coli) bacteria with (**a**) uncoated VICRYL® Plus, (**b**) PCL/Ag-MBG, (**c**) chitosan/Ag-MBG, (**d**) chitosan/Zn-BG, and (**e**) PCL/Zn-BG.

3. Discussion

The aim of this work was the development of composite coatings based on Zn-BG or Ag-MBG particles embedded in a polymer matrix of either chitosan or PCL for surgical sutures. Commercially available Vicryl® (Polyglactin 910) sutures were coated with chitosan/silver-doped ordered mesoporous bioactive glass (Ag-MBG), PCL/Ag-MBG, chitosan/Zn-BG, and PCL/Zn-BG, using a slurry dipping method previously applied in other studies [1,5,7]. The dipping process was optimized and uniform and homogeneous coatings were obtained in case of chitosan/BG coated sutures. On the contrary, sutures coated with PCL/BG were characterized by the presence of a lower amount of BG particles, as a consequence of the lower BG concentration used for the preparation of the slurry. The acellular in vitro bioactive behavior of the coated sutures was investigated by immersing the samples in SBF for three days. SEM observations revealed that the chitosan/Ag-MBG coated suture was able to form HCA on the surface after 3 days of immersion in SBF. Similar results were obtained by Blaker et al. [5], who coated commercially available Mersilk® and Vicryl® sutures with silver-doped BG. The authors observed the formation of crystalline HA after three days in SBF. In another work, Stamboulis et al. [7] corroborated that the coating of commercially available sutures with bioactive glass particles could act as a protective "shield", affecting the extent and rate of degradation of the sutures. However, also considering the present results, further investigations are necessary to understand the effect of the thickness and microstructure of the bioactive glass coating on the overall degradation rate and strength retention of the sutures. The adhesion and stability of the coatings were tested qualitatively, performing the knot test. Both coatings revealed limited adhesion and low mechanical stability on the substrate. However, the PCL/BG coated sutures showed better adhesion, which is probably due to the presence of fewer BG particles on the surface or due to the viscosity of the PCL solution produced in this work.

The antibacterial properties of the coated sutures were tested by means of agar diffusion tests. BG-coated sutures showed clear inhibition zones, which demonstrate their antibacterial effect. However, it should be noted that also pristine Vicryl® Plus sutures showed an inhibition zone. This can be explained by the fact that this type of resorbable suture is subject to a surface treatment with triclosan, a widespread antiseptic. Nowadays, the use of triclosan is being critically discussed because some studies have indicated that this compound could promote liver tumors and provoke muscle weakness [12–14]. For this reason, in the last years, researchers have focused their attention on the development of new materials doped with biologically active ions (e.g., Fe, Cu, Ag, or Ga) able to limit bacteria adhesion and proliferation [23–25]. Nevertheless, it should be noted that the sutures coated with PCL/BG did not show any antibacterial effect against both Gram-positive and Gram-negative bacteria. It might therefore be possible that the coating isolates the triclosan layer, hindering its antibacterial effect. For this reason, it can be ascertained that the antibacterial effect observed for chitosan coatings is due to the well-known bactericidal effect of silver and zinc ions and not to the presence of triclosan. The antibacterial activity of β-chitin/ZnO nanostructured composites was already investigated by Wysokowski et al. [26]. The authors corroborated that the antibacterial activity of chitin/ZnO composite could be caused by the formation of reactive oxygen species in the presence of light. The formation of these active radicals (e.g., OH^- or H_2O_2) can provoke the oxidative stress in the bacteria, leading to perturbation of the cell membrane and damage of cell proteins and DNA [26]. The antibacterial activity of silver ions has been extensively investigated. Studies [8,24,27] have shown that Ag^+ can cause the detachment of the cytoplasm membrane of bacteria from the cell wall, compromising the bacteria's ability to replicate [27,28]. Moreover, bioactive glasses doped with Zn and their possible application as scaffolds for bone tissue engineering, bone filling granules, bone cements, and coatings for orthopedic applications have been widely investigated [29]. Studies [30,31] have also shown the great antimicrobial properties of Zn against both Gram (+) and Gram (-) bacteria.

However, further tests should be carried out to better understand the antibacterial mechanism of the present antibacterial sutures, in particular the possible combined effects of BGs and triclosan, which may lead to a reduction or even elimination of triclosan to achieve the desired antibacterial effect.

4. Materials and Methods

Undyed braided 4-0 Vicryl® Plus sutures (Ethicon Inc., Edinburgh, Scotland) were used. The material is braided from fine filaments of Polyglactin 910, a copolymer of glycolide and lactide at a ratio of 90:10. The braided sutures are additionally coated with Irgacare® MP (Triclosan). The sutures were cut into pieces of 1 cm length (the dimension was measured by means of a caliper) and they were coated by either zinc-doped melt derived BG (ZnBG, composition: 46.13 mol.% SiO_2, 24.35 mol.% Na_2O, 2.60 mol.% P_2O_5, 20.91 mol.% CaO, 6 mol.% ZnO) or silver-doped mesoporous ordered BG (Ag-MBG, composition: 78.00 mol.% SiO_2, 1.20 mol.% P_2O_5, 20 mol.% CaO, 0.8 mol.% AgO), developed in previous works [22,27]. Chitosan (medium molecular weight 190-310 kDa, with 75–85% deacetylation degree) and poly(ε-caprolactone) (PCL, average M_n 8000) were purchased from Sigma-Aldrich (Schnelldorf, Germany) and used without further purification. A two-step coating was used to improve the adhesion between suture and coating; the first layer consists of a polymeric layer (chitosan or PCL) and the second one was a mixture of chitosan or PCL and Ag-MBG or Zn-BG particles. A slurry dipping method was performed manually in a glass beaker placed on a magnetic stirrer. An optimization process based on trial-and-error was used to determine the optimal composition of the slurry required to obtain uniform and homogeneous coatings.

The first layer was obtained by dipping the suture for 1 min in an aqueous chitosan (with 4% *v/v* acetic acid and 2% *w/v* chitosan) or PCL solution (with 50% *v/v* formic acid, 50% *v/v* acetic acid and 15% *w/v* PCL) and letting them dry for 24 h at room temperature. The second layer was applied by dipping the suture for 2 min into a suspension made of BG powder and chitosan/PCL. The second layer of chitosan/BG was prepared as follows: the chitosan solution mentioned beforehand was mixed with an aqueous slurry containing 40 wt.% BG (mixed for 2 h) and then stirred for 4 days. On the other hand, the second layer of PCL/BG was prepared by adding to the afore-mentioned PCL solution 30% *w/v* (with respect to PCL) of BG particles and the resultant suspension was stirred for 2 h. All slurries were produced by using benign solvents, which led to an increase in preparation time but enabled a safer work environment and may lead to a better biological compatibility of the coatings.

The microstructure and uniformity of the coatings were investigated using a light microscope (Leica M50 and IC80, Application Suite LAS V3.8 software, Leica Microsystems GmbH, Wetzlar, Germany) and scanning electron microscopy (SEM) (Gemini, Auriga, Carl Zeiss AG, Jena, Germany). The ability of coated and non-coated Vicryl® sutures to form hydroxyl-carbonate-apatite (HCA) once in contact with biological fluids was assessed by immersion in simulated body fluid (SBF) for different time periods. The standard procedure described by Kokubo et al. [32] was used to carry out these experiments. Samples were placed on CellCrowns™ (Scaffdex Ltd., Tampere, Finland) inserts and immersed in 6 mL of SBF for up to 3 days. Once removed from incubation, the samples were rinsed with deionized water and left to dry at room temperature. The adhesion and stability of the coating were qualitatively evaluated by performing a knot test. The following operations were performed: threading through the eyes of surgical needles, tying a surgical knot, and bending the extremes of the sutures. After these operations, the surface of the samples was observed by SEM. The antibacterial properties of the coated sutures were evaluated using agar diffusion tests against *Escherichia coli* (Gram-negative) and *Staphylococcus carnosus* (Gram-positive). These bacteria were chosen because they are common bacteria responsible for infections [33] and they enable the direct comparison between Gram-positive and Gram-negative strains. The bacteria were obtained from the Microbiology Department of the University of Erlangen-Nuremberg, where they were routinely isolated and characterized. The bacteria population was suspended in LB (lysogeny broth) medium and its optical density (O.D.) was adjusted (at 600 mm, Biophotometer Plus, Eppendorf AG, Hamburg, Germany) to reach the value of 0.015. Then, 20 µl of the prepared medium was deposited and spread homogeneously onto a Petri dish of 10 cm diameter, which was previously covered with a uniform layer of LB-Agar. The samples (sutures) of 1.5 cm length) were placed on top and incubated overnight at 37 °C and at high relative humidity (~80 °C). The next day, the halo of the bacterial growth inhibition zone was evaluated optically and computed.

5. Conclusions

Surgical sutures were successfully coated by a two-step coating process to improve the adhesion between suture and coating: The first layer consisted of a polymeric layer (chitosan or PCL), and the second one formed by a mixture of chitosan or PCL and Ag-MBG or Zn-BG particles. Ag-MBG coated sutures showed a high reactivity once in contact with simulating body fluid, developing a layer of HCA after three days of immersion, while Zn-BG did not lead to HCA formation. Moreover, the chitosan coated samples showed promising results in terms of antibacterial properties against both Gram-positive and Gram-negative strains. Coatings with PCL did not show any antibacterial properties, which might be due to the low glass concentration present in the outer layer of the coating. Future investigations to determine the mechanical properties of coated sutures should be performed, e.g., by combining both polymers, in different layer structures and by optimizing the BG content.

Author Contributions: Conceptualization, F.E.C. and A.R.B.; Methodology, F.E.C., A.R.B. and W.H.G.; Software, F.E.C. and K.S.; Validation, F.E.C., A.R.B. and W.H.G.; Formal Analysis, F.E.C. and K.S.; Investigation, F.E.C. and K.S.; Resources, K.S. and W.H.G.; Data curation, F.E.C.; Writing-Original Draft Preparation, F.E.C., and K.S.; Writing-Review & Editing, W.H.G. and A.R.B.; Visualization, F.E.C. and K.S.; Supervision, W.H.G. and A.R.B.; Project Administration, A.R.B.; Funding Acquisition, A.R.B.

Funding: This research was carried out within the EU Horizon 2020 framework project COACH (ITN-ETN, Grant agreement 642557) Ms Francesca Elisa Ciraldo acknowledges the financial support. W.H. Goldmann thanks the German Science Foundation (DFG Go598) for financial support.

Acknowledgments: The authors would like to thank Astrid Mainka (Biophysics group, FAU) for her technical support and A. Arkudas (University Hospital, Erlangen) for providing the sutures.

Conflicts of Interest: The authors declare no conflict of interest.

References

1. Pratten, J.; Nazhat, S.N.; Blaker, J.J.; Boccaccini, A.R. In Vitro Attachment of Staphylococcus Epidermidis to Surgical Sutures with and without Ag-Containing Bioactive Glass Coating. *J. Biomater. Appl.* **2004**, *19*, 47–57. [CrossRef] [PubMed]

2. Capperauld, I. Suture materials: A review. *Clin. Mater.* **1989**, *4*, 3–12. [CrossRef]

3. Qin, Y. Surgical sutures. In *Medical Textile Materials*; Qin, Y., Ed.; Woodhead Publishing: Cambridge, UK, 2016; pp. 123–132.

4. Alshomer, F.; Madhavan, A.; Pathan, O.; Song, W. Bioactive Sutures: A Review of Advances in Surgical Suture Functionalisation. *Curr. Med. Chem.* **2017**, *24*, 215–223. [CrossRef] [PubMed]

5. Blaker, J.J.; Nazhat, S.N.; Boccaccini, A.R. Development and characterisation of silver-doped bioactive glass-coated sutures for tissue engineering and wound healing applications. *Biomaterials* **2004**, *25*, 1319–1329. [CrossRef] [PubMed]

6. Boccaccini, A.R.; Stamboulis, A.G.; Rashid, A.; Roether, J.A. Composite surgical sutures with bioactive glass coating. *J. Biomed. Mater. Res.* **2003**, *67B*, 618–626. [CrossRef] [PubMed]

7. Stamboulis, A.; Hench, L.L.; Boccaccini, A.R. Mechanical properties of biodegradable polymer sutures coated with bioactive glass. *J. Mater. Sci. Mater. Med.* **2002**, *13*, 843–848. [CrossRef] [PubMed]

8. Blaker, J.J.; Boccaccini, A.R.; Nazhat, S.N. Thermal Characterizations of Silver-containing Bioactive Glass-coated Sutures. *J. Biomater. Appl.* **2005**, *20*, 81–98. [CrossRef]

9. Miguez-Pacheco, V.; Hench, L.L.; Boccaccini, A.R. Bioactive glasses beyond bone and teeth: Emerging applications in contact with soft tissues. *Acta Biomater.* **2015**, *13*, 1–15. [CrossRef]

10. Jones, J.R. Review of bioactive glass: From Hench to hybrids. *Acta Biomater.* **2013**, *9*, 4457–4486. [CrossRef]

11. Catauro, M.; Tranquillo, E.; Risoluti, R.; Vecchio Ciprioti, S. Sol-Gel Synthesis, Spectroscopic and Thermal Behavior Study of SiO2/PEG Composites Containing Different Amount of Chlorogenic Acid. *Polymers (Basel).* **2018**, *10*, 682. [CrossRef]

12. Cherednichenko, G.; Zhang, R.; Bannister, R.A.; Timofeyev, V.; Li, N.; Fritsch, E.B.; Feng, W.; Barrientos, G.C.; Schebb, N.H.; Hammock, B.D.; et al. Triclosan impairs excitation-contraction coupling and Ca2+ dynamics in striated muscle. *Proc. Natl. Acad. Sci.* **2012**, *109*, 14158–14163. [CrossRef] [PubMed]

13. Yueh, M.-F.; Taniguchi, K.; Chen, S.; Evans, R.M.; Hammock, B.D.; Karin, M.; Tukey, R.H. The commonly used antimicrobial additive triclosan is a liver tumor promoter. *Proc. Natl. Acad. Sci.* **2014**, *111*, 17200–17205. [CrossRef] [PubMed]

14. James, M.O.; Li, W.; Summerlot, D.P.; Rowland-Faux, L.; Wood, C.E. Triclosan is a potent inhibitor of estradiol and estrone sulfonation in sheep placenta. *Environ. Int.* **2010**, *36*, 942–949. [CrossRef] [PubMed]

15. Dennis, C.; Sethu, S.; Nayak, S.; Mohan, L.; Morsi, Y.Y.; Manivasagam, G. Suture materials—Current and emerging trends. *J. Biomed. Mater. Res. Part A* **2016**, *104*, 1544–1559. [CrossRef]

16. Yan, X.; Yu, C.; Zhou, X.; Tang, J.; Zhao, D. Highly ordered mesoporous bioactive glasses with superior in vitro bone-forming bioactivities. *Angew. Chemie Int. Ed.* **2004**, *43*, 5980–5984. [CrossRef]

17. Wu, C.; Chang, J. Multifunctional mesoporous bioactive glasses for effective delivery of therapeutic ions and drug/growth factors. *J. Control. Release* **2014**, *193*, 282–295. [CrossRef]

18. Vallet-Regí, M.; Manzano Garcia, M.; Colilla, M. *Biomedical Application of Mesopous Ceramics Drug Delivery, Smart Materials and Bone Tissue Engineering*; Vallet-Regí, M., Manzano Garcia, M., Colilla, M., Eds.; CRC Press: New York, NY, USA, 2013; pp. 1–231.

19. Wu, C.; Chang, J. Mesoporous bioactive glasses: Structure characteristics, drug/growth factor delivery and bone regeneration application. *Interface Focus* **2012**, *2*, 292–306. [CrossRef]

20. Baino, F.; Fiorilli, S.; Vitale-Brovarone, C. Bioactive glass-based materials with hierarchical porosity for medical applications: Review of recent advances. *Acta Biomater.* **2016**, *42*, 18–32. [CrossRef]

21. Salinas, A.J.; Vallet-Regí, M. Evolution of ceramics with medical applications. *Zeitschrift fur Anorg. und Allg. Chemie* **2007**, *633*, 1762–1773. [CrossRef]

22. Miola, M.; Verné, E.; Ciraldo, F.E.; Cordero-Arias, L.; Boccaccini, A.R. Electrophoretic Deposition of Chitosan/45S5 Bioactive Glass Composite Coatings Doped with Zn and Sr. *Front. Bioeng. Biotechnol.* **2015**, *3*, 1–13. [CrossRef]

23. Catauro, M.; Tranquillo, E.; Barrino, F.; Blanco, I.; Dal Poggetto, F.; Naviglio, D. Drug Release of Hybrid Materials Containing Fe(II)Citrate Synthesized by Sol-Gel Technique. *Materials (Basel).* **2018**, *11*, 2270. [CrossRef] [PubMed]

24. Jung, W.K.; Koo, H.C.; Kim, K.W.; Shin, S.; Kim, S.H.; Park, Y.H. Antibacterial Activity and Mechanism of Action of the Silver Ion in Staphylococcus aureus and Escherichia coli. *Appl. Environ. Microbiol.* **2008**, *74*, 2171–2178. [CrossRef] [PubMed]

25. Gritsch, L.; Lovell, C.; Goldmann, W.H.; Boccaccini, A.R. Fabrication and characterization of copper(II)-chitosan complexes as antibiotic-free antibacterial biomaterial. *Carbohydr. Polym.* **2018**, *179*, 370–378. [CrossRef] [PubMed]

26. Wysokowski, M.; Motylenko, M.; Stöcker, H.; Bazhenov, V.V.; Langer, E.; Dobrowolska, A.; Czaczyk, K.; Galli, R.; Stelling, A.L.; Behm, T.; et al. An extreme biomimetic approach: Hydrothermal synthesis of β-chitin/ZnO nanostructured composites. *J. Mater. Chem. B* **2013**, *1*, 6469–6476. [CrossRef]

27. Ciraldo, F.; Liverani, L.; Gritsch, L.; Goldmann, W.H.; Boccaccini, A.R. Synthesis and Characterization of Silver-Doped Mesoporous Bioactive Glass and Its Applications in Conjunction with Electrospinning. *Materials (Basel).* **2018**, *11*, 692. [CrossRef] [PubMed]

28. Goudouri, O.; Kontonasaki, E.; Lohbauer, U.; Boccaccini, A.R. Acta Biomaterialia Antibacterial properties of metal and metalloid ions in chronic periodontitis and peri-implantitis therapy. *Acta Biomater.* **2014**, *10*, 3795–3810. [CrossRef]

29. Balasubramanian, P.; Strobel, L.A.; Kneser, U.; Boccaccini, A.R. Zinc-containing bioactive glasses for bone regeneration, dental and orthopedic applications. *Biomed. Glas.* **2015**, *1*, 51–69. [CrossRef]

30. Aydin Sevinç, B.; Hanley, L. Antibacterial activity of dental composites containing zinc oxide nanoparticles. *J. Biomed. Mater. Res. Part B Appl. Biomater.* **2010**, *94*, 22–31. [CrossRef]

31. Esteban-Tejeda, L.; Díaz, L.; Prado, C.; Cabal, B.; Torrecillas, R.; Moya, J. Calcium and Zinc Containing Bactericidal Glass Coatings for Biomedical Metallic Substrates. *Int. J. Mol. Sci.* **2014**, *15*, 13030–13044. [CrossRef]

32. Kokubo, T.; Takadama, H. How useful is SBF in predicting in vivo bone bioactivity? *Biomaterials* **2006**, *27*, 2907–2915. [CrossRef]

33. Bellantone, M.; Williams, H.D.; Hench, L.L. Broad-spectrum bactericidal activity of Ag_2O-doped bioactive glass. *Antimicrob. Agents Chemother.* **2002**, *46*, 1940–1945. [CrossRef] [PubMed]

materials

MDPI

Article

Studies on Cell Compatibility, Antibacterial Behavior, and Zeta Potential of Ag-Containing Polydopamine-Coated Bioactive Glass-Ceramic

Rocío Tejido-Rastrilla [1,2], Sara Ferraris [3], Wolfgang H. Goldmann [4], Alina Grünewald [1], Rainer Detsch [1], Giovanni Baldi [2,*], Silvia Spriano [3] and Aldo R. Boccaccini [1,*]

[1]   Institute of Biomaterials, University of Erlangen-Nuremberg, 91058 Erlangen, Germany;
    rocio.tejido@fau.de (R.T.-R.); alina.gruenewald@fau.de (A.G.); rainer.detsch@fau.de (R.D.)
[2]   Colorobbia Consulting s.r.l., 50053 Sovigliana Vinci, Florence, Italy
[3]   Department of Applied Science and Technology, Politecnico di Torino, 10129 Torino, Italy;
    sara.ferraris@polito.it (S.F.); silvia.spriano@polito.it (S.S.)
[4]   Centre for Medical Physics and Technology, University of Erlangen-Nuremberg, 91052 Erlangen, Germany;
    wgoldmann@biomed.uni-erlangen.de
*   Correspondence: baldig@colorobbia.it (G.B.); aldo.boccaccini@ww.uni-erlangen.de (A.R.B.)

Received: 23 December 2018; Accepted: 29 January 2019; Published: 6 February 2019

Abstract: Dopamine is a small molecule that mimics the adhesive component (L-DOPA) of marine mussels with a catecholamine structure. Dopamine can spontaneously polymerize to form polydopamine (PDA) in a mild basic environment. PDA binds, in principle, to all types of surfaces and offers a platform for post-modification of surfaces. In this work, a novel Ag-containing polydopamine coating has been developed for the functionalization of bioactive glass-ceramics. In order to study the interactions between the surface of uncoated and coated samples and the environment, we have measured the surface zeta potential. Results confirmed that PDA can interact with the substrate through different chemical groups. A strongly negative surface zeta potential was measured, which is desirable for biocompatibility. The dual function of the material, namely the capability to exhibit bioactive behavior while being antibacterial and not harmful to mammalian cells, was assessed. The biocompatibility of the samples with MG-63 (osteoblast-like) cells was determined, as well as the antibacterial behavior against Gram-positive *Staphylococcus carnosus* and Gram-negative *Escherichia coli* bacteria. During cell biology tests, uncoated and PDA-coated samples showed biocompatibility, while cell viability on Ag-containing PDA-coated samples was reduced. On the other hand, antibacterial tests confirmed the strong antimicrobial properties of Ag-containing PDA-coated samples, although tailoring of the silver release will be necessary to modulate the dual effect of PDA and silver.

Keywords: polydopamine; silver; antibacterial; biocompatibility; bioactive glass-ceramic; coatings

---

1. Introduction

The average life expectancy of humans is increasing worldwide thanks to advances in medicine and science. As a consequence, there has been an increase in osteoarthritis and other pathologies in elderly people, who consequently require orthopedic and dental implants [1]. The introduction of implants has the intrinsic risk of microbial infection, which can lead to implant failure. Bacterial infections are becoming more difficult to treat due to antimicrobial resistance (AMR). In the European Union, which has a population of ~500 million people, there are about 25,000 deaths per year due to bacterial infections [2]. Not only does this represent an unacceptable loss of lives, it carries an economic burden; if AMR is not controlled, the annual global gross domestic product will lose

3.8% by 2050 [3]. New approaches must therefore be put forward to prevent bacterial colonization of different surfaces, thus limiting the spread of infections.

Numerous biomaterials have been developed over the years for the application as orthopedic or dental implants, including stainless steel 316 L, cobalt-based alloys, titanium and its alloys, polymers, glass-ceramics and bioactive glasses, and their composites [1]. Among these, bioactive glasses and glass-ceramics have received much attention, because of their desirable mechanical properties and their capability to interact with hard [4,5] and soft tissues [6].

Surface functionalization offers the possibility to tailor the properties of a material's surface to obtain an optimal interface between the substrate and the biological environment [7]. Since the discovery of the outstanding properties of polydopamine [8], several materials have been coated with this oligomer, which mimics the molecules found near the plaque–substrate interface of mussels [8–10]. Thanks to the presence of catechol, amine, and imine moieties in its structure, polydopamine can undergo further functionalization by electroless metallization with copper or silver [11,12]. Silver nanoparticles and nanocomposites are well known for their strong antibacterial capabilities [13,14]. Silver has been also shown to be non-cytotoxic to human cells [13–15].

The incorporation of silver in bioactive glasses has been investigated for many years [16–18]. This is usually achieved by the sol-gel technique; another option is to coat the surfaces with silver using different techniques such as plasma spraying, molten salt ion exchange, or sputtering [16,17,19,20].

Coatings, based on polydopamine incorporating silver particles, represent an attractive approach to obtain dual-functional composite layers on the surface of biomaterials. Such composites should facilitate interactions with biological tissues without showing cytotoxicity and also exhibit antibacterial capability [21–25]. In our previous study [26], Ag-containing polydopamine coatings on bioactive glass-ceramic surfaces were developed; however, only a limited characterization of the coatings has been reported. In this work, such coatings have been characterized by testing their biocompatibility. We have carried out antibacterial tests against Gram-positive *Staphylococcus carnosus* and Gram-negative *Escherichia coli* bacterial strains. We have further conducted studies regarding the surface zeta potential of the samples and hypothesize that due to the presence of several chemical moieties, polydopamine can bind differently to different biomaterials depending on the characteristics of the substrate.

2. Materials and Methods

2.1. Fabrication of the Samples and Structural Characterization

All tests were performed on sintered glass-ceramic (labelled BGC1) pellets, uncoated and coated with PDA and PDA@Ag (BGC1@PDA, BGC1@PDA@Ag), of which the details are described elsewhere [26]. Briefly, we produced a melt-derived glass of the nominal composition BGC1 (in wt%): 47.8% SiO_2, 4.9% Na_2O, 0.4% K_2O, 30.6% CaO, 2.9% MgO, 11.8% P_2O_5, and 1.6% CaF_2. Ground BGC1 powder was then cold-pressed at 0.7 MPa for 30 s to form cylindrical pellets with a diameter of 11 mm and a height of 3 mm. The pellets were sintered at 860 °C for 30 min. In order to coat the pellets with polydopamine, sintered pellets were immersed for 24 h in aqueous solution of dopamine hydrochloride (Sigma-Aldrich, Schnelldorf, Germany) in Tris-buffer solution (Tris(hydroxymethyl) aminomethane, Sigma-Aldrich, Schnelldorf, Germany). The solution's pH value was adjusted to pH 8.5 using HCl 1 M. The pellets were then thoroughly rinsed with deionized water. Additionally, the PDA-coated pellets (BGC1@PDA) were further functionalized by immersing them for 24 h in an aqueous solution of $AgNO_3$ 5×10^{-3} M (Sigma-Aldrich, Schnelldorf, Germany). BGC1@PDA@Ag pellets were then rinsed with deionized water and allowed to dry in normal air.

The surface roughness was measured for each type of coating by using a laser profilometer (UBM Microfocus Expert, ISC-2). A measurement length of 5 mm was used with a scanning velocity of 400 points per second. The mean roughness (R_a) and maximum roughness (R_{max}) were calculated using the LMT Surface View UBM software (UBM Messtechnik GmbH, Ettlingen, Germany). R_a is calculated as the average roughness of the coating's surfaces by measuring microscopic peaks and valleys, while R_{max} is calculated by measuring the vertical distance from the highest peak to the lowest valley. Three samples were measured for each type of coating, and mean values of R_a and R_{max} were reported with standard deviation.

Morphological analysis of the samples, prior to and after coating, was performed by means of scanning electron microscopy (SEM) (Auriga SEM instrument Zeiss, Oberkochen, Germany). Samples were sputter-coated with gold in vacuum.

2.2. Surface Zeta Potential

The zeta potential of the samples was measured by means of the streaming potential technique, using an electrokinetic analyser (SurPASS, Anton Paar, Graz, Austria) equipped with an adjustable gap cell. All measurements were performed in simulated body fluid (SBF) [27], which was diluted in ultrapure water up to a pH of about 7.4 and had a conductivity close to 16 mS·m^{-1} (measured by the monitoring function of the instrument), without pH titration during measurements. For the analyses, two cylindrical samples (11 mm diameter, 3 mm height) were prepared for each type (bare BGC1, BGC1@PDA and BGC1@PDA@Ag) and mounted parallel to each other on the sample holder of the cell. The gap was adjusted close to 100 μm and the electrolyte flow to approximately 100 mL·min^{-1}. The zeta potential values were obtained from measured streaming currents using the Helmholtz–Smoluchowski Equation (1):

$$\zeta = \frac{dU}{dp} \times \frac{\eta}{\epsilon \times \epsilon_0} \times \kappa_B, \tag{1}$$

where dU/dp is the slope of streaming potential vs. differential pressure, η is electrolyte viscosity, ϵ is dielectric coefficient of electrolyte, ϵ_0 is permittivity, and κ_B is electrolyte conductivity.

2.3. Preconditioning of the Samples

Uncoated BGC1 pellets were sterilized by dry heat (160 °C for 2 h). The samples were placed in a 24-well plate and covered with 1 mL DMEM (+10% FBS, +1% penicillin/streptomycin (PS)). The pH of the medium was measured after 1, 2, 3, 4, 7, and 8 days. Measurements were made in duplicate and the average was calculated. After the measurements, the medium was removed and fresh medium was added. The preconditioning of the samples was carried out under sterile conditions.

2.4. Cell Biology Studies

2.4.1. Cell Seeding and Culture

MG-63 (osteoblast-like) cells (Sigma-Aldrich, Schnelldorf, Germany) were used. The cells were cultured at 37 °C in an atmosphere of 95% humidified air and 5% CO$_2$, in Dulbecco's modified Eagle's medium (DMEM; Gibco, ThermoFisher Scientific, Waltham, MA, USA) supplemented with 10 vol % fetal bovine serum (FBS; Gibco, Germany) and 1% of penicillin/streptomycin (PS; Gibco, ThermoFisher Scientific, Waltham, MA, USA). Cells were grown to 80% confluence in 75 cm^2 culture flasks, washed with phosphate buffered saline and detached using trypsin/EDTA (Sigma-Aldrich, Schnelldorf, Germany). Cells were counted by a hemocytometer (Neubauer improved) and diluted with culture medium to a final concentration of 1×10^5 cells/mL. Subsequently, 1 mL of cell solution was seeded in direct contact in a 24-well cell culture plate. To ensure statistical significance, eight replicates of each sample type were performed.

2.4.2. Cell Viability

The viability of MG-63 cells was assessed using the WST-8 assay kit (Sigma-Aldrich, Schnelldorf, Germany). WST-8 (2-(2-methoxy-4-nitrophenyl)-3-(4-nitrophenyl)-5-(2,4-disulfophenyl)-2H-tetrazolium, monosodium salt) is reduced by cellular dehydrogenase to a formazan product, which is directly proportional to the number of living cells.

After 48 h of incubation, the culture medium was completely removed, and samples were washed with PBS. Subsequently, 0.25 mL of WST-8 medium (containing 1 vol % of WST-8 reagent and 99 vol % of DMEM medium) was added and incubated for 2 h. Afterwards, 100 µL of supernatant from each sample was transferred into a well of a 96-well plate, and the absorbance at 450 nm was measured with a microplate reader (PHOmo, Autobio Labtec Instruments Co. Ltd., Zhengzhou, China). From the obtained absorbance, the cell viability was calculated by taking the absorbance of each specimen (A_i) and of the respective control (A_0) as follows:

$$\text{Cell viability}(\%) = \frac{A_i}{A_0} \times 100, \tag{2}$$

2.4.3. Cell Morphology

To visualize the adhered cells on the samples, green Calcein AM (calcein acetoxymethyl ester, Invitrogen, ThermoFisher Scientific, Waltham, MA, USA) cell-labelling solution was used to stain the cytoplasm of the cells. Cell culture medium was removed and 0.25 mL of staining solution containing 0.5 vol % of dye-labelling solution and 99.5 vol % of PBS was added to the samples and incubated for 30 min. Afterwards, the solution was removed and the samples were washed with 0.5 mL PBS. Cells on the surface were fixed by 3.7 vol % paraformaldehyde. Samples were washed with PBS again and blue fluorescent DAPI (4′,6-diamidino-2-phenylindole dihydrochloride, Roche, Basel, Switzerland) was added to stain the nucleus. The samples were incubated for 5 min, and the solution was removed. The samples were left in PBS for microscopic viewing, using a fluorescence microscope (Axio Scope, ZEISS, Jena, Germany).

2.5. Statistical Analysis

The differences in analytical parameters between the different samples were analyzed by one-way analysis of variance (ANOVA). The significance level was set as * $p < 0.05$, ** $p < 0.01$ and *** $p < 0.001$. For comparison of the mean values, the Tukey post hoc test was used (N = 8).

2.6. Antibacterial Tests

A direct contact bacterial assay was performed on uncoated and coated BGC1 pellets. The dimensions of the pellets were 10 mm (diameter) and 2 mm (height). The samples were tested before and after a pre-incubation time of 8 days in DMEM (Dulbecco's modified Eagle's medium) to maintain the same conditions as in the case of cell viability tests.

Prior to coating the samples, they were dry-sterilized in an oven at 160 °C for 2 h. Afterwards, the samples were coated as described elsewhere [26]. After the coating of the samples, they were sterilized by UV irradiation for 1 h.

Gram-positive *Staphylococcus carnosus* and Gram-negative *Escherichia coli* were chosen as test bacterial strains. Isolated colonies of both Gram-positive and Gram-negative bacteria were suspended in 10 mL of lysogeny broth (LB #968.1, Carl Roth GmbH) and grown overnight in an orbital shaker at 100 rpm at 37 °C.

The next day, the fresh bacteria suspension was diluted to an optical density of 0.015 at 600 nm (OD600) (Biophotometer Plus, Eppendorf AG, Hamburg, Germany).

Uncoated and coated samples were placed in a 24 well-plate with 2 mL of LB, and fresh bacteria suspension was inoculated into the samples and put in the incubator at 37 °C. Optical density (OD600) was measured after 1, 4, 8, 24, and 48 hours. To ensure statistical significance, three replicates of each sample type were performed. In addition, bacterial cultures were done in duplicate, on different days.

3. Results and Discussion

3.1. Structural Characterization

The surface roughness was measured, and results are summarized in Table 1. It can be seen that for all samples the roughness is quite homogeneous. In this case, the roughness measurements are related to the substrate only, since polydopamine coating has been determined to be ~50 nm in thickness, as reported in literature [9,28]. In this way, for the roughness measurements, the influence of the substrate would be more significant than the polydopamine film itself.

Table 1. Roughness measurements of the different samples investigated.

-	R_a (µm)	R_{max} (µm)
Sintered ceramic-glass pellets, uncoated (BGC1)	1.4 ± 0.3	12 ± 5
BGC1 coated with polydopamine (BGC1@PDA)	1.0 ± 0.2	8 ± 2
BGC1 coated with polydopamine and Ag (BGC1@PDA@Ag)	0.8 ± 0.1	6.1 ± 0.3

In order to study the morphology of the surface, SEM micrographs were obtained (Figure 1). In the case of BGC1, it is possible to observe sharpened microcrystals (Figure 1A,B), although the overall surface is quite homogeneous. For BGC1@PDA, the surface is smoother, probably due to the deposition of polydopamine (Figure 1C,D), which forms spherical aggregates. After the deposition of silver onto the surface of the coated bioactive glass-ceramic, some silver aggregates are visible, being well distributed on the surface (Figure 1E,F).

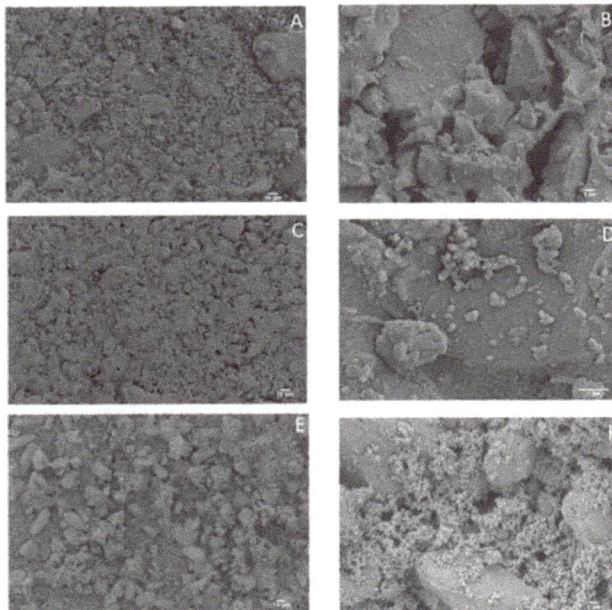

Figure 1. Scanning electron microscopy (SEM) micrographs of uncoated BGC1 (**A**,**B**), BGC1@PDA (**C**,**D**), and BGC1@PDA@Ag (**E**,**F**).

3.2. Surface Zeta Potential

The contact between a solid surface and a water-based medium leads to the development of a surface charge at the interface. This charge is one of the surface characteristics, which could affect the interaction between the material and the biological environment (e.g., protein adsorption, cellular, and bacterial adhesion) [29]. In this context, zeta potential measurements were carried out to determine the surface charge at physiological pH (about 7.4). The pH of an aqueous solution is the driving force for an acid-base reaction, meaning that at high pH values, the dissociation of acidic groups will be enhanced while the protonation of basic groups will be suppressed, and vice versa.

The surface charge on polydopamine films is probably due to quinone imine and catechol groups [30]. More specifically, the positive or negative surface charges may arise from the reversible dissociation and deprotonation/protonation of amine and catechol groups, featuring PDA zwitterionicity [10]. It has been reported in the literature that the overall charge of PDA coatings is negative, although there is no general agreement regarding the zeta potential value; values reported vary between -4.58 mV (at pH = 7) and -39 mV (in Tris-buffer at pH = 8.5) [30–33], which can strongly depend on the measurement conditions (pH, electrolyte). Obtained results on surface zeta potential measurements are summarized in Table 2.

For uncoated BGC1, we obtained a highly negative zeta potential value (-120 ± 9 mV). Such negative surface charge at physiological pH is in accordance with the acidic isoelectric point of silica-based bioactive glasses [34]. Moreover, the negative surface charge of BGC1 is likely the result of the presence of siloxane (Si–OH) groups and hydroxyapatite ($Ca_5(PO_4)_3(OH)$) on the surface (Figure 2), which in the presence of an aqueous solution at pH ~7.4 remains negatively charged. The negative surface charge of the glass-ceramic BGC1 could enhance the affinity and adhesion of cells. It has been reported that strongly negative surface charges promote the adsorption of specific proteins, leading to increased cell adhesion [35].

When BGC1 was modified with PDA, the value of the zeta potential at physiological pH remained negative, but its absolute value was reduced (-83 ± 1 mV) with respect to uncoated BGC1. It is possible that, thanks to the presence of PDA, positive ions were attracted from the solution (for example Ca^{2+}), increasing the positive charge of the surface; this is in accordance with the literature [36–38]. However, this hypothesis is unlikely in the present case because, due to the high dilution of SBF, the availability of Ca^{2+} is relatively low.

It has been proposed in the literature that polydopamine could interact with aqueous solutions through various mechanisms due to its molecular structure [39]. If PDA binds differently, it means that exposed chemical groups would be different. Therefore, it is possible that for BGC1, polydopamine binds through OH^- groups from quinone, leaving more amine groups exposed to the aqueous environment and thus provoking the coated surface to become less negative.

With the introduction of silver onto the surfaces of PDA-modified samples (BGC1@PDA@Ag), the value of the zeta potential becomes slightly more negative (-98 ± 1 mV) than for BGC1@PDA. This behavior could be explained by the presence of silver particles which possess negative zeta potential, as described in the literature for silver nanoparticles [40–42] and for coatings containing silver nanoparticles at the investigated pH [43].

In summary, the measurements reported here evidence a strong negative charge on all tested surfaces in physiological conditions. The absolute value of these charges can depend on the typology and distribution of the surface functional groups as well as on their acidity strength and on their effect on surface wettability. However, a clear attribution of these differences cannot be obtained by measurements at a fixed pH; indeed, a zeta potential titration as a function of pH should be performed in future works to clarify this point.

Table 2. Surface zeta potential of the different samples investigated (measurements in simulated body fluid (SBF)).

-	Initial Conditions		Measurements in SBF	
-	pH	Conductivity (mS·m^{-1})	pH	Z (mV)
BGC1	7.32	16.14	7.34 ± 0.01	-120 ± 9
BGC1@PDA	7.33	16.84	7.33 ± 0.00	-83 ± 1
BGC1@PDA@Ag	7.38	16.20	7.35 ± 0.00	-98 ± 1

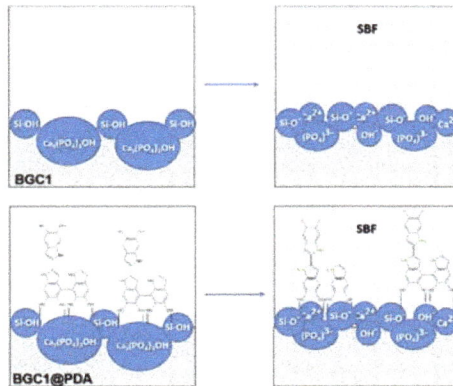

Figure 2. Schematic representations of the origin of negative charge on the uncoated (BGC1) and coated (BGC1@PDA) surface in diluted SBF at pH ≈ 7.4.

3.3. Preconditioning of the Samples

Figure 3 shows the comparison of pH variation in DMEM and in SBF in contact with uncoated BGC1 pellets as a function of time. It must be taken into account that in the case of DMEM, the medium was changed every day, while for SBF the medium was not changed. Both measurements were carried out under static conditions. Observing the behavior in SBF, it is quite clear that a chemical process is occurring at the surface of BGC1 as already proved elsewhere [26]. In this work, the selected medium for pre-treatment was DMEM, since it is the medium used in cell biology tests. The preconditioning time needed for the pH to be lower than 7.75 was selected as 8 days, which is in good agreement with the work of Verné et al. [44]. The limit for the pH value, set to be lower than pH 7.75, has been established as an optimal value for the adhesion of osteoblasts [45]. The duration of the preconditioning treatment, namely 8 days, was relatively short. DMEM does not seem to provoke an extreme reaction at the surface of BGC1. Since BGC1 has a low content of Na_2O (4.9%), a rapid exchange of sodium ions is not expected, which would cause a detrimental burst of local pH increase. Finally, the morphology of the samples also plays an important role in determining their bioreactivity [46]. It should be noted that in this study we have used BGC1 in the form of dense pellets, which exhibit a slower rate of ion exchange (bioreactivity) compared, for example, with porous materials or powders, which have much higher surface areas.

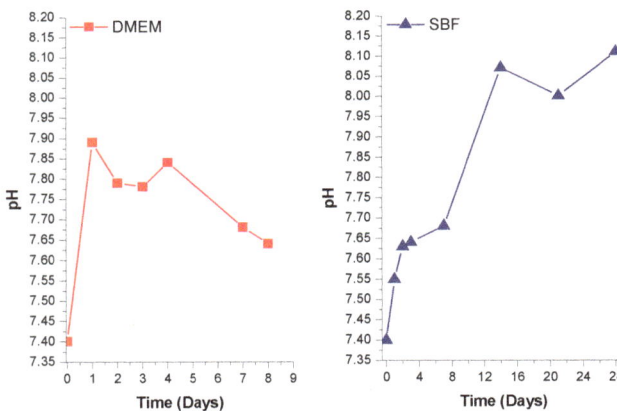

Figure 3. pH variation of DMEM (**left**) and SBF (**right**) containing uncoated BGC1 pellets.

3.4. Cell Biology Studies

The viability of MG-63 cells cultured onto BGC1, BGC1@PDA, and BGC1@PDA@Ag samples for 48 hours was determined by WST-8 assay, and the results are shown in Figure 4. The bioactive glass-ceramic BGC1 showed, as expected, good compatibility with mammalian cells. Sintered BGC1 was determined to have a glassy matrix (15.8%) and the following crystalline phases: hydroxyapatite (47.6%), wollastonite (28.7%), and cristobalite (7.9%) [26]. An important phase in BGC1 is wollastonite. This crystalline phase has compatibility with bone tissue, as described in literature [47]. In the same way, hydroxyapatite and glass-ceramic implants have shown favorable interactions with marrow stromal cells [48–50]. Figure 5 shows fluorescence microscope images of MG-63 cells after 48 hours of direct culture. For BGC1, it was difficult to determine the morphology of the cells because of the brightness of the surface. However, the surface is seen to be completely covered by cells.

Polydopamine has been proven to have good cell compatibility, mostly due to (among other properties) its hydrophilicity, stiffness, and surface charge [9,10,51,52]. Figure 4 shows a slight suppression of cell viability for BGC1@PDA, which can be related to a modification of the surface roughness during the coating process. It is important to highlight that there is a strong influence of the substrate, since the polydopamine coating on BGC1 does not form a continuous film, reaching only 50 nm of thickness. Ryu et al. determined that polydopamine surfaces lead to mammalian cell proliferation without toxicity [36]; in addition, Chien and Tsai described that polydopamine does not support the adhesion of all types of cells [53]. On this basis, the lower number of cells on BGC1@PDA samples compared to BGC1 may not be related to cell death, but to a lower rate of proliferation. On the other hand, it can be seen in Figure 5 that cells spread well on BGC1@PDA surfaces and were interconnected to each other, which is an expected behaviour of bioactive materials.

BGC1@PDA@Ag samples exhibited a strong decrease in cell viability (52%). Fluorescence microscope images (Figure 5) show that cells are round shaped, which indicates cell stress. It is possible to observe calcein-stained cells, which means that cells are alive but under high stress. Forte et al. [24] used polydopamine as an interface between calcium phosphate and silver. It was also found that for certain samples coated with silver, cell toxicity occurred [24]. Therefore, more research efforts should be undertaken to optimize the silver dose to obtain an effective dual-function material.

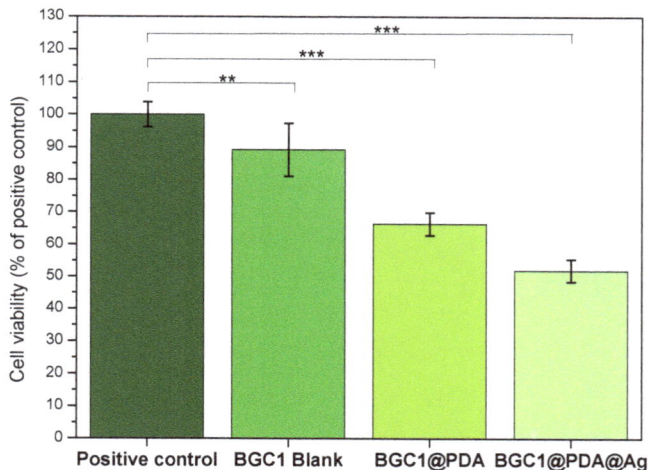

Figure 4. Viability of MG-63 cells cultured onto BGC1, BGC1@PDA and BGC1@PDA@Ag samples for 48 hours. Significant differences are indicated in comparison to control: * $p < 0.05$, ** $p < 0.01$ and *** $p < 0.001$ (Tukey's posthoc test).

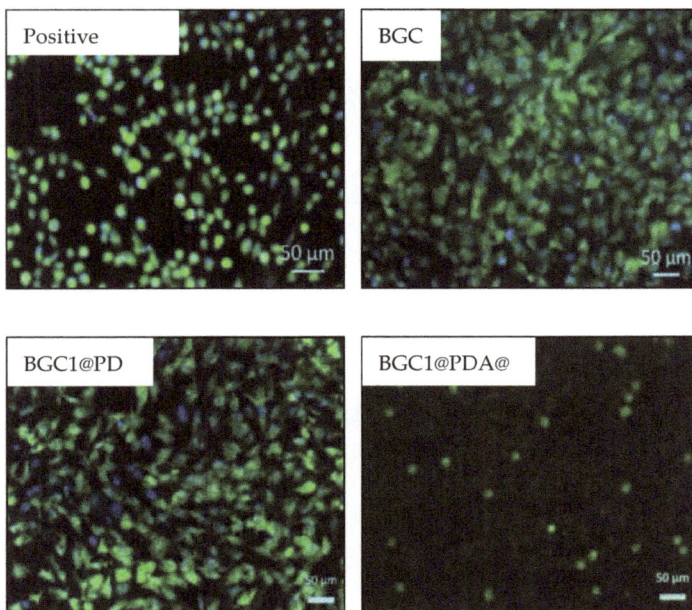

Figure 5. Fluorescence microscope images on different representative samples showing the results of calcein-DAPI staining of MG-63 cells after 48 hours of direct culture.

3.5. Antibacterial Test

Figure 6 shows antibacterial test results against *S. carnosus* (Gram-positive) and *E. coli* (Gram-negative) bacteria strains after 48 hours of incubation in LB medium. For the test carried out against *S. carnosus*, BGC1 and BGC1@PDA did not show a significant effect on the growth of bacteria suspended in the medium. The preconditioning of the samples was important to confirm that there was no increase of local pH for BGC1, because of the lack of any dissolution process within the measuring time period, avoiding bacterial death. In the case of BGC1@PDA, since polydopamine confers hydrophilicity to the samples and does not have an active antimicrobial component, no significant antibacterial effect was observed. This result is in good agreement with the literature, as reports have shown that polydopamine coating itself has no antimicrobial effect against Gram-positive bacteria strains [25].

For the test carried out against *E. coli*, it seems that both BGC1 and BGC1@PDA inhibited the growth of bacteria after 24 hours of incubation. There is a controversy in literature, as some authors have described that polydopamine has an intrinsic antimicrobial effect, which is relatively weak against *E. coli* [32,54]. Meanwhile, some other works have determined that polydopamine itself does not show an antibacterial effect against *E. coli* [55]. In Figure 6 it is possible to observe a similar trend for both samples (BGC1 and BGC1@PDA), rejecting the hypothesis that polydopamine itself presents antimicrobial properties.

The tests carried out on BGC1@PDA@Ag samples against both *S. carnosus* and *E. coli* showed that the material is able to strongly reduce the growth of both chosen bacterial strains, as expected. A similar effect has been reported in the literature, in which several materials have been surface-modified with polydopamine and silver particles [10,21,25,32,56,57].

4. Conclusions

A bioactive glass-ceramic, BGC1, and the functionalized samples, BGC1@PDA and BGC1@PDA@Ag, showed a strongly negative surface zeta potential, which is desirable for in vitro biocompatibility. Tests carried out with the MG-63 cell line demonstrated the non-toxicity of BGC1 and BGC1@PDA. BGC1@PDA@Ag showed a moderate biocompatibility. Antibacterial tests indicated that BGC1@PDA@Ag possess a strong antimicrobial effect against both Gram-positive and Gram-negative bacterial strains. An antibacterial effect of PDA was not observed in the present experiments.

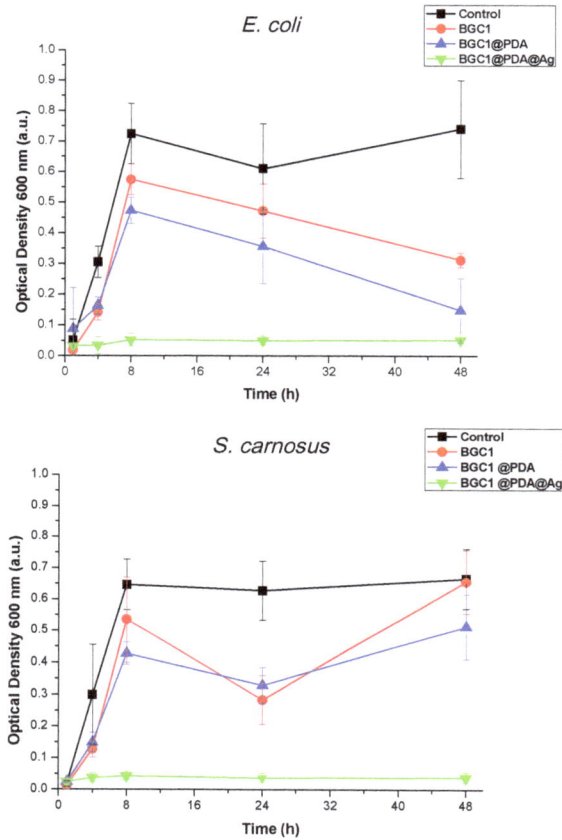

Figure 6. Turbidity measurements on suspensions of both tested *Staphylococcus carnosus* and *Escherichia coli* strains in lysogeny broth (LB) medium on different samples, showing their different antibacterial effects.

These promising findings encourage further investigations, which should lead to the tailoring of silver content in PDA-modified bioactive glass-ceramic to obtain a dual-function composite with strong antibacterial activity and non-toxicity for mammalian cells.

Author Contributions: R.T.-R. conceived the initial idea with A.R.B. and G.B.; R.T.-R. carried out the experiments and wrote the paper; A.G. and R.D. provided support during cell culture and later during the characterization, as well as analysis of the data; the antibacterial tests were carried out in the laboratory of W.H.G., who provided data evaluation; the surface zeta potential measurements were carried out in the laboratories of the Department of Applied Science and Technology, Politecnico di Torino, Italy, under the supervision of S.F. and S.S., who provided data evaluation. All authors checked the manuscript, provided necessary corrections and gave intellectual input.

Funding: This research was carried out within the EU Horizon 2020 framework project COACH (ITN-ETN, Grant agreement 642557).

Conflicts of Interest: The authors declare no conflict of interest.

References

1. Mantripragada, V.P.; Lecka-Czernik, B.; Ebraheim, N.A.; Jayasuriya, A.C. An overview of recent advances in designing orthopedic and craniofacial implants. *J. Biomed. Mater. Res. Part A* **2013**, *101*, 3349–3364. [CrossRef] [PubMed]

2. European Centre for Disease Prevention and Control; European Medicines Agency. *The Bacterial Challenge: Time to React*; European Centre for Disease Prevention and Control: Stockholm, Sweden, 2009. [CrossRef]

3. World Bank Group. *Drug-Resistant Infections: A Threat to Our Economic Future*; World Bank Group: Washington, DC, USA, 2017. [CrossRef]

4. Kokubo, T. Bioactive glass ceramics: Properties and applications. *Biomaterials* **1991**, *12*, 155–163. [CrossRef]

5. Peitl, O.; Zanotto, E.D.; Hench, L.L. Highly bioactive P_2O_5–Na_2O–CaO–SiO_2 glass-ceramics. *J. Non Cryst. Solids* **2001**, *292*, 115–126. [CrossRef]

6. Miguez-Pacheco, V.; Hench, L.L.; Boccaccini, A.R. Bioactive glasses beyond bone and teeth: Emerging applications in contact with soft tissues. *Acta Biomater.* **2015**, *13*, 1–15. [CrossRef] [PubMed]

7. Tallawi, M.; Rosellini, E.; Barbani, N.; Cascone, M.G.; Rai, R.; Saint-Pierre, G.; Boccaccini, A.R. Strategies for the chemical and biological functionalization of scaffolds for cardiac tissue engineering: A review. *J. R. Soc. Interface* **2015**, *12*, 20150254. [CrossRef] [PubMed]

8. Lee, H.; Dellatore, S.M.; Miller, W.M.; Messersmith, P.B. Mussel-inspired surface chemistry for multifunctional coatings. *Science* **2007**, *318*, 426–430. [CrossRef] [PubMed]

9. Liu, Y.; Ai, K.; Lu, L. Polydopamine and its derivative materials: Synthesis and promising applications in energy, environmental, and biomedical fields. *Chem. Rev.* **2014**, *114*, 5057–5115. [CrossRef] [PubMed]

10. Ho, C.C.; Ding, S.J. Structure, properties and applications of mussel-inspired polydopamine. *J. Biomed. Nanotechnol.* **2014**, *10*, 3063–3084. [CrossRef] [PubMed]

11. Mondin, G.; Wisser, F.M.; Leifert, A.; Mohamed-Noriega, N.; Grothe, J.; Dörfler, S.; Kaskel, S. Metal deposition by electroless plating on polydopamine functionalized micro- and nanoparticles. *J. Colloid Interface Sci.* **2013**, *411*, 187–193. [CrossRef] [PubMed]

12. Schaubroeck, D.; Mader, L.; Dubruel, P.; Vanfleteren, J. Surface modification of an epoxy resin with polyamines and polydopamine: Adhesion toward electroless deposited copper. *Appl. Surf. Sci.* **2015**, *353*, 238–244. [CrossRef]

13. Morones, J.R.; Elechiguerra, J.L.; Camacho, A.; Holt, K.; Kouri, J.B.; Ramírez, J.T.; Yacaman, M.J. The bactericidal effect of silver nanoparticles. *Nanotechnology* **2005**, *16*, 2346–2353. [CrossRef] [PubMed]

14. Lara, H.H.; Ayala-Núñez, N.V.; del Turrent, L.C.I.; Padilla, C.R. Bactericidal effect of silver nanoparticles against multidrug-resistant bacteria. *World J. Microbiol. Biotechnol.* **2010**, *26*, 615–621. [CrossRef]

15. Panácek, A.; Smékalová, M.; Kilianová, M.; Prucek, R.; Bogdanová, K.; Věcěrová, R.; Kolár, M.; Havrdová, M.; Płaza, G.A.; Chojniak, J.; et al. Strong and nonspecific synergistic antibacterial efficiency of antibiotics combined with silver nanoparticles at very low concentrations showing no cytotoxic effect. *Molecules* **2016**, *21*, 26. [CrossRef] [PubMed]

16. Blaker, J.J.; Nazhat, S.N.; Boccaccini, A.R. Development and characterisation of silver-doped bioactive glass-coated sutures for tissue engineering and wound healing applications. *Biomaterials* **2004**, *25*, 1319–1329. [CrossRef] [PubMed]

17. Miola, M.; Ferraris, S.; di Nunzio, S.; Robotti, P.F.; Bianchi, G.; Fucale, G.; Maina, G.; Cannas, M.; Gatti, S.; Massé, A.; et al. Surface silver-doping of biocompatible glasses to induce antibacterial properties. Part II: Plasma sprayed glass-coatings. *J. Mater. Sci. Mater. Med.* **2009**, *20*, 741–749. [CrossRef] [PubMed]

18. Ciraldo, F.E.; Liverani, L.; Gritsch, L.; Goldmann, W.H.; Boccaccini, A.R. Synthesis and characterization of silver-doped mesoporous bioactive glass and its applications in conjunction with electrospinning. *Materials* **2018**, *11*, 692. [CrossRef] [PubMed]

19. Kawashita, M.; Tsuneyama, S.; Miyaji, F.; Kokubo, T.; Kozuka, H.; Yamamoto, K. Antibacterial silver-containing silica glass prepared by sol-gel method. *Biomaterials* **2000**, *21*, 393–398. [CrossRef]

20. Di Nunzio, S.; Brovarone, C.V.; Spriano, S.; Milanese, D.; Verné, E.; Bergo, V.; Maina, G.; Spinelli, P. Silver containing bioactive glasses prepared by molten salt ion-exchange. *J. Eur. Ceram. Soc.* **2004**, *24*, 2935–2942. [CrossRef]

21. Saidin, S.; Chevallier, P.; Kadir, M.R.A.; Hermawan, H.; Mantovani, D. Polydopamine as an intermediate layer for silver and hydroxyapatite immobilisation on metallic biomaterials surface. *Mater. Sci. Eng. C* **2013**, *33*, 4715–4724. [CrossRef]

22. Li, M.; Liu, Q.; Jia, Z.; Xu, X.; Shi, Y.; Cheng, Y.; Zheng, Y. Polydopamine-induced nanocomposite Ag/CaP coatings on the surface of titania nanotubes for antibacterial and osteointegration functions. *J. Mater. Chem. B* **2015**, *3*, 8796–8805. [CrossRef]

23. Wang, Z.; Ou, J.; Wang, Y.; Xue, M.; Wang, F.; Pan, B.; Li, C.; Li, W. Anti-bacterial superhydrophobic silver on diverse substrates based on the mussel-inspired polydopamine. *Surf. Coat. Technol.* **2015**, *280*, 378–383. [CrossRef]

24. Forte, L.; Torricelli, P.; Bonvicini, F.; Boanini, E.; Gentilomi, G.A.; Lusvardi, G.; della Bella, E.; Fini, M.; Nepita, E.V.; Bigi, A. Biomimetic fabrication of antibacterial calcium phosphates mediated by polydopamine. *J. Inorg. Biochem.* **2018**, *178*, 43–53. [CrossRef] [PubMed]

25. Sileika, T.S.; Kim, H.D.; Maniak, P.; Messersmith, P.B. Antibacterial performance of polydopamine-modified polymer surfaces containing passive and active components. *ACS Appl. Mater. Interfaces* **2011**, *3*, 4602–4610. [CrossRef] [PubMed]

26. Tejido-Rastrilla, R.; Baldi, G.; Boccaccini, A.R. Ag containing polydopamine coating on a melt-derived bioactive glass-ceramic: Effect on surface reactivity. *Ceram. Int.* **2018**, *44*, 16083–16087. [CrossRef]

27. Kokubo, T.; Takadama, H. How useful is SBF in predicting in vivo bone bioactivity? *Biomaterials* **2006**, *27*, 2907–2915. [CrossRef] [PubMed]

28. Zangmeister, R.A.; Morris, T.A.; Tarlov, M.J. Characterization of polydopamine thin films deposited at short times by autoxidation of dopamine. *Langmuir* **2013**, *29*, 8619–8628. [CrossRef] [PubMed]

29. Spriano, S.; Chandra, V.S.; Cochis, A.; Uberti, F.; Rimondini, L.; Bertone, E.; Vitale, A.; Scolaro, C.; Ferrari, M.; Cirisano, F.; et al. How do wettability, zeta potential and hydroxylation degree affect the biological response of biomaterials? *Mater. Sci. Eng. C* **2017**, *74*, 542–555. [CrossRef] [PubMed]

30. Bernsmann, F.; Frisch, B.; Ringwald, C.; Ball, V. Protein adsorption on dopamine-melanin films: Role of electrostatic interactions inferred from ζ-potential measurements versus chemisorption. *J. Colloid Interface Sci.* **2010**, *344*, 54–60. [CrossRef]

31. Fu, J.; Chen, Z.; Wang, M.; Liu, S.; Zhang, J.; Zhang, J.; Han, R.; Xu, Q. Adsorption of methylene blue by a high-efficiency adsorbent (*Polydopamine microspheres*): Kinetics, isotherm, thermodynamics and mechanism analysis. *Chem. Eng. J.* **2015**, *259*, 53–61. [CrossRef]

32. Su, L.; Yu, Y.; Zhao, Y.; Liang, F.; Zhang, X. Strong Antibacterial Polydopamine Coatings Prepared by a Shaking-assisted Method. *Sci. Rep.* **2016**, *6*, 24420. [CrossRef] [PubMed]

33. Ball, V. Impedance spectroscopy and zeta potential titration of dopa-melanin films produced by oxidation of dopamine. *Colloids Surf. A Phys. Eng. Asp.* **2010**, *363*, 92–97. [CrossRef]

34. Cazzola, M.; Corazzari, I.; Prenesti, E.; Bertone, E.; Vernè, E.; Ferraris, S. Bioactive glass coupling with natural polyphenols: Surface modification, bioactivity and anti-oxidant ability. *Appl. Surf. Sci.* **2016**, *367*, 237–248. [CrossRef]

35. Suzuki, T.; Nishizawa, K.; Yokogawa, Y.; Nagata, F.; Kawamoto, Y.; Kameyama, T. Time-dependent variation of the surface structure of bioceramics in tissue culture medium and the effect on adhesiveness of cells. *J. Ferment. Bioeng.* **1996**, *81*, 226–232. [CrossRef]

36. Ryu, J.; Ku, S.H.; Lee, H.; Park, C.B. Mussel-inspired polydopamine coating as a universal route to hydroxyapatite crystallization. *Adv. Funct. Mater.* **2010**, *20*, 2132–2139. [CrossRef]

37. El-Ghannam, A.; Hamazawy, E.; Yehia, A. Effect of thermal treatment on bioactive glass microstructure, corrosion behavior ζ potential, and protein adsorption. *J. Biomed. Mater. Res.* **2001**, *55*, 387–395. [CrossRef]

38. Lu, H.H.; Pollack, S.R.; Ducheyne, P. 45S5 Bioactive glass surface charge variations and the formation of a surface calcium phosphate layer in a solution containing fibronectin. *J. Biomed. Mater. Res.* **2001**, *54*, 454–461. [CrossRef]

39. Zhang, C.; Gong, L.; Xiang, L.; Du, Y.; Hu, W.; Zeng, H.; Xu, Z.K. Deposition and Adhesion of Polydopamine on the Surfaces of Varying Wettability. *ACS Appl. Mater. Interfaces* **2017**, *9*, 30943–30950. [CrossRef]

40. Sathishkumar, M.; Sneha, K.; Won, S.W.; Cho, C.W.; Kim, S.; Yun, Y.S. Cinnamon zeylanicum bark extract and powder mediated green synthesis of nano-crystalline silver particles and its bactericidal activity. *Colloids Surf. B Biointerfaces* **2009**, *73*, 332–338. [CrossRef]
41. Roy, A.K.; Park, B.; Lee, K.S.; Park, S.Y.; In, I. Boron nitride nanosheets decorated with silver nanoparticles through mussel-inspired chemistry of dopamine. *Nanotechnology* **2014**, *25*, 445603. [CrossRef]
42. Wang, J.; Rahman, M.F.; Duhart, H.M.; Newport, G.D.; Patterson, T.A.; Murdock, R.C.; Hussain, S.M.; Schlager, J.J.; Ali, S.F. Expression changes of dopaminergic system-related genes in PC12 cells induced by manganese, silver, or copper nanoparticles. *Neurotoxicology* **2009**, *30*, 926–933. [CrossRef]
43. Rehman, M.A.U.; Ferraris, S.; Goldmann, W.H.; Perero, S.; Bastan, F.E.; Nawaz, Q.; di Confiengo, G.G.; Ferraris, M.; Boccaccini, A.R. Antibacterial and Bioactive Coatings Based on Radio Frequency Co-Sputtering of Silver Nanocluster-Silica Coatings on PEEK/Bioactive Glass Layers Obtained by Electrophoretic Deposition. *ACS Appl. Mater. Interfaces* **2017**, *9*, 32489–32497. [CrossRef] [PubMed]
44. Verné, E.; Bretcanu, O.; Balagna, C.; Bianchi, C.L.; Cannas, M.; Gatti, S.; Vitale-Brovarone, C. Early stage reactivity and in vitro behavior of silica-based bioactive glasses and glass-ceramics. *J. Mater. Sci. Mater. Med.* **2009**, *20*, 75–87. [CrossRef] [PubMed]
45. Vitale-Brovarone, C.; Verné, E.; Robiglio, L.; Martinasso, G.; Canuto, R.A.; Muzio, G. Biocompatible glass-ceramic materials for bone substitution. *J. Mater. Sci. Mater. Med.* **2008**, *19*, 471–478. [CrossRef] [PubMed]
46. Ciraldo, F.E.; Boccardi, E.; Melli, V.; Westhauser, F.; Boccaccini, A.R. Tackling bioactive glass excessive in vitro bioreactivity: Preconditioning approaches for cell culture tests. *Acta Biomater.* **2018**, *75*, 3–10. [CrossRef] [PubMed]
47. Nakamura, T.; Yamamuro, T.; Higashi, S.; Kokubo, T.; Itoo, S. A new glass-ceramic for bone replacement: Evaluation of its bonding to bone tissue. *J. Biomed. Mater. Res.* **1985**, *19*, 685–698. [CrossRef] [PubMed]
48. Ozawa, S.; Kasugai, S. Evaluation of implant materials (hydroxyapatite, glass-ceramics, titanium) in rat bone marrow stromal cell culture. *Biomaterials* **1996**, *17*, 23–29. [CrossRef]
49. Hench, L.L.; Paschall, H.A. Direct chemical bond of bioactive glass-ceramic materials to bone and muscle. *J. Biomed. Mater. Res.* **1973**, *7*, 25–42. [CrossRef]
50. Deligianni, D.D.; Katsala, N.D.; Koutsoukos, P.G.; Missirlis, Y.F. Effect of surface roughness of hydroxyapatite on human bone marrow cell adhesion, proliferation, differentiation and detachment strength. *Biomaterials* **2001**, *22*, 87–96. [CrossRef]
51. Shin, Y.M.; Lee, Y.B.; Shin, H. Time-dependent mussel-inspired functionalization of poly(l-lactide-co-e{open}-caprolactone) substrates for tunable cell behaviors. *Colloids Surf. B Biointerfaces* **2011**, *87*, 79–87. [CrossRef]
52. Lynge, M.E. Recent developments in poly (dopamine)—Based coatings for biomedical applications. *Nanomedicine* **2015**, *10*, 2725–2742. [CrossRef]
53. Chien, H.W.; Tsai, W.B. Fabrication of tunable micropatterned substrates for cell patterning via microcontact printing of polydopamine with poly(ethylene imine)-grafted copolymers. *Acta Biomater.* **2012**, *8*, 3678–3686. [CrossRef]
54. Iqbal, Z.; Lai, E.P.C.; Avis, T.J. Antimicrobial effect of polydopamine coating on *Escherichia coli*. *J. Mater. Chem.* **2012**, *22*, 21608–21612. [CrossRef]
55. Lim, K.; Chua, R.R.Y.; Ho, B.; Tambyah, P.A.; Hadinoto, K.; Leong, S.S.J. Development of a catheter functionalized by a polydopamine peptide coating with antimicrobial and antibiofilm properties. *Acta Biomater.* **2015**, *15*, 127–138. [CrossRef] [PubMed]
56. He, S.; Zhou, P.; Wang, L.; Xiong, X.; Zhang, Y.; Deng, Y.; Wei, S. Antibiotic-decorated titanium with enhanced antibacterial activity through adhesive polydopamine for dental/bone implant Antibiotic-decorated titanium with enhanced antibacterial activity through adhesive polydopamine for dental/bone implant. *J. R. Soc. Interface* **2014**, *11*, 20140169. [CrossRef] [PubMed]
57. Zhou, P.; Deng, Y.; Lyu, B.; Zhang, R.; Zhang, H.; Ma, H.; Lyu, Y.; Wei, S. Rapidly-Deposited Polydopamine Coating via High Temperature and Vigorous Stirring: Formation, Characterization and Biofunctional Evaluation. *PLoS ONE* **2014**, *9*, e113087. [CrossRef] [PubMed]

materials

MDPI

Article

Oxidation Protective Hybrid Coating for Thermoelectric Materials

Francesco Gucci [1,2,]*[©], Fabiana D'Isanto [3], Ruizhi Zhang [1,2], Michael J. Reece [1,2], Federico Smeacetto [4] and Milena Salvo [3][©]

1   School of Engineering and Material Science, Queen Mary University of London, London E1 4NS, UK; ruizhi.zhang@qmul.ac.uk (R.Z.); m.j.reece@qmul.ac.uk (M.J.R.)
2   Nanoforce Technology Limited, London E1 4NS, UK
3   Department of Applied Science and Technology, Politecnico di Torino, 10129 Turin, Italy; fabiana.disanto@polito.it (F.D.); milena.salvo@polito.it (M.S.)
4   Department of Energy, Politecnico di Torino, 10129 Turin, Italy; federico.smeacetto@polito.it
*   Correspondence: f.f.gucci@qmul.ac.uk; Tel.: +44-20-7882-2773

Received: 14 December 2018; Accepted: 6 February 2019; Published: 14 February 2019

Abstract: Two commercial hybrid coatings, cured at temperatures lower than 300 °C, were successfully used to protect magnesium silicide stannide and zinc-doped tetrahedrite thermoelectrics. The oxidation rate of magnesium silicide at 500 °C in air was substantially reduced after 120 h with the application of the solvent-based coating and a slight increase in power factor was observed. The water-based coating was effective in preventing an increase in electrical resistivity for a coated tethtraedrite, preserving its power factor after 48 h at 350 °C.

Keywords: Thermoelectrics; oxidation resistance; hybrid-coating

1. Introduction

Thermoelectric materials are able to convert thermal gradient into electricity and recover energy from waste heat [1,2]. They are, usually, semiconductors or intermetallics, often containing elements such as Mg, Pb, Te, Bi, Mn, Ge, Si, Sb, Co or In [3–7].

A considerable effort has been made to produce materials with high figure of merit, and several compounds have been identified and improved using different approaches: doping elements; composites; nanostructuring; and mesostructuring [3,4,8]. The effect of conventional synthesis and sintering techniques have been evaluated [9] and innovative methods are constantly under development [10–14]. One of the main challenges in the thermoelectrics field is the identification of efficient materials that are inexpensive, easy to be produced, and formed of earth-abundant and environmentally friendly elements. In this respect, magnesium silicide [15] and tetrahedrite [16] are considered attractive and sustainable candidates for n and p-type thermoelectrics, respectively. One important aspect for the development of high temperature thermoelectric generators is their long-term stability in air at high temperature [17].

Magnesium silicide is a semiconductor of the Mg_2X (X = Si, Ge, Sn and Pb) compounds family. It possess an anti-fluorite structure with a bandgap of 0.784 eV [18]. It can be doped to achieve good thermoelectric properties (ZT of 0.86 at 862 K with Bi-doping [19]) but it is limited by its relatively high thermal conductivity. N-type solid solutions of Mg_2Si with Mg_2Sn [20–23] and/or Mg_2Ge [24,25] have been produced in an attempt to reduce their thermal conductivity. One of the best values has been reported for $Mg_{2.08}Si_{0.364}Sn_{0.6}Sb_{0.036}$ (ZT of 1.5 at 723 K [26]).

Magnesium silicide and its solid solutions are prone to oxidation above ~400 °C; Skomedal et al. [27] reported breakaway oxidation behavior for $Mg_2Si_{1-x}Sn_x$ (0.1 < x < 0.6) at temperatures above 430 °C. Sondergard et al. [28] showed the substantial stability in air of $Mg_2Si_{0.4}Sn_{0.6}$

and $Mg_2Si_{0.6}Sn_{0.4}$ up to 400 °C when the material has a high relative density. Yin et al. [29] reported the oxidation behavior of Sb-doped $Mg_2Si_{0.3}Sn_{0.7}$ (360–720 h) to prevent the decomposition they proposed and studied the effect of BN spray-coating which was effective up to 500 °C. Tani et al. [30] studied the effect of magnetron sputtered β-FeSi on Mg_2Si, observing that it improved the oxidation resistance up to 600 °C.

Tetrahedrite ($Cu_{12}Sb_4S_{13}$) is a ternary I-V-VI semiconductor, which has a complex crystal structure with a large number of atoms per unit cell, helpful in providing low thermal conductivity, and a high band degeneracy (1.7 eV) [31,32] due to its highly symmetric crystal structure, which is useful for improving the power factor [33]. It has a sphalerite-like structure with 58 atoms arranged in a high symmetry cubic cell (I43m) made of CuS_4 tetrahedra, CuS_3 triangles and SbS_3 pyramids [34]. This structure, with lone-pair electrons on Sb sites is the origin of the low lattice thermal conductivity [32], which is shared by other compounds in the Cu-Sb-S system [35] such as chalcostibite [36] ($CuSbS_2$), famatinite [37,38] (Cu_3SbS_4) and skinnerite (Cu_3SbS_3) [39]. Naturally, tetrahedrite occurs with the composition $Cu_{12-x}M_x(Sb,As)_4S_{13}$, which is a solid solution of As rich tennantite ($Cu_{12}As_4S_{13}$) and Sb rich tetrahedrite ($Cu_{12}Sb_4S_{13}$). The literature related to this material shows the best Thermoelectric properties are achieved by replacing Cu^{2+} atoms with Zn: ZT is 0.6 (at 400 ° C) for an un-doped sample, and increases up to 0.9 (at 447 °C) with Zn substitution due to a reduction of the thermal conductivity [40].

Tetrahedrite has limited thermal stability and is subjected to sulphur loss. Braga et al. [41] reported the phase decomposition of $Cu_{12}SbS_{13}$ at around 795 K. Barbier et al. [42] confirmed the same phase transformation at 803 K corresponding to a weight loss due to Sulphur volatilization. Nevertheless, Chetty et al. [16] reported that tetrahedrite is usually stable only up to 600 K, and that the overall stability may increase or decrease depending on the dopants. The oxidation behavior of a tetrahedrite was tested only by Gonçalves et al. [43]. They discovered the formation of a $Cu_{2-x}S$ surface barrier, which decreases the corrosion rate at 275 °C acting as a weak passivation layer. Nevertheless, this layer was not effective at 350 °C and 375 °C because of the simultaneous action of sulphur sublimation.

Second phases were present in all of the samples at the end of the test, evidencing the need of an effective protective coating, but no studies have been carried out to identify a suitable one for this thermoelectric material. In general, there is a relatively small body of literature that is concerned with oxidation protective coatings for TE modules.

The oxidation of the surface of the thermoelectric degrades the power generation and significantly limits the long-term reliability and efficiency of TE modules. For this reason the application of an oxidation resistant coating is needed to improve thermoelectric properties.

In this work, we investigated the potential of two commercial hybrid (ceramic-polymer) coatings with nominal temperature resistance up to 590 °C. To evaluate their effectiveness, we tested the properties of $Mg_2Si_{0.487}Sn_{0.55}Sb_{0.13}$ and $Cu_{11.5}Zn_{0.5}Sb_4S_{13}$ as sintered and after aging in air, with and without the coatings. The low curing temperature (250 °C) of these resins is a great advantage; in fact, glass-ceramics coatings would require deposition temperatures too high for tetrahedrites (for example, a Higher Manganese Silicides coating was prepared at 700 °C [44]). Moreover, the coating procedure does not require the need of expensive equipment (such as magnetron sputtering) making it more appealing for actual device production.

2. Materials and Methods

$Mg_{2.1}Si_{0.487}Sn_{0.5}Sb_{0.13}$ (Mg-silicide) powders were provided by European Thermodynamics Ltd (Leicester, UK). Powders were then sintered into 30 mm diameter discs using a Spark Plasma Sintering furnace (FCT HPD 25; FCT Systeme GmbH, Rauenstein, Germany). The sintering of Mg-silicide was carried out at a temperature of 750 °C with a heating and cooling rate of 100 °C/min, a dwell time of 5 min and a pressure of 50 MPa.

$Cu_{11.5}Zn_{0.5}Sb_4S_{13}$ (THD) was prepared starting from single elements powders: Cu (Alpha Aesar, 150 mesh, purity 99.5%), Sb (Alpha Aesar, 100 mesh, purity 99.5%), S (Sigma Aldrich, 100 mesh, purity reagent grade) and Zn (Sigma Aldrich, $\geq$ 99%). They were weighted in the appropriate stoichiometry and sealed in a stainless steel jar in an argon filled glove box, processed in a ball milling machine (QM-3SP2, Nanjiing University, China) employing stainless steel balls at 360 rpm for 96 h, with a ball to powder ratio of 30:1. The sintering of tetrahedrite was carried out at a temperature of 400 °C with a heating and cooling rate of 50 °C/min, a dwell time of 5 min and a pressure of 50 MPa. The density of the pellets was measured using the Archimede's method. Each pellet was cut into bars having square base of 3 mm per side and 10 mm height.

After preliminary tests, a solvent-based resin (CP4040-S1) was chosen for Mg-Silicide and a water-based resin (CP4040) for tetrahedrite, both purchased from AREMCO SCIENTIFIC COMPANY (Los Angeles, USA). They were applied using a foam brush and subsequently cured in a tubular furnace (Carbolite Gero STF/180, Neuhausen, Germany) for 45 min at 250 °C with a heating and cooling rate of 1.6 °C/min. Aging tests was performed in a muffle oven (Manfredi OVMAT 2009, Pinerolo, Italy) in air at a temperature of 500 °C for 120 h (for Mg-silicide) and at 350°C for 48h (for THD) with a heating rate of 1.2 °C/min. The choice of oxidation temperatures was guided by previous literature and the potential operating temperatures of the materials. Both Mg-Silicide and tetrahedrite are oxidized in air at these temperatures without being subjected to any phase transformations and their properties are near their optimum values. The tests would provide an initial benchmark for the tested hybrid coatings.

XRD data were collected using X'Pert Pro MRD <u>diffractometer with Cu Kα radiation</u> (PANalytical X'Pert Pro, Philips, Almelo, The Netherlands, with the aid of the X-Pert HighScore software) and the different phases were identified using the JCPDS data base. Field emission scanning electron microscope (FE-SEM, Merlin electron microscope, ZEISS, Oberkochen, Germany) and energy dispersive X-ray Spectroscopy (EDS, ZEISS Supra TM 40, Oberkochen, Germany)were used to characterize the microstructure morphology and chemical composition of uncoated and coated samples, before and after ageing. The measurements of the electrical properties were carried out using a Linseis LSR-3 (Linseis Messgeraete GmbH, Selb, Germany) with Pt thermocouples and electrodes. The oxide layer of the aged samples was removed before measuring their properties.

3. Results and Discussion

3.1. Mg-Silicide (Solvent-Based Coating)

Ball milling of the elemental powders effectively produced a single phase solid solution ($Mg_2Si_{0.4}Sn_{0.6}$) of Mg_2Si and Mg_2Sn, and no peak splitting was observed in the XRD pattern (Figure 1a). After sintering, no phase separation was visible in the XRD pattern and the peaks simply became sharper (Figure 1b).

The density of the as sintered sample was about 96% of the theoretical one.

The aging at 500 °C for 120 h in air had a very clear effect on the uncoated sample; it was completely burned and turned into powder (Figure 2). The coated sample did not suffer such a catastrophic effect despite the fact that the applied coating was damaged at the edges.

The XRD pattern of the uncoated sample after aging (Figure 1c) shows the decomposition of $Mg_2Si_{0.4}Sn_{0.6}$ into a mixture of compounds (MgO, SiO, SnO_2 and Sn), as already observed by Skomedal et al. [27].

Figure 3 shows the cross-section of a coated sample after the curing. The interface between Mg-silicide and the hybrid coating is continuous, without cracks. However, the coating shows a few cracks parallel to the surface; this was likely due to a mismatch in CTE between the matrix and ceramic filler or more likely an effect due to solvent evaporation during curing, with consequent shrinkage effects. The thickness of the layer was found to be about 30–100 µm, being thinner at the edges.

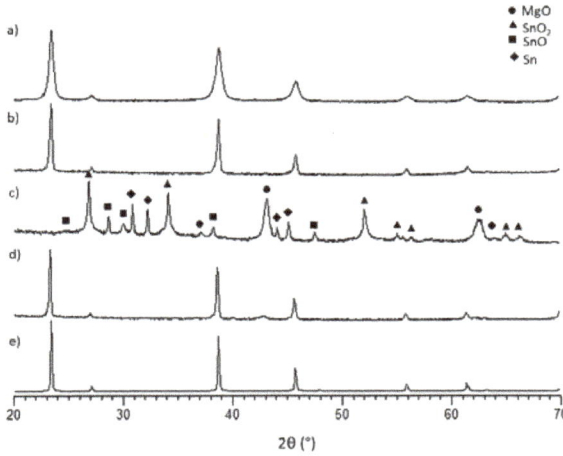

Figure 1. XRD spectra for (**a**) Mg-silicide powders (**b**) as sintered sample (**c**) aged Mg-silicide at 500 °C for 120 h without coating (**d**) aged Mg-silicide at 500 °C for 120 h with coating and (**e**) PDF card of $Mg_2Si_{0.4}Sn_{0.6}$ (n. 01-089-4254).

Figure 2. Mg-silicide uncoated after aging at 500 °C for 120 h, in air.

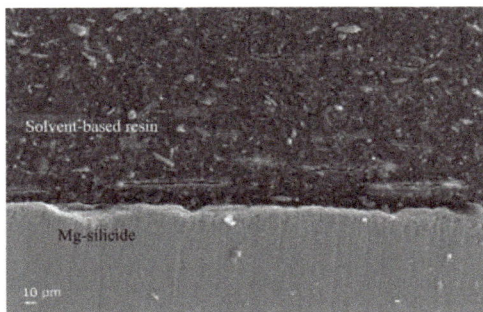

Figure 3. Cross sectional SEM image of coated sample after curing at 250 °C for 45 min under flowing Ar.

After aging for 120 h at 500 °C in air, the coated sample did not experience significant oxidation, and was still mainly composed of a single phase (Figure 1d). However, a small amount of MgO was present and the coating showed cracks on the edges and peeled off in some areas. Figure 4 shows the cross-section of a coated Mg-silicide after the oxidation test: an uneven oxide scale formed on the TE surface, thus indicating that the coating effectively reduced the oxidation reaction rate compared to the uncoated sample but did not prevent it as an oxide layer growing at the coating/sample interface.

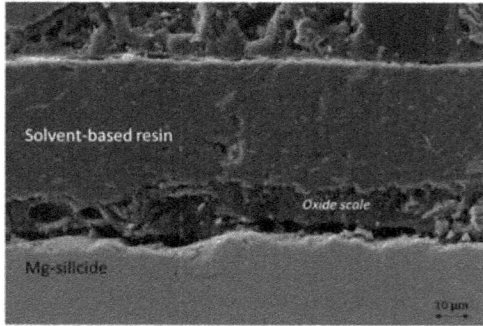

Figure 4. Cross sectional SEM image of solvent-based resin coated Mg-silicide after aging at 500 °C for 120 h, in air.

The comparison between the electrical properties of the as-sintered sample and coated sample after ageing is a useful tool to understand the effectiveness of the coating (Figure 5).

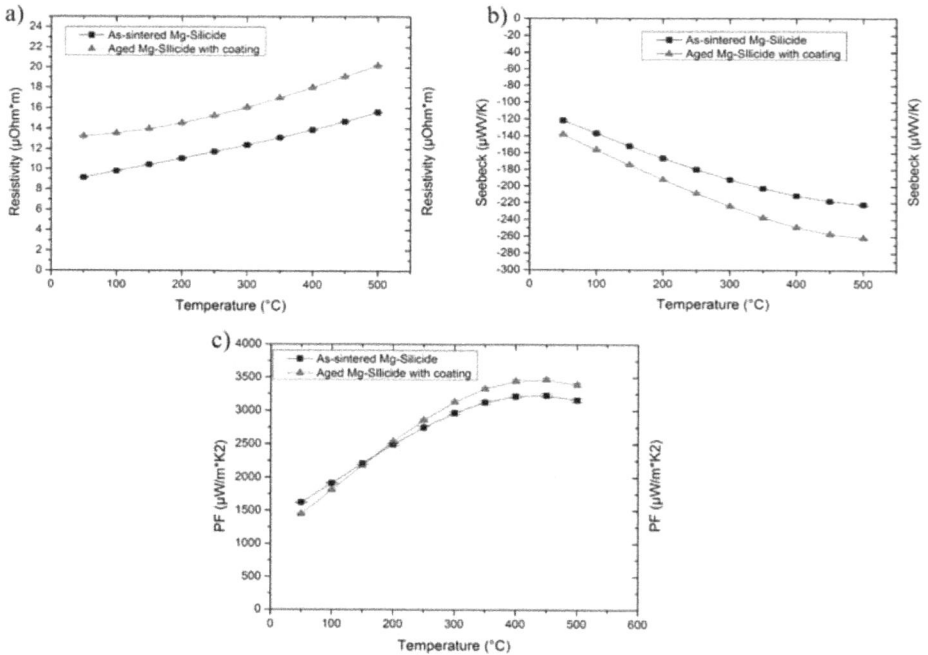

Figure 5. Temperature dependence of the: (**a**) Electrical resistivity; (**b**) Seebeck coefficient; and (**c**) Power Factor for Mg-silicide as sintered and coated after ageing at 500 °C for 120 h, uncoated could not be measured. (For all electrical measurements, the outer layers were removed coated sample to reveal pristine silicide onto which electrical contacts have been made).

It was clearly impossible to measure the properties of the uncoated sample after the oxidation test since it was completely destroyed (Figure 2). The initial properties of the sample are comparable to those of a similar composition reported in the literature [20,22]. The electrical resistivity (ρ) of the coated sample after aging was increased by about 50%, while the Seebeck coefficient (S) increased by about 10%. Different stoichiometric ratio of Mg_2Si and Mg_2Sn as well as Mg vacancies or interstitial (due to the fact that Mg diffuses towards the surface) can influence the electronic properties [15,22,45,46].

The coating provides a very good degree of protection of Mg-silicide, against a complete burning of the Mg-silicide substrate.

From the data, the power factor (S^2/ρ) of the coated samples seems to be slightly increased. On the other hand, it is also clear that the effectiveness is time-limited and longer ageing times may likely determine the total oxidation of the sample, as occurred for the uncoated sample.

Further work is needed to fully understand the implications of defects or non-homogeneous areas through the coating.

Further work should focus on the production of a homogeneous coating with a controlled optimal thickness, which should prevent the surface damage and remove easy path for oxygen diffusion. Due to the sample shape, it was not possible to evaluate thermal conductivity and, therefore, ZT.

3.2. Tetrahedrite (Water-Based Coating)

The Ball milled powders produced starting from single elements consisted of single phase $Cu_{12}Sb_4S_{13}$ (PDF Card n.00-024-1318) (Figure 6a). The density of the as sintered THD measured with the Archimede's method was found to be about 98% of the theoretical density. The XRD pattern of the as-sintered THD (Figure 6b) confirmed that the main phase is $Cu_{12}Sb_4S_{13}$ with a minor amount of Cu_3SbS_4 (Famatinite PDF Card n. 01-071-0555).

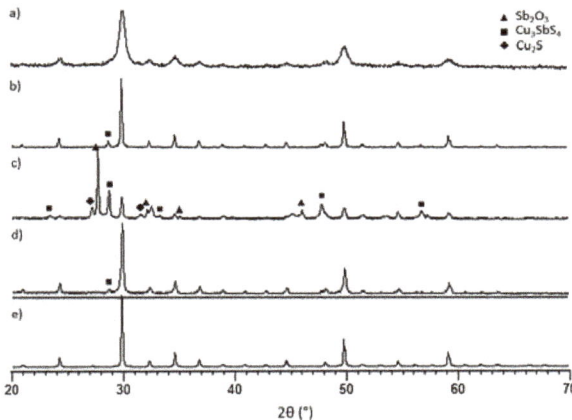

Figure 6. XRD spectra for (**a**) Tetrahdrite powders (**b**) as sintered sample (**c**) aged THD at 350 °C for 48 h without coating (**d**) aged THD at 350 °C for 48 h with coating and (**e**) $Cu_{12}Sb_4S_{13}$ PDF Card n.00-024-1318.

After the ageing at 350 °C for 48 h in air, the uncoated THD was oxidised with a darker surface than the as-sintered sample. Cross sectional SEM images of the uncoated THD (Figure 7) shows the formation of an inhomogeneous layer (around 3–5 μm) on the whole surface of the thermoelectric. The point indicated with the black arrow can be attributed to antimony oxide.

The XRD analysis of the uncoated sample surface after ageing (Figure 6c) shows that the main phase was Sb_2O_3 (PDF Card n. 00-043-1071), confirming the SEM analysis, with the presence of Cu_3SbS_4 and Cu_2S (PDF Card n. 01-072-1071) as secondary phases, as also reported by Chetty et al. and Harish at al. [16,47]. The cross-section of the water-based resin coated THD after curing at 250 °C for 45 min (Figure 8) shows crystals of different shape and composition well dispersed in the silicone resin matrix, and no pores, cracks or other defects are visible at the coating/THD interface.

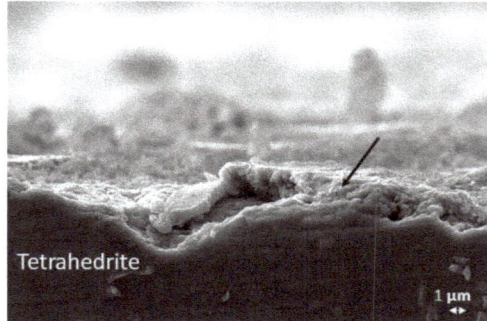

Figure 7. SEM image of cross-section of uncoated THD after ageing at 350 °C, dwelling time 48 h, in air.

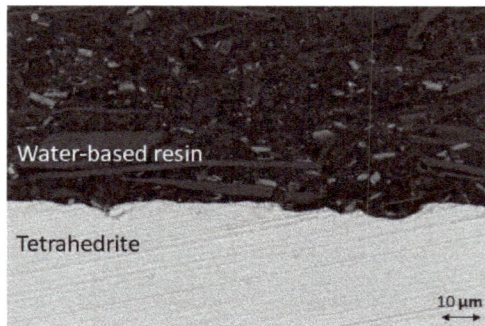

Figure 8. SEM image of cross-section of water-based resin THD coated after curing at 250 °C for 45 min under flowing Ar.

The cross-sectional image of the water-based coated THD after ageing at 350 °C for 48 h (Figure 9) evidenced the absence of cracks within the coating, which is still well-adhered to the substrate. As can be seen in the SEM image, no evidence for the formation of oxidation layers was found at the coating/THD interface. Furthermore, XRD analysis (Figure 6d) shows that after the ageing at 350 °C for 48 h in air there were no apparent changes in the THD compared to the as-sintered sample, confirming that the hybrid coating provided an effective protection, inhibiting the oxidation of THD under thermal ageing.

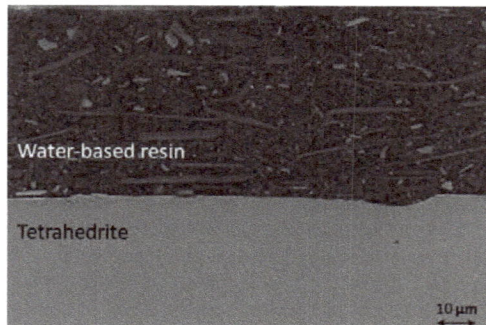

Figure 9. SEM image of cross-section of water-based resin coated THD after aging at 350 °C, dwelling time 48 h, in air.

A comparison between the properties of the as-sintered sample and the uncoated and coated counterparts after ageing at 350 °C for 48 h in air (Figure 10), further confirm the effectiveness of the coating. The Seebeck coefficients of the three samples did not show any differences, but the values of the aged THD with coating were slightly higher than those without coating, at least starting from 150 °C. The coating is able to prevent the increase in electrical resistivity noticed in the uncoated sample, as it maintains the original chemical composition. Consequently, the power factor of the uncoated sample suffers a significant reduction, while the coated sample maintains a similar value.

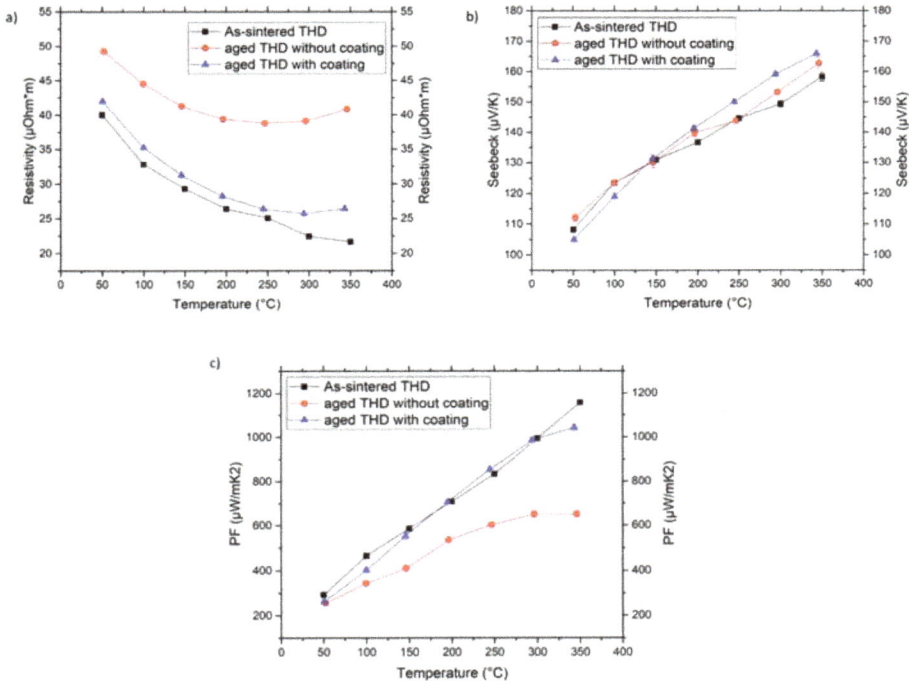

Figure 10. Temperature dependence of the (**a**) Electrical resistivity (**b**) Seebeck coefficient and (**c**) Power Factor of as-sintered THD and aged THD, with and without coating after ageing at 350 °C for 48 h.

4. Conclusions

The effectiveness of two hybrid protective coatings for $Mg_{2.1}Si_{0.487}Sn_{0.5}Sb_{0.13}$ and $Cu_{11.5}Zn_{0.5}Sb_4S_{13}$ thermoelectric materials was reviewed and discussed.

The solvent-based resin significantly reduced the oxidation rate of magnesium silicide at 500 °C in air. Even with some imperfections (incomplete adhesion, few cracks within the resin after curing), the coated sample was not significantly oxidised after 120 h and the electrical properties were not severely modified.

The water-based hybrid coating was effective at providing a barrier coating to avoid the oxidation. Consequently, the values of the power factor did not decrease in the presence of the hybrid coating, indicating that it is a promising candidate for protecting THD against high temperature oxidation in air.

Author Contributions: Conceptualization, F.S. and M.S.; Data curation, F.G. and F.D.; Funding acquisition, M.J.R. and M.S.; Inestigation, F.G., F.D. and R.Z.; Resources, M.J.R. and M.S.; Supervision M.J.R.; F.S. and M.S.; Visualization F.G. and F.D.; Writing—original draft, F.G. and F.D.; Writing—review and editing, M.J.R.; F.S. and M.S.

Funding: This research was funded by the European Union's Horizon 2020 Programme through a Marie Skłodowska-Curie Innovative Training Network ('CoACH-ETN", http://www.coach-etn.eu/, g.a. no. 642557).

Conflicts of Interest: The authors declare no conflict of interest.

References

1. Vineis, C.J.; Shakouri, A.; Majumdar, A.; Kanatzidis, M.G. Nanostructured thermoelectrics: big efficiency gains from small features. *Adv. Mater.* **2010**, *22*, 3970–3980. [CrossRef] [PubMed]
2. Snyder, G.J.; Toberer, E.S. Complex thermoelectric materials. *Nat. Mater.* **2008**, *7*, 105–114. [CrossRef] [PubMed]
3. Sootsman, J.R.; Chung, D.Y.; Kanatzidis, M.G. New and old concepts in thermoelectric materials. *Angew. Chem. Int. Edit.* **2009**, *48*, 8616–8639. [CrossRef]
4. Alam, H.; Ramakrishna, S. A review on the enhancement of figure of merit from bulk to nano-thermoelectric materials. *Nano Energy* **2013**, *2*, 190–212. [CrossRef]
5. Srinivasan, B.; Gucci, F.; Boussard-Pledel, C.; Chevire, F.; Reece, M.J.; Tricot, S.; Calvez, L.; Bureau, B. Enhancement in thermoelectric performance of n-type Pb-deficit Pb-Sb-Te alloys. *J. Alloy. Compd.* **2017**, *729*, 198–202. [CrossRef]
6. Srinivasan, B.; Boussard-Pledel, C.; Bureau, B. Thermoelectric performance of codoped (Bi, In)-GeTe and (Ag, In, Sb)-SnTe materials processed by Spark Plasma Sintering. *Mater. Lett.* **2018**, *230*, 191–194. [CrossRef]
7. Srinivasan, B.; Gautier, R.; Gucci, F.; Fontaine, B.; Halet, J.F.; Chevire, F.; Boussard-Pledel, C.; Reece, M.J.; Bureau, B. Impact of Coinage Metal Insertion on the Thermoelectric Properties of GeTe Solid-State Solutions. *J. Phys. Chem. C* **2018**, *122*, 227–235. [CrossRef]
8. He, J.Q.; Kanatzidis, M.G.; Dravid, V.P. High performance bulk thermoelectrics via a panoscopic approach. *Mater. Today* **2013**, *16*, 166–176. [CrossRef]
9. Li, J.F.; Pan, Y.; Wu, C.F.; Sun, F.H.; Wei, T.R. Processing of advanced thermoelectric materials. *Sci. China Technol. Sci.* **2017**, *60*, 1347–1364. [CrossRef]
10. Gucci, F.; Saunders, T.G.; Reece, M.J. In-situ synthesis of n-type unfilled skutterudite with reduced thermal conductivity by hybrid flash-spark plasma sintering. *Scripta Mater.* **2018**, *157*, 58–61. [CrossRef]
11. Srinivasan, B.; Gellé, A.; Gucci, F.; Boussard-Pledel, C.; Fontaine, B.; Gautier, R.; Halet, J.-F.; Reece, M.; Bureau, B. Realizing a Stable High Thermoelectric zT ~ 2 over a Broad Temperature Range in $Ge_{1-x-y}Ga_xSb_yTe$ via Band Engineering and Hybrid Flash-SPS Processing. *Inorg. Chem. Front.* **2018**, 63–73. [CrossRef]
12. Xie, H.Y.; Su, X.L.; Yan, Y.G.; Liu, W.; Chen, L.J.; Fu, J.F.; Yang, J.H.; Uher, C.; Tang, X.F. Thermoelectric performance of $CuFeS_{2+2x}$ composites prepared by rapid thermal explosion. *Npg Asia Mater.* **2017**, *9*, e390. [CrossRef]
13. Yang, F.; Fan, X.A.; Rong, Z.Z.; Cai, X.Z.; Li, G.Q. Lattice thermal conductivity reduction due to in situ-generated nano-phase in $Bi_{0.4}Sb_{1.6}Te_3$ alloys by microwave-activated hot pressing. *J. Electron. Mater.* **2014**, *43*, 4327–4334. [CrossRef]
14. Srinivasan, B.; Fontaine, B.; Gucci, F.; Dorcet, V.; Saunders, T.G.; Yu, M.; Chevire, F.; Boussard-Pledel, C.; Halet, J.F.; Gautier, R.; et al. Effect of the Processing Route on the Thermoelectric Performance of Nanostructured $CuPb_{18}SbTe_{20}$. *Inorg. Chem.* **2018**, *57*, 12976–12986. [CrossRef] [PubMed]
15. Fedorov, M.I.; Isachenko, G.N. Silicides: Materials for thermoelectric energy conversion. *Jpn. J. Appl. Phys.* **2015**, *54*. [CrossRef]
16. Chetty, R.; Bali, A.; Mallik, R.C. Tetrahedrites as thermoelectric materials: an overview. *J. Mater. Chem. C* **2015**, *3*, 12364–12378. [CrossRef]
17. LeBlanc, S. Thermoelectric generators: Linking material properties and systems engineering for waste heat recovery applications. *Sustain. Mater. Technol.* **2014**, *1–2*, 26–35. [CrossRef]
18. Morris, R.G.; Redin, R.D.; Danielson, G.C. Semiconducting Properties of Mg_2Si Single Crystals. *Phys. Rev.* **1958**, *109*, 1909–1915. [CrossRef]
19. Tani, J.; Kido, H. Thermoelectric properties of Bi-doped Mg_2Si semiconductors. *Phys. B* **2005**, *364*, 218–224. [CrossRef]

20. Liu, W.; Tang, X.F.; Li, H.; Yin, K.; Sharp, J.; Zhou, X.Y.; Uher, C. Enhanced thermoelectric properties of n-type $Mg_{2.16}(Si_{0.4}Sn_{0.6})_{1-y}Sb_y$ due to nano-sized Sn-rich precipitates and an optimized electron concentration. *J. Mater. Chem.* **2012**, *22*, 13653–13661. [CrossRef]

21. Liu, W.; Zhang, Q.; Yin, K.; Chi, H.; Zhou, X.Y.; Tang, X.F.; Uher, C. High figure of merit and thermoelectric properties of Bi-doped $Mg_2Si_{0.4}Sn_{0.6}$ solid solutions. *J. Solid State Chem.* **2013**, *203*, 333–339. [CrossRef]

22. Zaitsev, V.K.; Fedorov, M.I.; Gurieva, E.A.; Eremin, I.S.; Konstantinov, P.P.; Samunin, A.Y.; Vedernikov, M.V. Highly effective $Mg_2Si_{1-x}Sn_x$ thermoelectrics. *Phys. Rev. B* **2006**, *74*. [CrossRef]

23. Chen, L.X.; Jiang, G.Y.; Chen, Y.; Du, Z.L.; Zhao, X.B.; Zhu, T.J.; He, J.; Tritt, T.M. Miscibility gap and thermoelectric properties of ecofriendly $Mg_2Si_{1-x}Sn_x$ ($0.1 \leq x \leq 0.8$) solid solutions by flux method. *J. Mater. Res.* **2011**, *26*, 3038–3043. [CrossRef]

24. You, S.W.; Shin, D.K.; Kim, I.H. The effects of Sb on the thermoelectric properties of $Mg_2Si_{1-x}Ge_x$ prepared by using solid-state synthesis. *J. Korean. Phys. Soc.* **2014**, *64*, 1346–1350. [CrossRef]

25. Khan, A.U.; Vlachos, N.; Kyratsi, T. High thermoelectric figure of merit of $Mg_2Si_{0.55}Sn_{0.4}Ge_{0.05}$ materials doped with Bi and Sb. *Scripta Mater.* **2013**, *69*, 606–609. [CrossRef]

26. Gao, P.; Berkun, I.; Schmidt, R.D.; Luzenski, M.F.; Lu, X.; Sarac, P.B.; Case, E.D.; Hogan, T.P. Transport and mechanical properties of high-ZT $Mg_{2.08}Si_{0.4-x}$ $Sn_{0.6}Sb_x$ thermoelectric materials. *J. Electron. Mater.* **2014**, *43*, 1790–1803. [CrossRef]

27. Skomedal, G.; Burkov, A.; Samunin, A.; Haugsrudd, R.; Middletona, H. High temperature oxidation of Mg_2(Si-Sn). *J. Corros. Sci.* **2016**, *111*, 325–333. [CrossRef]

28. Sondergaard, M.; Christensen, M.; Borup, K.A.; Yin, H.; Iversen, B.B. Thermal stability and thermoelectric properties of $Mg_2Si_{0.4}Sn_{0.6}$ and $Mg_2Si_{0.6}Sn_{0.4}$. *J. Mater. Sci.* **2013**, *48*, 2002–2008. [CrossRef]

29. Yin, K.; Zhang, Q.; Zheng, Y.; Su, X.; Tang, X.; Uher, C. Thermal stability of $Mg_2Si_{0.3}Sn_{0.7}$ under different heat treatment conditions. *J. Mater. Chem. C* **2015**, *3*, 10381–10387. [CrossRef]

30. Tani, J.; Takahashi, M.; Kido, H. Thermoelectric properties and oxidation behaviour of magnesium silicide. In *IOP Conference Series: Materials Science and Engineering*; IOP Publishing: Bristol, UK, 2011; Volume 18, p. 142013. [CrossRef]

31. Jeanloz, R.; Johnson, M.L. A Note on the bonding, optical-spectrum and composition of tetrahedrite. *Phys. Chem. Miner.* **1984**, *11*, 52–54. [CrossRef]

32. Du, B.L.; Chen, K.; Yan, H.X.; Reece, M.J. Efficacy of lone-pair electrons to engender ultralow thermal conductivity. *Scripta Mater.* **2016**, *111*, 49–53. [CrossRef]

33. Suekuni, K.; Tsuruta, K.; Ariga, T.; Koyano, M. Thermoelectric properties of mineral tetrahedrites $Cu_{10}Tr_2Sb_4S_{13}$ with low thermal conductivity. *Appl. Phys. Express* **2012**, *5*. [CrossRef]

34. Wuensch, B.J. The crystal structure of tetrahedrite, $Cu_{12}Sb_4S_{13}$. *Z. Krist. Cryst. Mater.* **1964**, *119*, 437–453. [CrossRef]

35. Van Embden, J.; Tachibana, Y. Synthesis and characterisation of famatinite copper antimony sulfide nanocrystals. *J. Mater. Chem.* **2012**, *22*, 11466–11469. [CrossRef]

36. Du, B.L.; Zhang, R.Z.; Chen, K.; Mahajan, A.; Reece, M.J. The impact of lone-pair electrons on the lattice thermal conductivity of the thermoelectric compound $CuSbS_2$. *J. Mater. Chem. A* **2017**, *5*, 3249–3259. [CrossRef]

37. Chen, K.; Du, B.L.; Bonini, N.; Weber, C.; Yan, H.X.; Reece, M.J. Theory-guided synthesis of an eco-friendly and low-cost copper based sulfide thermoelectric material. *J. Phys. Chem. C* **2016**, *120*, 27135–27140. [CrossRef]

38. Chen, K.; Di Paola, C.; Du, B.L.; Zhang, R.Z.; Laricchia, S.; Bonini, N.; Weber, C.; Abrahams, I.; Yan, H.X.; Reece, M. Enhanced thermoelectric performance of Sn-doped Cu_3SbS_4. *J. Mater. Chem. C* **2018**, *6*, 8546–8552. [CrossRef]

39. Du, B.; Zhang, R.; Liu, M.; Chen, K.; Zhang, H.; Reece, M.J. Crystal structure and improved thermoelectric performance of iron stabilized cubic Cu_3SbS_3 compound. *J. Mater. Chem. C* **2019**, *7*, 394–404. [CrossRef]

40. Lu, X.; Morelli, D.T.; Xia, Y.; Zhou, F.; Ozolins, V.; Chi, H.; Zhou, X.Y.; Uher, C. High performance thermoelectricity in earth-abundant compounds based on natural mineral tetrahedrites. *Adv. Energy Mater.* **2013**, *3*, 342–348. [CrossRef]

41. Braga, M.H.; Ferreira, J.A.; Lopes, C.; Malheiros, L.F. Phase transitions in the Cu-Sb-S system. *Mater. Sci. Forum* **2008**, *587–588*, 435–439. [CrossRef]

42. Barbier, T.; Lemoine, P.; Gascoin, S.; Lebedev, O.I.; Kaltzoglou, A.; Vaqueiro, P.; Powell, A.V.; Smith, R.I.; Guilmeau, E. Structural stability of the synthetic thermoelectric ternary and nickel-substituted tetrahedrite phases. *J. Alloy. Compd.* **2015**, *634*, 253–262. [CrossRef]

43. Goncalves, A.P.; Lopes, E.B.; Montemor, M.F.; Monnier, J.; Lenoir, B. Oxidation Studies of $Cu_{12}Sb_{3.9}Bi_{0.1}S_{10}Se_3$ tetrahedrite. *J. Electron. Mater.* **2018**, *47*, 2880–2889. [CrossRef]

44. Salvo, M.; Smeacetto, F.; D'Isanto, F.; Viola, G.; Demitri, P.; Gucci, F.; Reece, M.J. Glass-ceramic oxidation protection of higher manganese silicide thermoelectrics. *J. Eur. Ceram. Soc.* **2018**, *39*, 66–71. [CrossRef]

45. Liu, W.; Tang, X.F.; Li, H.; Sharp, J.; Zhou, X.Y.; Uher, C. Optimized thermoelectric properties of Sb-doped $Mg_{2(1+z)}Si_{0.5-y}Sn_{0.5}Sb_y$ through Adjustment of the Mg Content. *Chem. Mater.* **2011**, *23*, 5256–5263. [CrossRef]

46. Du, Z.L.; Zhu, T.J.; Chen, Y.; He, J.; Gao, H.L.; Jiang, G.Y.; Tritt, T.M.; Zhao, X.B. Roles of interstitial Mg in improving thermoelectric properties of Sb-doped $Mg_2Si_{0.4}Sn_{0.6}$ solid solutions. *J. Mater. Chem.* **2012**, *22*, 6838–6844. [CrossRef]

47. Harish, S.; Sivaprahasam, D.; Battabyal, M.; Gopalan, R. Phase stability and thermoelectric properties of $Cu_{10.5}Zn_{1.5}Sb_4S_{13}$ tetrahedrite. *J. Alloy. Compd.* **2016**, *667*, 323–328. [CrossRef]

materials

MDPI

Article

Glass-Ceramic Foams from 'Weak Alkali Activation' and Gel-Casting of Waste Glass/Fly Ash Mixtures

Acacio Rincón Romero [1], Nicoletta Toniolo [2], Aldo R. Boccaccini [2] and Enrico Bernardo [1,*]

[1] Department of Industrial Engineering, University of Padova, Via Marzolo 9, Padova 35131, Italy; acacio.rinconromero@unipd.it

[2] Institute of Biomaterials, Department of Materials Science and Engineering, University of Erlangen-Nuremberg, Cauerstraße 6, 91058 Erlangen, Germany; nicoletta.t.toniolo@fau.de (N.T.); aldo.boccaccini@ww.uni-erlangen.de (A.R.B.)

* Correspondence: enrico.bernardo@unipd.it; Tel.: +39-049-827-5510; Fax: +39-049-827-5505

Received: 10 December 2018; Accepted: 12 February 2019; Published: 15 February 2019

Abstract: A 'weak alkali activation' was applied to aqueous suspensions based on soda lime glass and coal fly ash. Unlike in actual geopolymers, an extensive formation of zeolite-like gels was not expected, due to the low molarity of the alkali activator (NaOH) used. In any case, the suspension underwent gelation and presented a marked pseudoplastic behavior. A significant foaming could be achieved by air incorporation, in turn resulting from intensive mechanical stirring (with the help of a surfactant), before complete hardening. Dried foams were later subjected to heat treatment at 700–900 °C. The interactions between glass and fly ash, upon firing, determined the formation of new crystal phases, particularly nepheline (sodium alumino–silicate), with remarkable crushing strength (~6 MPa, with a porosity of about 70%). The fired materials, finally, demonstrated a successful stabilization of pollutants from fly ash and a low thermal conductivity that could be exploited for building applications.

Keywords: alkali activation; inorganic gel casting; glass–ceramic foams; waste glass; fly ash

1. Introduction

Coal fly ash is a fundamental waste produced by power stations; it consists of a fine particulate material with fluctuating chemical and phase compositions, depending on the original coal and burning conditions, configuring a significant environmental issue. In fact, despite the exploitation of renewable energy resources, the amounts of waste generated worldwide are increasing (the currently produced amount, of about 900 million tons, is expected to increase up to 2 billion tons in 2020) [1]. Coal fly ash is mainly landfilled, producing significant dangers, such as the potential leaching of heavy metals or polycyclic aromatic hydrocarbons [2–4]. As a consequence, the use of coal fly ash in new valuable materials is a critical issue for a sustainable society.

Significant amounts of fly ash are used in the building industry, due to their pozzolanic properties: mixed with cement, they are well known to improve concrete durability [3,5,6]. Fly ash valorization has also been realized by the production of dense glass–ceramic materials, to be used as an alternative to natural stones or traditional ceramic tiles [7,8].

Compared to dense glass–ceramics, lightweight glass–ceramic foams, to be used for thermal and acoustic insulation, may represent a more valuable product. Waste glass/fly ash mixtures have been variously foamed, by viscous flow sintering of glass, determining a pyroplastic mass (in turn incorporating fly ash), with concurrent gas evolution, by the addition of different foaming agents, such as SiC [9]. The amount and nature of waste glass are significant in reducing the processing temperature [10,11]. As an example, carbonates, in the form of dolomite ($CaMg(CO_3)_2$), or sludge from a marble cutting–polishing plant (containing mainly calcite, $CaCO_3$) may lead to ceramic foams with low apparent density (0.36–0.41 g/cm^3) and relatively high compressive strength values (2.4–2.8 MPa). However, the highest amount that could be incorporated for the foam production does not exceed 20 wt% [3]. Coal fly ash may be increased up to 40 wt%, in foams generated by means $CaCO_3$ decomposition, but 30 wt% borax must be considered as extra fluxing agent [11]. Some attempts to produce lightweight aggregates have been made even avoiding the use of any foaming agents, e.g., by mixing the fly ash (75 wt%) and waste window panes (25 wt%) exploiting the bloating of the fly ash at high temperature [12].

According to the fact that fly ash is rich in amorphous alumina and silica, another promising application concerns the formulation of waste-derived geopolymers [13]. Geopolymer materials are a class of inorganic polymers synthesized through the reaction of a solid alumino–silicate precursor with a highly concentrated alkali solution. When immersed in strongly alkaline solutions, the alumino–silicate-reactive materials are almost entirely dissolved, leading to the formation of hydrated alumino–silicate oligomers, including [SiO_4] and [AlO_4] tetrahedral units. The following polymerization of the oligomers creates a highly stable three-dimensional network structure where the [SiO_4] and [AlO_4] tetrahedra are linked together by sharing oxygen atoms. This structure ('zeolite-like gel') resembles that of zeolites, but it mostly reveals an amorphous nature [14].

The use of fly ash in the geopolymer formulation, alone or mixed with other natural sources or industrial wastes, has been widely studied [15–17], with the perspective of obtaining inorganic binders with low CO_2 emissions (compared to ordinary Portland cement) and reduced use of natural raw materials [18,19]. These advantages, however, are somewhat counterbalanced by the costs of the standard reagents used as alkali activators in fly ash geopolymer-based materials, such as sodium silicate (also known as water glass) and sodium or potassium hydroxide [20,21]. In fact, it should be noted that significant energy demand and CO_2 emissions are associated with sodium silicate production, where temperatures around 1300 °C are required, in order to melt sodium carbonate and silica mixtures [22]. Some efforts have recently been made to substitute the conventional activators with low cost and more environmentally friendly alternatives, such as industrial residues of discarded cleaning solutions for aluminum molds [23] and, above all, waste glass [24].

In recent investigations, we proposed the synthesis of dense fly ash [25] or red mud [26] based geopolymers using waste glass from municipal waste collection, replacing water glass as main silica source in the geopolymeric formulation. It was established that, despite the alternative formulation, stable and chemically resistant geopolymeric gels could be obtained by mixing fly ash and waste glass, in an appropriate ratio, and using NaOH in relatively low molarities (NaOH 8M) [25]. The present investigation was conceived as a further extension of the approach, in the development of inorganic gels with an even weaker alkali activation of waste glass/fly ash aqueous suspensions. The suspensions were subjected to extensive foaming (by mechanical stirring) at the early stage of gelation, leading to highly porous bodies later stabilized by means of a low temperature firing treatment.

2. Materials and Methods

The initial raw materials used in this study were low calcium fly ash (FA) class F (ASTM C 618) [27], with a mean particle size of 20 μm, supplied by Steag Power Minerals (Dinslaken, Germany), and soda–lime glass waste (SLG), produced by SASIL S.r.l. (Brusnengo, Biella, Italy) as fine powders with a particle size under 30 μm. In particular, we considered the finest fraction produced during the purification process of glass cullet carried out in the company, with limited industrial application. Table 1 summarizes the chemical composition of the two basic raw materials determined by means of X-ray fluorescence [25].

FA/SLG were prepared by progressive addition of powders into an NaOH aqueous solution, for a liquid/solid of 0.45. According to a methodology already presented in previous studies [25,26], we varied the theoretical molar ratio between SiO_2 and Al_2O_3 in the final product by changing the FA and SLG proportions, being 76/24, 64/36, 54/46, corresponding to a SiO_2/Al_2O_3 theoretical molar ratio of 5, 6, and 7, respectively. However, in this case, a lower concentration (3M) of NaOH was considered. As in previous studies, the mixtures were kept under low-speed mechanical stirring (500 rpm) for 4 h, to ensure a good dissolution of the starting materials and the proper dispersion of the remaining undissolved particles in the slurries.

Table 1. Chemical composition (expressed in wt%) of the starting materials.

Oxide (wt%)	SiO_2	Al_2O_3	Na_2O	K_2O	CaO	MgO	Fe_2O_3	TiO_2
FA	54.36	24.84	0.83	3.03	2.56	2.06	8.28	1.07
SLG	70.5	3.2	12	1	10	2.3	0.42	0.07

The suspensions were foamed by the addition of 4 wt% of an aqueous solution of sodium lauryl sulphate (SLS) ($CH_3(CH_2)_{11}OSO_3Na$, supplied by Carlo Erba, Cornaredo, Milan, Italy), previously prepared with a SLS/water = 1/10 and then by application of vigorous mechanical stirring (2000 rpm) for 10 min. The prepared wet foams were poured into cylindrical molds (6 cm diameter) and kept at 60 °C for 48 h in sealed conditions. Dried samples were finally demolded and subjected to thermal treatment at 800, 900, and 1000 °C for 1 h (10 °C/min heating rate). The overall manufacturing process is represented by Figure 1.

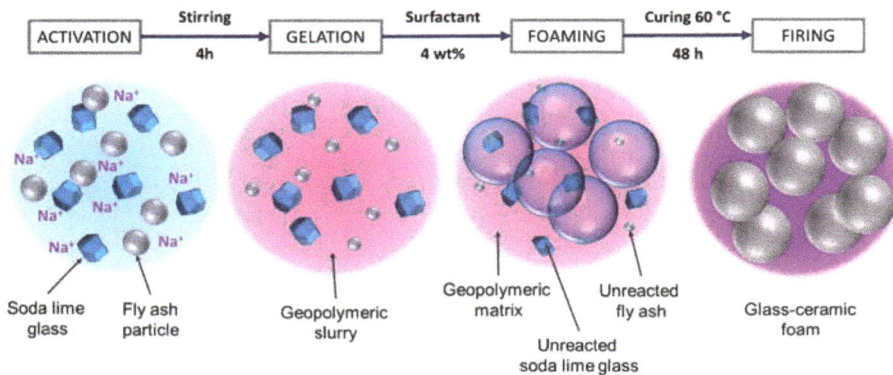

Figure 1. Processing scheme for the production of glass–ceramic foams.

X-ray diffraction analyses (XRD) were carried out on powdered samples (Bruker D8 Advance, Karlsruhe, Germany) using CuKα radiation (0.15418 mm), 40 kV-40mA, and 2θ angles between 10–70°. A step size of 0.02° and 2 s counting time was set for the analysis of the starting materials and of the hardened foams. A step size of 0.02° and a 0.5 s counting time was adopted for the fired samples. The phase identification was performed using the Match!® software suite (Crystal Impact GbR, Bonn, Germany), supported by data from PDF-2 database (ICDD- International Centre for Diffraction Data, Newtown Square, PA, USA).

Samples form larger specimens were cut to about 10 × 10 × 10 mm. The bulk density was evaluated considering the mass measured with an analytical balance and the volume carefully measured with a digital caliper. The apparent and true densities were determined using a helium pycnometer (Micromeritics AccuPyc 1330, Norcross, GA, USA), working on foamed samples or on fine powders from crushed samples, respectively. The total, open, and closed porosity were computed using the three density values. Compression tests were done using an Instron 1121 UTS (Instron, Danvers, MA, USA) machine, with a crosshead speed of 0.5 mm/min, with each data point corresponding to 9–10 samples.

The European Standard for the compliance test for leaching of granular waste materials and sludge (EN 12457-2) was followed to evaluate the release of heavy metals in the initial raw materials and in selected fired samples. Fragments below 4 mm were placed in distilled water with a pH value of ~7 to a liquid/solid ratio of 10, softly stirred at 25 °C for 24 h. The resulting eluates were filtered through a 0.6 μm filter and analyzed using inductively coupled plasma (ICP; SPECTRO Analytical Instruments GmbH, Kleve, Germany).

The semi-industrial production of lightweight panels was developed for selected samples. The alkali activation process was conducted in the same way. However, a relatively significant amount of activated suspension was prepared for each batch (around 5 kg), and the process was carried out with technical grade reagents. The hardened foamed panels were obtained after casting the wet foams in bigger molds (15 cm × 20 cm). Finally, hardened foamed gels were fired at 800 °C for 1 h, using a pilot scale tunnel furnace (Nanetti ER-15S) at SASIL S. p. a. (Brusnengo, Biella, Italy).

Thermal conductivity tests on the lightweight panels were performed using a Fox 50 Heat Flow Meter by TA Instruments (New Castle, DE, USA) operating at 25 °C. The measure was conducted on cylindrical samples (50 mm diameter and 10 mm thickness) cut from the panels; three replicated tests were performed on samples taken from different panels.

3. Results and Discussion

Figure 2 shows the microstructure for the geopolymer foam after 24 h of curing for the mixtures 5S, 6S, and 7S (Figure 2a–c, respectively). The hardened foams reveal a high microstructural uniformity, presenting pores in a range from 10 to 30 μm diameter. No significant changes in the pore distribution between samples with different compositions can be detected, even if the viscosity of the initial slurries was different. High amounts of glass were expected to determine a slight viscosity increase [28], but the spherical morphology of the FA ensured good workability in all conditions; the observed great homogeneity presented can be explained thanks to the rapid setting of the wet foams, preventing any coalescence effect [29].

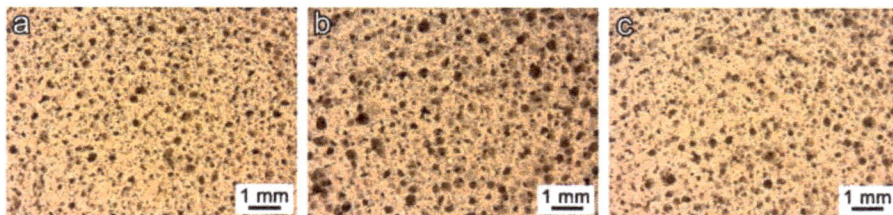

Figure 2. Microstructural details of the soda–lime glass waste/fly ash (SLG/FA foams in the hardened state; (**a**) 5S; (**b**) 6S, and (**c**) 7S.

The X-ray diffraction patterns of the starting materials are illustrated in Figure 3; a different intensity scale was selected for each material in order to highlight the crystalline phases present (the intensity of the strongest peak in Figure 3b is ~20 times higher than the intensity of the strongest peak in Figure 3a). SLG (Figure 3a) presents the typical 'amorphous halo' of silicate glasses centered approximately at $2\theta = 24$–$26°$, along with weak peaks, attributed to hydrated phases, such as calcium aluminum silica hydrate (gismondine $CaAl_2Si_2O_8 \cdot 4H_2O$, PDF#020-0452), calcium silica hydrate ($Ca_{1.5}SiO_{3.5} \cdot xH_2O$, PDF#033-0306), and sodium aluminum silica hydrate ($K_2NaAl_3Si_9O_{24} \cdot 7H_2O$, PDF#022-0773). These hydrated phases had formed probably according to surface reaction of fine glass particles with environmental humidity during storage.

The original fly ash (Figure 3b) contained quartz (SiO_2, PDF#083-0539) and mullite (Al_4SiO_8, PDF#079-1275) as the main crystal phases; moreover, minor traces of hematite (Fe_2O_3, PDF#033-0664) were also detected. In the pattern, it is still possible to detect an amorphous halo at around 24–26°.

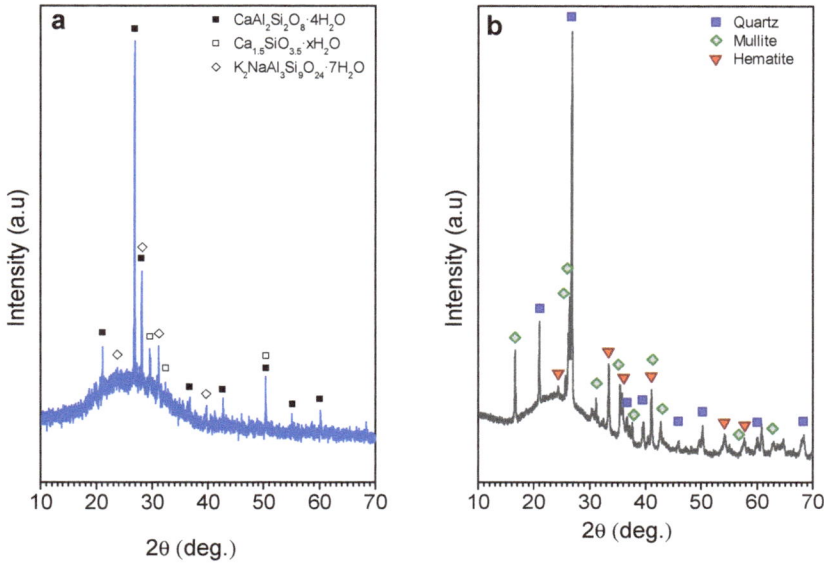

Figure 3. X-ray diffraction patterns of the initial materials (**a**) waste soda lime glass and (**b**) fly ash.

The X-ray diffraction patterns of the hardened foams represented in Figure 4 show that quartz and mullite from the initial fly ash remained practically unaltered. The intensity of these peaks simply decreased with a higher amount of glass in the initial formulation; as a result of this 'dilution' effect, hematite (Fe_2O_3, PDF#033-0664) is no longer visible.

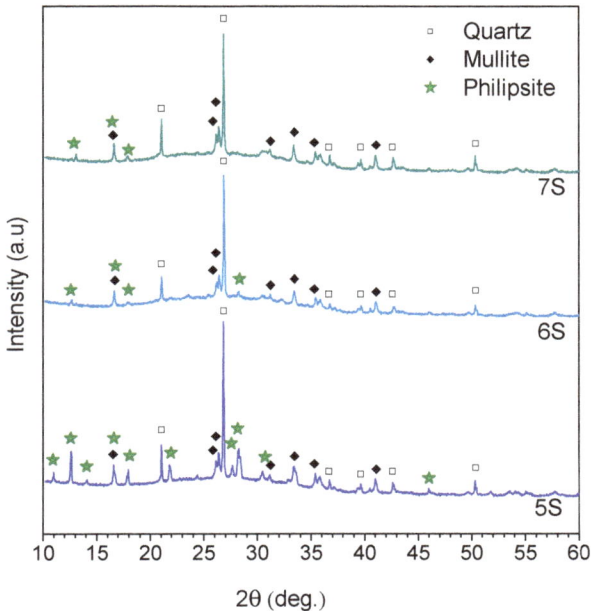

Figure 4. X-rays diffraction patterns of the hardened geopolymeric foams.

The center of the amorphous halo moved from 2θ~24°, in the original raw materials, to 2θ~30°, after alkali activation. This shift is consistent with the literature on geopolymers, concerning the development of an alkaline alumino–silicate hydrate (N–A–S–H) gel as a reaction product [30–32].

A second change, with alkali activation, concerns the presence of philipsite ($Na_4KAl_5Si_{11}O_{32}(H_2O)_{10}$, PDF#073-1419), as newly formed crystal phase, for all compositions. This phase, a sodium potassium aluminum silicate hydrate with a zeolite-type structure, has already been detected in geopolymeric gels [33,34]. Although quite rough, both shift of amorphous halo and formation of philipsite could be considered as a proof of the reaction between starting materials and development of a semi-crystalline 'geopolymer-like' gel.

Microstructural details of the cell struts from hardened sample 5S were studied using SEM (Figure 5). The low magnification image (Figure 5a) reveals the spherical shape of the fly ash mixed with irregular shape particles of SLG. As could be noticed, the low activator molarity did not allow a significant dissolution of the initial raw materials. However, the reaction products detected in the X-ray analysis effectively promoted the binding of adjacent particles in the hardened samples. At higher magnification, zeolite crystals were well visible on the surface of the unreacted particles (Figure 5b). The particular morphology of deposits on spherical fly ash particles is typically associated with the formation of zeolite compounds [35,36].

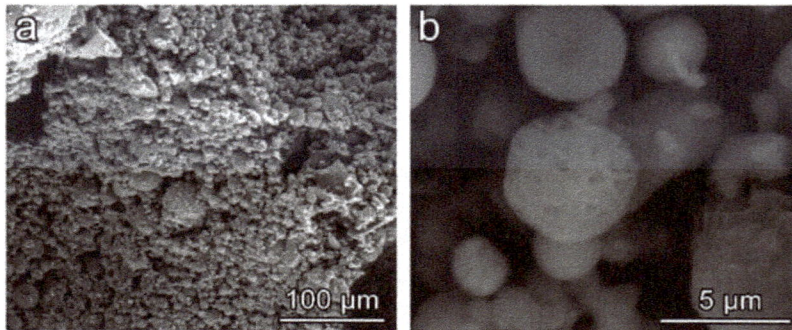

Figure 5. High magnification morphology of 5S hardened state foam: (**a**) Low magnification; (**b**) high magnification.

The hardened foams kept their structural integrity after drying. However, after immersion in boiling water, the water became turbid, and pH rapidly approached 13. This fact revealed the poor chemical stability of the hardened samples when activated at low molarity and justified the application of a firing treatment for further stabilization.

The weight losses of the initial SLG waste and FA are presented in Figure 6a. The waste glass presents a negligible loss of weight around 1 wt% at 700 °C, which is attributed to the burn-out of the plastic impurities inside the waste glass fraction. The weight losses of the FA are approximately 4 wt% in the temperature range studied (20–1200 °C). Losses between 200 and 800 °C are associated with the combustion of carbon present in the FA, and marginal loss of near 1 wt% at 1000 °C is associated with sulphate decomposition [37].

The TGA results of the hardened samples shown in Figure 6b reveal a higher weight loss with the higher content of fly ash in the initial formulation. The surfactant influence is not taken into account as it is added in an aqueous solution 1/10, so the total weight loss attributed to it could not exceed 0.4%.

It is quite challenging to identify the weight losses in the hardened foams, as they are the result of decomposition reaction of several compounds. Two principal changes, however, could be reasonably explained. The gradual weight loss up to 500 °C was attributed mainly to the evaporation of physical bonded and combined water in the gel, whereas the loss at 500–700 °C is thought to correspond mainly to the carbon combustion and secondly to the final dehydration of the C–S–H and N–A–S–H compounds [38]. Beyond 700 °C, the weight losses were stabilized in all the samples, and only the sulphate decomposition from the initial fly ash could be noticed [37].

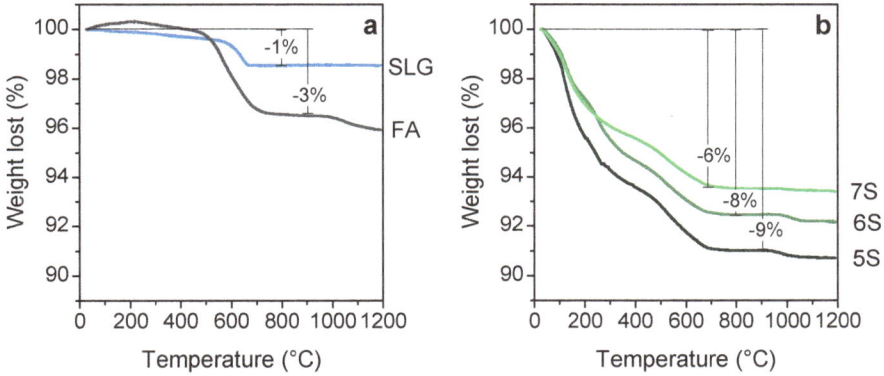

Figure 6. Differential thermal analysis plots; (**a**) initial soda lime glass cullet and fly ash; (**b**) hardened foams.

After the heat treatment, all the samples resulted in well-sintered bodies, illustrated in Figure 7. Homogeneous well distributed round pores, with diameters from 10–30 μm, remained from the initial structure (after hardening). Even if the treatment temperature between 700–900 °C was well beyond the softening point of glass (around 650 °C), no coalescence of the cells, by viscous flow, could be observed, as an effect of the presence of crystal inclusions evolved from FA as well as from glass/FA interactions, and that increased the apparent viscosity on the melt.

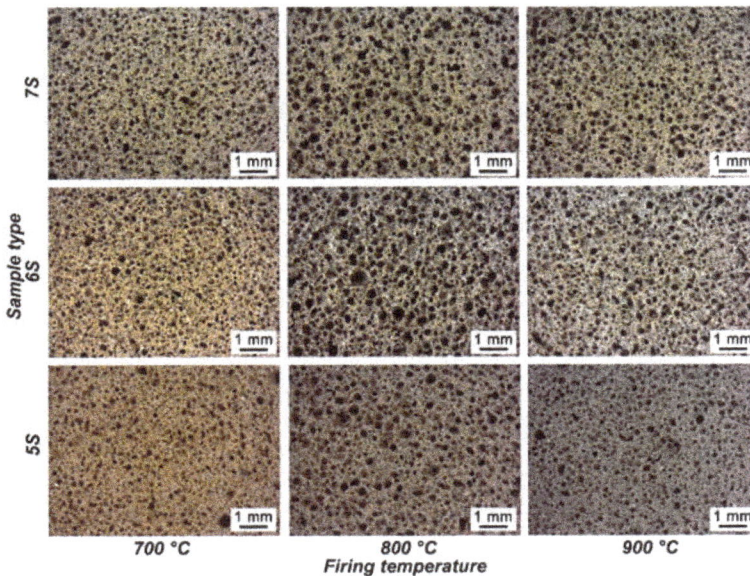

Figure 7. Morphological characterization of the fired foams with different SiO_2/Al_2O_3 molar ratio, after firing.

The evolution of the crystalline phases upon firing is illustrated by the X-ray diffraction patterns in Figure 8. The samples treated at 700 °C (Figure 8a) show that quartz (SiO_2; PDF#083-0539) and mullite (Al_4SiO_8; PDF#079-1275) remained as the only identifiable phases. At this temperature, just above the softening point, SLG could just 'glue' FA particles, with limited chemical interaction. By contrast, softened SLG dissolved some quartz, especially for low FA/glass ratio (sample 7S), starting from 800 °C (Figure 8b). At 800 °C and, above all, at 900 °C, the SLG/FA interaction was substantial enough to lead to the precipitation of nepheline ($Na_{6.65}Al_{6.24}Si_{9.76}O_{32}$, PDF#083-2372), a quite typical crystalline phase formed upon transformation of geopolymers at high temperature [39–41].

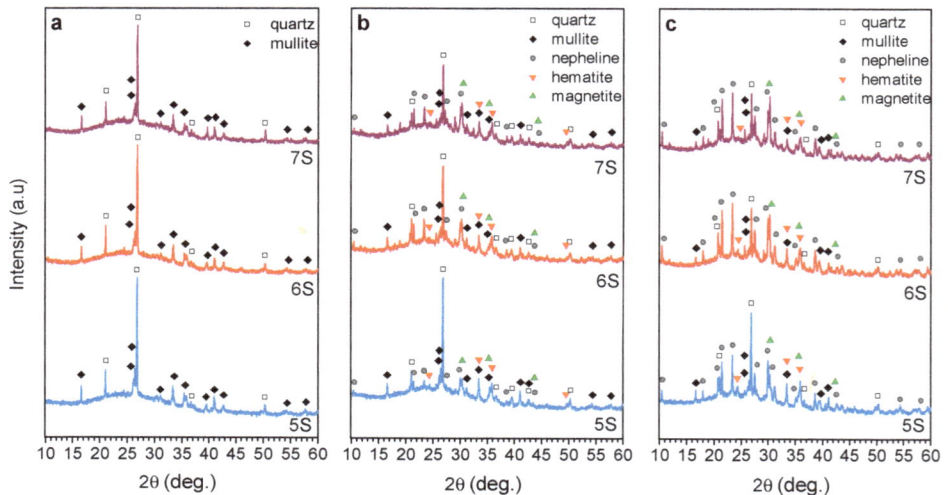

Figure 8. X-ray diffraction patterns of the fired foams with different compositions; (**a**) 700 °C; (**b**) 800 °C; and (**c**) 900 °C.

The increase of temperature also determines the precipitation of iron oxides, in the form of hematite (Fe_2O_3, PDF#89-0691) and magnetite (Fe_3O_4, PDF#89-0691). Its relative abundance is difficult to quantify. However, a significant increase of the magnetite can be noticed after the heat treatment at 900 °C.

Density and porosity data of fired samples are shown in Table 2. High total porosity values are achieved, and it may be noted that the porosity remained mostly open for samples of the 5S and 6S series. Samples with higher glass content (7S) exhibited a more substantial reduction of both overall porosity and open porosity by increasing firing temperature, as an effect of viscous flow. The viscous flow is evident from the microstructural details in Figure 9: Especially samples with relatively low glass content (5S and 6S) were evidently poorly sintered at 700 °C; passing to 800 and 900 °C effectively led to the mixing of the constituents and to the formation of a uniform solid phase.

The densification with increasing glass content is even more evident considering the morphology of cell struts, shown in Figure 10, for samples fired at 800 °C. It can be appreciated that in the sample with lower glass amount, 5S (Figure 10a), some spherical particles, recognized as fly ash, remained visible; by contrast, they were completely incorporated with higher glass content, with evidence of formation of fibrous crystals.

Table 2. Density, porosity, and mechanical properties of the fired treated foams at different heating temperatures.

Sample	Firing T^a (°C)	Density, ρ (g/cm^3)			Porosity (%)			σ_{comp} (MPa)
		Geometric	Apparent	True	Total	Open	Closed	
5	700	0.71 ± 0.01	2.28 ± 0.03	2.46 ± 0.01	71 ± 1	68 ± 2	2 ± 1	1.8 ± 0.2
	800	0.82 ± 0.01	2.32 ± 0.01	2.49 ± 0.01	67 ± 1	64 ± 1	2 ± 1	**5.1 ± 0.3**
	900	0.87 ± 0.01	2.36 ± 0.01	2.51 ± 0.01	65 ± 1	62 ± 1	2 ± 1	**7.4 ± 0.3**
6	700	0.54 ± 0.08	2.25 ± 0.02	2.42 ± 0.01	78 ± 2	75 ± 3	2 ± 1	1.9 ± 0.2
	800	0.66 ± 0.01	2.26 ± 0.02	2.43 ± 0.01	72 ± 1	70 ± 2	2 ± 1	**4.5 ± 0.2**
	900	0.71 ± 0.01	2.26 ± 0.01	2.45 ± 0.01	70 ± 3	68 ± 3	2 ± 1	**5.4 ± 0.4**
7	700	0.71 ± 0.01	2.31 ± 0.06	2.46 ± 0.01	71 ± 1	69 ± 2	2 ± 1	**3.8 ± 0.2**
	800	0.83 ± 0.01	2.13 ± 0.05	2.49 ± 0.01	66 ± 2	61 ± 2	6 ± 1	5.3 ± 0.4
	900	1.02 ± 0.3	2.08 ± 0.02	2.44 ± 0.01	58 ± 3	50 ± 4	7 ± 2	8.7 ± 0.6

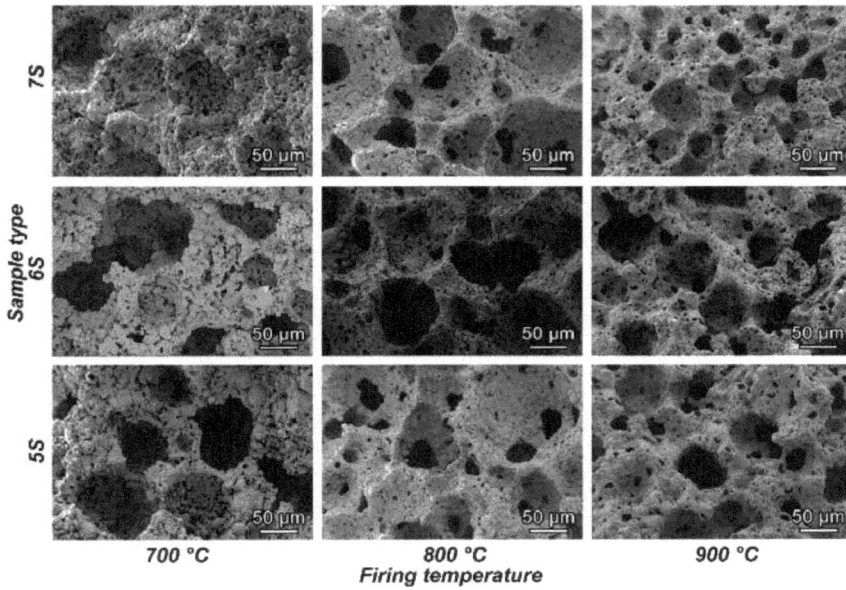

Figure 9. Microstructural details of glass–ceramic foams with different $SiO_2/Al2O_3$ molar ratio, after firing.

Figure 10. High magnification details of the foams struts after firing at 800 °C; (**a**) 5S, (**b**) 6S, and (**c**) 7S.

The strength data reported in Table 2 are interesting, above all, for the correlation with density data. In fact, any serious comparison should take into account the downscaling of mechanical properties operated by porosity. For samples possessing an abundant overall porosity, with very limited closed porosity (samples of the 5S and 6S series, 7S only fired at 900 °C), we may consider a simplified expression of the model proposed by Gibson and Ahsby [42], for the crushing strength (σ_c) of bending-dominated cellular structures, as follows:

$$\sigma_c \approx \sigma_{bend} \cdot C \cdot (\rho_{rel})^{3/2} \tag{1}$$

where the relative density (ρ_{rel}) is defined as $\rho_{rel} = 1 - P/100$ (P is the total porosity), C is a dimensionless constant (~0.2), and σ_{bend} is the bending strength of the solid phase. If we consider the relative densities of samples 5S and 6S fired at 800–900 °C, and of sample 7S fired at 700 °C, ranging from 0.3 to 0.35, the observed crushing strength values (marked in bold character in Table 2) are remarkable. In fact, these values correspond to a bending strength of the solid phase well above 100 MPa, in turn exceeding the bending strength of most glasses [43].

Table 3 shows the chemical analysis of the leachate of the glass–ceramic foams after the heat treatment at 800 and 900 °C and the initial raw materials. According to EN-12457, the leachates of the selected samples are well below the thresholds for inert materials in all the analyzed metal ions, with just one exception. The 5S sample treated at 800 °C exhibited a release of molybdenum ions slightly above the limit, but it should be considered as safe in any case, observing that the leaching tests were applied on highly porous samples, featuring a high surface-to-volume ratio, i.e., in particularly severe conditions.

Table 3. Leachate chemical analysis of selected samples and initial materials.

Element (ppm)	5S3M		6S3M		7S3M		Initial Materials		Limits [UE] (ppm)	
	800	900	800	900	800	900	FA	SLG	Inert material	Non-hazardous material
As	0.0316	0.0635	0.068	0.0503	0.0491	0.0795	<0.0049	<0.0049	0.5	2
Ba	>al	>al	>al	>al	0.0672	0.1108	<0.000	>al	20	100
Cd	<0.0002	<0.0002	<0.0002	<0.0002	<0.0002	<0.0002	<0.0002	0.001	0.04	1
Cr	0.3406	0.0805	0.0255	0.0598	0.0689	0.3001	0.4672	0.0043	0.5	10
Cu	0.0029	0.0183	0.0024	0.0065	0.0053	0.0207	0.0282	0.0036	2	50
Hg	0.0032	<0.0004	0.0006	<0.0004	0.0020	0.0017,	0.8983	<0.0004	0.01	0.2
Mo	0.5324	0.0472	0.2107	0.0087	0.1973	0.2435	<0.0004	0.007	0.5	10
Ni	<0.0014	<0.0014	<0.0014	<0.0014	<0.0014	<0.0014	<0.0014	<0.0014	0.4	10
Pb	<0.0047	<0.0047	<0.0047	<0.0047	<0.0047	<0.0047	<0.0047	0.018	0.5	10
Se	0.0133	<0.0122	0.0255	<0.0122	0.0226	<0.0122	<0.0122	0.018	0.1	0.5
Zn	<0.0203	<0.0203	<0.0203	<0.0203	<0.0203	<0.0203	<0.0203	0.088	4	50
pH	9.4	8.6	8.2	7.7	8.0	7.6				

Abbreviation: >al, above detection limit

The stabilization of the potential pollutants present in the initial waste materials supports the possible use of the waste-derived glass–ceramic foams as environmentally friendly materials for thermal and acoustic insulation. For this reason, selected samples were produced on a semi-industrial scale, given that the proposed method with alkali activation and foaming is beneficial in the manufacturing of large panels. In addition, low-temperature foaming does not imply any geometrical limitation. The firing temperature (800 °C) is far below that adopted in the case of cheapest ceramics for building applications, such as clay bricks.

The overall aspect of the 5S hardened lightweight panels produced with a big mold, after 24 h of post-foaming, is illustrated in Figure 11a (thickness of about 20 mm). The panels show good consistency with no cracks and acceptable mechanical properties, which make them easy to handle. After firing at 800 °C (Figure 11b), the overall structure remained unaltered, as observed on laboratory scale (the sample shown here was not cut or rectified). The faster heat treatment applied in a semi-industrial furnace did not cause the formation of any visible cracks (no preheating was applied to the samples, since they were inserted directly into the furnace).

Figure 11. General view of lightweight panels; (**a**) 6S hardened panel; (**b**) heat treated panel at 800 °C.

The thermal conductivity measured at 25 °C in the 5S samples treated at 800 °C was 0.163 ± 0.005 W·m^{-1}·K^{-1}. Such a relatively low value could be explained due to the low density and the particular cellular structure developed in the panels. Despite the open porosity, the reduced size and the homogeneous distribution probably had an advantageous effect, by reducing air convection [44]. These optimized foam panels could find applications in the building industry, as thermal insulators; moreover, the open-celled morphology presented by the ceramic foams along with the abundant content of iron oxide could support additional applications, e.g., as catalytic supports.

4. Conclusions

We may conclude that:

- The technique ensures an excellent approach to produce glass–ceramic foams allowing the incorporation of high proportions of fly ash.
- This approach provides a recycling route to glass fraction currently landfilled, providing a solution to the landfill derived problems as well as a significant economic advantage.
- The possibility to use low alkali activator concentrations to produce a geopolymer-like gel, which acts as a binding phase, is demonstrated.
- The decomposition of the gel and the SLG/FA interactions upon firing promote the formation of the glass–ceramic foams.
- The developed glass–ceramic foams have high porosity, low thermal conductivity, and reasonable mechanical properties to be applied as thermal insulation materials.
- The chemical stability of the glass ceramic foams was assessed by leaching tests; the release of heavy metals remained below the threshold specification for inert materials.

Author Contributions: For this paper, E.B. and A.R.B. formulated research ideas and supervised the experiments. A.R.R. and N.T. performed the general experimentation. A.R.R. and E.B. have written and edited the article. All authors read, corrected, and approved the article.

Funding: The authors acknowledge the support of the European Community's Horizon 2020 Programme through a Marie Skłodowska-Curie Innovative Training Network ("CoACH-ETN", g.a. no. 642557).

Acknowledgments: The authors thank SASIL S.p.a. (Brusnengo, Biella, Italy), for the supply of the soda–lime glass and experiments on bigger samples, and Steag Power Minerals (Gladbeck, Germany), for the supply of the fly ash. The authors also thank Matteo Cavasin Element (Hitchin, United Kingdom) for the support with the thermal conductivity measurements.

Conflicts of Interest: The authors declare no conflict of interest.

References

1. Blissett, R.S.; Rowson, N.A. A review of the multi-component utilisation of coal fly ash. *Fuel* **2012**, *97*, 1–23. [CrossRef]
2. Carlson, C.L.; Adriano, D.C. Environmental impacts of coal combustion residues. *J. Environ. Qual.* **1993**, *22*, 227–247. [CrossRef]
3. Fernandes, H.R.; Tulyaganov, D.U.; Ferreira, J.M.F. Preparation and characterization of foams from sheet glass and fly ash using carbonates as foaming agents. *Ceram. Int.* **2009**, *35*, 229–235. [CrossRef]
4. Asokan, P.; Saxena, M.; Asolekar, S.R. Coal combustion residues–Environmental implications and recycling potentials. *Resour. Conserv. Recycl.* **2005**, *43*, 239–262. [CrossRef]
5. Kulasuriya, C.; Vimonsatit, V.; Dias, W.; De Silva, P. Design and development of Alkali Pozzolan Cement (APC). *Constr. Build. Mater.* **2014**, *68*, 426–433. [CrossRef]
6. Bui, P.T.; Ogawa, Y.; Nakarai, K.; Kawai, K. Effect of internal alkali activation on pozzolanic reaction of low-calcium fly ash cement paste. *Mater. Struct. Mater. Constr.* **2016**, *49*, 3039–3053. [CrossRef]
7. Leroy, C.; Ferro, M.C.; Monteiro, R.C.C.; Fernandes, M.H.V. Production of glass-ceramics from coal ashes. *J. Eur. Ceram. Soc.* **2001**, *21*, 195–202. [CrossRef]
8. Barbieri, L.; Lancellotti, I.; Manfredini, T.; Ignasi, Q.; Rincon, J.; Romero, M. Design, obtainment and properties of glasses and glass–ceramics from coal fly ash. *Fuel* **1999**, *78*, 271–276. [CrossRef]
9. Wu, J.P.; Boccaccini, A.R.; Lee, P.D.; Kershaw, M.J.; Rawlings, R.D. Glass ceramic foams from coal ash and waste glass: Production and characterisation. *Adv. Appl. Ceram.* **2006**, *105*, 32–39. [CrossRef]
10. Bai, J.; Yang, X.; Xu, S.; Jing, W.; Yang, J. Preparation of foam glass from waste glass and fly ash. *Mater. Lett.* **2014**, *136*, 52–54. [CrossRef]
11. Mi, H.; Yang, J.; Su, Z.; Wang, T.; Li, Z.; Huo, W.; Qu, Y. Preparation of ultra-light ceramic foams from waste glass and fly ash. *Adv. Appl. Ceram.* **2017**, *116*, 400–408. [CrossRef]
12. Wei, Y.-L.; Cheng, S.-H.; Ko, G.-W. Effect of waste glass addition on lightweight aggregates prepared from F-class coal fly ash. *Constr. Build. Mater.* **2016**, *112*, 773–782. [CrossRef]

13. Feng, J.; Zhang, R.; Gong, L.; Li, Y.; Cao, W.; Cheng, X. Development of porous fly ash-based geopolymer with low thermal conductivity. *Mater. Des.* **2015**, *65*, 529–533. [CrossRef]

14. Xu, H.; Van Deventer, J.S.J. The geopolymerisation of alumino-silicate minerals. *Int. J. Miner. Process.* **2000**, *59*, 247–266. [CrossRef]

15. Palomo, A.; Grutzeck, M.W.; Blanco, M.T. Alkali-activated fly ashes: A cement for the future. *Cem. Concr. Res.* **1999**, *29*, 1323–1329. [CrossRef]

16. Toniolo, N.; Boccaccini, A.R. Fly ash-based geopolymers containing added silicate waste. A review. *Ceram. Int.* **2017**, *43*, 14545–14551. [CrossRef]

17. Swanepoel, J.C.; Strydom, C.A. Utilisation of fly ash in a geopolymeric material. *Appl. Geochem.* **2002**, *17*, 1143–1148. [CrossRef]

18. Fernández-Jiménez, A.; Palomo, A.; Criado, M. Microstructure development of alkali-activated fly ash cement: A descriptive model. *Cem. Concr. Res.* **2005**, *35*, 1204–1209. [CrossRef]

19. McLellan, B.C.; Williams, R.P.; Lay, J.; van Riessen, A.; Corder, G.D. Costs and carbon emissions for geopolymer pastes in comparison to ordinary portland cement. *J. Clean. Prod.* **2011**, *19*, 1080–1090. [CrossRef]

20. Komljenović, M.; Baščarević, Z.; Bradić, V. Mechanical and microstructural properties of alkali-activated fly ash geopolymers. *J. Hazard. Mater.* **2010**, *181*, 35–42. [CrossRef]

21. Somna, K.; Jaturapitakkul, C.; Kajitvichyanukul, P.; Chindaprasirt, P. NaOH-activated ground fly ash geopolymer cured at ambient temperature. *Fuel* **2011**, *90*, 2118–2124. [CrossRef]

22. Puertas, F.; Torres-Carrasco, M. Use of glass waste as an activator in the preparation of alkali-activated slag. Mechanical strength and paste characterisation. *Cem. Concr. Res.* **2014**, *57*, 95–104. [CrossRef]

23. Fernández-Jiménez, A.; Cristelo, N.; Miranda, T.; Palomo, Á. Sustainable alkali activated materials: Precursor and activator derived from industrial wastes. *J. Clean. Prod.* **2017**, *162*, 1200–1209. [CrossRef]

24. Zhang, S.; Keulen, A.; Arbi, K.; Ye, G. Waste glass as partial mineral precursor in alkali-activated slag/fly ash system. *Cem. Concr. Res.* **2017**, *102*, 29–40. [CrossRef]

25. Toniolo, N.; Rincón, A.; Roether, J.A.; Ercole, P.; Bernardo, E.; Boccaccini, A.R. Extensive reuse of soda-lime waste glass in fly ash-based geopolymers. *Constr. Build. Mater.* **2018**, *188*, 1077–1084. [CrossRef]

26. Toniolo, N.; Rincón, A.; Avadhut, Y.S.; Hartmann, M.; Bernardo, E.; Boccaccini, A.R. Novel geopolymers incorporating red mud and waste glass cullet. *Mater. Lett.* **2018**, *219*, 152–154. [CrossRef]

27. *Standard Specification for Coal Fly Ash and Raw or Calcined Natural Pozzolan for Use in Concrete*; ASTM Committee C-09 on Concrete and Concrete Aggregates: West Conshohocken, PA, USA, 2013.

28. Ranjbar, N.; Mehrali, M.; Alengaram, U.J.; Metselaar, H.S.C.; Jumaat, M.Z. Compressive strength and microstructural analysis of fly ash/palm oil fuel ash based geopolymer mortar under elevated temperatures. *Constr. Build. Mater.* **2014**, *65*, 114–121. [CrossRef]

29. Studart, A.R.; Gonzenbach, U.T.; Tervoort, E.; Gauckler, L.J. Processing Routes to Macroporous Ceramics: A Review. *J. Am. Ceram. Soc.* **2006**, *89*, 1771–1789. [CrossRef]

30. Álvarez-Ayuso, E.; Querol, X.; Plana, F.; Alastuey, A.; Moreno, N.; Izquierdo, M.; Font, O.; Moreno, T.; Diez, S.; Vázquez, E.; et al. Environmental, physical and structural characterisation of geopolymer matrixes synthesised from coal (co-)combustion fly ashes. *J. Hazard. Mater.* **2008**, *154*, 175–183. [CrossRef]

31. Torres-Carrasco, M.; Puertas, F. Waste glass in the geopolymer preparation. Mechanical and microstructural characterisation. *J. Clean. Prod.* **2015**, *90*, 397–408. [CrossRef]

32. Rattanasak, U.; Chindaprasirt, P. Influence of NaOH solution on the synthesis of fly ash geopolymer. *Miner. Eng.* **2009**, *22*, 1073–1078. [CrossRef]

33. Desbats-Le Chequer, C.; Frizon, F. Impact of sulfate and nitrate incorporation on potassium- and sodium-based geopolymers: Geopolymerization and materials properties. *J. Mater. Sci.* **2011**, *46*, 5657–5664. [CrossRef]

34. Hu, N.; Bernsmeier, D.; Grathoff, G.H.; Warr, L.N. The influence of alkali activator type, curing temperature and gibbsite on the geopolymerization of an interstratified illite-smectite rich clay from Friedland. *Appl. Clay Sci.* **2017**, *135*, 386–393. [CrossRef]

35. Fernández-Jiménez, A.; Palomo, A. Composition and microstructure of alkali activated fly ash binder: Effect of the activator. *Cem. Concr. Res.* **2005**, *35*, 1984–1992. [CrossRef]

36. Katz, A. Microscopic Study of Alkali-Activated Fly Ash. *Cem. Concr. Res.* **1998**, *28*, 197–208. [CrossRef]

37. Li, J.; Zhuang, X.; Monfort, E.; Querol, X.; Llaudis, A.S.; Font, O.; Moreno, N.; Ten, F.J.G.; Izquierdo, M. Utilization of coal fly ash from a Chinese power plant for manufacturing highly insulating foam glass: Implications of physical, mechanical properties and environmental features. *Constr. Build. Mater.* **2018**, *175*, 64–76. [CrossRef]

38. ul Haq, E.; Kunjalukkal Padmanabhan, S.; Licciulli, A. Synthesis and characteristics of fly ash and bottom ash based geopolymers—A comparative study. *Ceram. Int.* **2014**, *40*, 2965–2971. [CrossRef]

39. Bakharev, T. Thermal behaviour of geopolymers prepared using class F fly ash and elevated temperature curing. *Cem. Concr. Res.* **2006**, *36*, 1134–1147. [CrossRef]

40. Dombrowski, K.; Buchwald, A.; Weil, M. The influence of calcium content on the structure and thermal performance of fly ash based geopolymers. *J. Mater. Sci.* **2007**, *42*, 3033–3043. [CrossRef]

41. Kuenzel, C.; Grover, L.M.; Vandeperre, L.; Boccaccini, A.R.; Cheeseman, C.R. Production of nepheline/quartz ceramics from geopolymer mortars. *J. Eur. Ceram. Soc.* **2013**, *33*, 251–258. [CrossRef]

42. Gibson, L.J.; Ashby, M.F. *Cellular Solids: Structure and Properties*, 2nd ed.; Cambridge University Press: Cambridge, UK, 2014; pp. 1–510.

43. Bernardo, E.; Scarinci, G.; Maddalena, A.; Hreglich, S. Development and mechanical properties of metal–particulate glass matrix composites from recycled glasses. *Compos. Part A Appl. Sci. Manuf.* **2004**, *35*, 17–22. [CrossRef]

44. Petersen, R.R.; König, J.; Yue, Y. The mechanism of foaming and thermal conductivity of glasses foamed with MnO_2. *J. Non-Cryst. Solids* **2015**, *425*, 74–82. [CrossRef]

materials

MDPI

Article

Exposure of Glass Fiber Reinforced Polymer Composites in Seawater and the Effect on Their Physical Performance

Matteo Cavasin [1,2,]*[ ], **Marco Sangermano** [1][ ], **Barry Thomson** [2] **and Stefanos Giannis** [3]

[1]   Dipartimento di Scienza Applicata e Tecnologia (DISAT), Politecnico di Torino,
     Corso Duca Degli Abruzzi 24, 10129 Torino, Italy; marco.sangermano@polito.it
[2]   Element Materials Technology Ltd., Wilbury Way, Hitchin SG4 0TW, UK; barry.thomson@element.com
[3]   National Physical Laboratory, Hampton Road, Teddington TW11 0LW, UK; stefanos.giannis@npl.co.uk
*   Correspondence: matteo.cavasin@polito.it

Received: 25 December 2018; Accepted: 1 March 2019; Published: 8 March 2019

Abstract: An innovative testing methodology to evaluate the effect of long-term exposure to a marine environment on Glass Fiber Reinforced Polymers (GFRPs) has been investigated and is presented in this paper. Up to one-year ageing was performed in seawater, to simulate the environment for offshore oil and gas applications. The performance of an epoxy and epoxy-based GFRP exposed at different temperatures from 25 to 80 °C was quantified. The materials were also aged in dry air, to de-couple the thermal effect from the seawater chemical action. Gravimetric testing and Dynamic Mechanical Analysis (DMA) were conducted in parallel on progressively aged specimens. The effect of specimen geometry and the anisotropic nature of diffusion are comprehensively discussed in this paper. For the quasi-infinite specimens, the results show an exponential increase in the seawater absorption rate with temperature. The methodology allowed for the prediction of the diffusivity at a temperature of 4 °C as 0.23 and 0.05×10^{-13} m^2/s for the epoxy and the epoxy-based composite, respectively. The glass transition temperature reduces as sea water is absorbed, yet the sea water effects appear to be reversible upon drying.

Keywords: PMCs; GFRPs; seawater exposure; diffusion; ageing; accelerated testing; gravimetric; DMA

1. Introduction

Polymer Matrix Composites (PMCs) are becoming widespread in the oil and gas industry. The potential to exploit their outstanding mechanical properties, along with their reduced density compared to metals, makes them a suitable candidate to overcome the technical limitations of traditional structural alloys for deep-water fossil fuels recovery.

The marine environment is aggressive towards building materials [1–4]. Both structural steel and concrete are severely attacked by the high saline content of seawater. Standard stainless-steel alloys cannot withstand the corrosive action without other protective means (e.g., passivation techniques, protective paints and liners) [5,6]. Even these approaches can fail to prevent localized corrosion, which will induce catastrophic failures in critical components, particularly if they are not periodically inspected. The oil and gas industry is interested in qualifying materials for eXtreme High Pressure High Temperature (XHPHT) operative conditions [7–9], which are usually met in deep-water offshore reservoirs. In these working scenarios, hydrostatic pressure can exceed 700 bar and temperatures can rise above 180 °C. The acidity of the extracted fossil fuels and fluids, which can be rich in CO_2 and H_2S, particularly for enhanced oil recovery, poses a further challenge to the material durability [10,11].

Duplex stainless alloys have been introduced for particularly demanding applications in corrosive environments, but their cost rarely makes them a viable option for large structures, such as pipelines in the petrochemical industry [10]. Polymers demonstrate a remarkable chemical inertness and they do not suffer from electrochemical corrosion due to their dielectric nature. However, this is not their only benefit: they are significantly lighter compared to metals. Considering their application in a submerged structure, on average, their relative weight is seven-fold lighter compared to the more common steel alloys. This means a significant reduction in the tensile stresses generated by the weight of pillars or pipelines with a very long vertical drop [8,11]. However, they cannot perform as well as structural materials, as they do not possess the required mechanical strength, in particular, with regards to tensile and creep stress. They are much more effective when combined with engineered fibers whose mechanical properties are far superior, resulting in a fiber reinforced Polymer Matrix Composite (PMC). In this form, not only are they excellent structural materials, but their very low relative density (about 2:1 compared to water) allows designers to exploit the hydrostatic pressure to relieve most of the tensile stress due to the weight of a high drop construction [11]. This is a particularly sensitive design constraint for components such as offshore risers, which are used to transfer the fossil fuel and working fluids from the wellhead, sitting at the seabed, to the surface production and storage facilities. Due to the exhaustion of many onshore and shallow water oilfields, the need to drill at greater depths is pushing the design of these structures to the technical limits of traditional building materials. There is an increasing need for composite and hybrid materials able to withstand XHPHT conditions and to enable the exploitation of deep-water reservoirs [1,11].

There is not yet extensive experience with PMCs when it comes to marine applications. The qualification of composite materials for harsh oil and gas services is less advanced than in the aerospace industry, where specific grades and testing routines have been developed over many decades [12]. Pipelines and petrochemical component manufacturers are finally ready to shift to fully composite structures [8]. The main ongoing issue is that the operative working life for those components should be in the order of tens of years in order for them to be economically viable. Due to the very inaccessible environment in which they are installed, no maintenance should be needed for the entire planned lifespan, lest to incur relevant additional costs, and their durability has to be consistently proven.

To introduce a new material grade in a commercial product, such as a flexible riser, the material has to be fully qualified, first at the specimen level and then as a component prototype, to certify that it is going to be safe and reliable for the whole working life. For long-term performance, testing cannot be carried out at a real timescale, as the technology would be obsolete by the time the product was ready to be introduced to the market. Therefore, there is a call for improved accelerated testing methodologies, which shorten the test length to reasonable timeframes, in order to successfully predict the evolution of the material properties over a few decades.

Many operational parameters can affect the results of these tests, so controlling or excluding the highest possible number will return a more robust methodology. Some of the variables are also interacting in a highly nonlinear way, such as the temperature, diffusivity coefficient, and chemical reactions rate. Even when there are theoretical models available to describe these interactions, they still rely on experimental data to be fit to. Hence, there is a need to map the material properties for certain conditions and their evolution with time. The more representative the simulating environment is of the operative scenario, the more accurate the data that is returned. The understanding of the diffusion kinetics at different temperatures provides useful information for foreseeing how the material will behave when exposed to the operative conditions. Ideally, if the data available is extensive enough, it can be possible to define a time-temperature superposition relation, which will allow data from short-term high temperature testing to be extrapolated to predict long-term performance at lower temperatures. Holding this information, the materials, and component designers can result in making an informed decision in a conservative manner.

In this paper, we investigated suitable approaches to measure and quantify the diffusion of seawater in epoxy-based composites, and attempted to use the glass transition temperature as a means of quantifying the physical and chemical ageing of the materials. To achieve that, exposures at a range of temperatures have been conducted and the effect of the coupon size in obtaining a reliable measure of the diffusivity is discussed.

2. Materials and Methods

2.1. Materials

The materials used in the experimental work are all commercially available. They were selected as representative for use in a marine environment involving exposure to seawater and it should be noted that the present research work does not aim to formally evaluate the performance of these commercial products.

A two-component Ampreg 26 epoxy resin with amine hardener commercial grade thermoset epoxy was selected. This material is suitable for vacuum assisted infusion and was obtained from Gurit (Newport, UK). The epoxy resin consists of a blend of bisphenol A, bisphenol F, and 1,6 hexanedioldiglycidylether, while the hardener is a blend of amines (polyoxyalkyleneamine, 2,2'-dimethyl-4,4'methylenebis (cyclohexylamine), 4,4'-methylenebis (cyclohexylamine), and 2,2'-iminodiethylamine) [13].

The stitched unidirectional glass fiber fabric (1200 g/m^2, from Gamma Tensor, Alcoi, Spain) comprises 3B Advantex® SE 2020 Direct Roving made of boron free E-CR glass, which is designed for the production of non-crimped fabrics and has a proprietary sizing specifically designed for excellent adhesion with epoxy resin systems.

Neat epoxy plates were manufactured by compression molding to reduce the formation of voids. A hydraulic press (Mackey Bowley, now Wessex Hydraulic Services, Trowbridge, UK) was used to apply a pressure of about 2 MPa at incremental steps to the viscous resin, hence ejecting residual air from the mould cavity. A rubber seal placed between the mold plates allowed the plates to be adjusted and let the air evacuate. The compression mold also enabled us to control the plate's planarity. The resin was left to cure for 24 h at room temperature, followed by a post-cure of 5 h at 80 °C. Once the curing was completed, the solidified plate was carefully removed from the mold. The plates were machined by a CNC milling machine to obtain the specimens at the required geometries.

The Glass Fiber Reinforced Polymer (GFRP) composite was manufactured by infusing the unidirectional glass fiber fabric with the same Ampreg 26 epoxy. A vacuum bag infusion process was used. After 24 h curing at room temperature, the GFRP plate was removed from the bag and placed in a forced convection oven (Genlab Ltd., Widnes, UK) at 80 °C to post-cure for 5 h. Two machining methods were used to section the GFRP plates and obtain the required specimens, namely diamond blade cutting (for initial cut) and waterjet (for finishing of the coupons). There was not any effect of the cutting method on the specimens obtained for this test program.

2.2. Methods

The evolution of the material properties was monitored by different techniques while the specimens were exposed in simulated working environments, being soaked in substitute seawater. The coupons were immersed in plastic containers and kept at different exposure temperatures (25, 55, and 80 °C) in forced convection ovens. At scheduled intervals, gravimetric measurements were performed following the ASTM D5229M [14] standard procedure, allowing the evaluation of fluid absorption progress. In the first stages, the fluid diffusion proceeded at a higher rate, requiring more frequent sampling, particularly at the higher temperatures, even at just a few hours, to effectively capture the absorption through the weight increase. After saturation, the sampling intervals were prolonged up to weeks. In parallel, Dynamic Mechanical Analysis (DMA) runs were performed to evaluate the glass transition temperature (T_g).

2.2.1. Conditioning in Simulated Environments

The gravimetric and DMA test specimens were exposed to substitute (or synthetic) seawater, described by the ASTM D1141 [15] standard procedure and purchased by ReAgent, (Runcorn, UK), in order to simulate the marine environment. The composition of the seawater is presented in Table 1.

Table 1. Standard substitute seawater as per ASTM D1141 [15].

Compound	Concentration (g/L)
NaCl	24.53
$MgCl_2$	5.20
Na_2SO_4	4.09
$CaCl_2$	1.16
KCl	0.695
$NaHCO_3$	0.201
KBr	0.101
H_3BO_3	0.027
$SrCl_2$	0.025
NaF	0.003
Other metal nitrates/nitrites	<0.1 mg/L
H_2O	balance

The seawater containers where placed in forced convection ovens to reach the equilibrium temperature. Different temperatures were selected to accelerate the diffusion and ageing processes. The selected temperatures were 25, 55, and 80 °C, as it was decided not to exceed this higher temperature, which was already very close to the glass transition temperature of the pristine epoxy. The oven's temperature was constantly recorded by K-type thermocouples, logging to the lab server.

To evaluate the possible scale and edge effect due to the nominal coupons' dimensions, five different geometries were chosen and are listed in Table 2.

Table 2. Diffusion test coupons' nominal geometries for both neat Ampreg 26 and GFRP.

Code	$h \times w \times l$ (mm $\times$ mm $\times$ mm)
α	$2 \times 19 \times 19$
β	$2 \times 40 \times 40$
γ—quasi-infinite	$2 \times 100 \times 100$
δ [1]—fibers along l direction	$2 \times 10 \times 100$
ε [1]—fibers along w direction	$2 \times 100 \times 10$

[1] there is no physical difference between δ and ε for the neat Ampreg 26 (there is no fibre reinforcement); the 10×100 samples were unequivocally named as δ.

The specimens, following an initial pre-conditioning to eliminate any possible moisture content (14 days at 50 °C), were introduced in the seawater: this moment is considered the beginning of the conditioning/ageing process. At regular times, the gravimetric specimens were extracted from the containers and were weighed, to monitor the progress of the seawater diffusion, and reintroduced in the simulated environment, in an effort to reduce the exposure to the external atmosphere (5 min at most out of the seawater) to a minimum. Two scales with a 0.1 mg precision were used for weighing the coupons. Complete absorption curves were generated in this way. The results were recorded to monitor the evolution of material properties. At the scheduled end of the conditioning, after about one-year exposure, the remaining specimens were removed from the exposure and either tested "as-is" or re-dried in an oven set at 50 °C, to evaluate the desorption curves and explore any recovery of the elastic properties. As a reference control, some coupons were exposed at the same temperatures, but kept in dry air within sealed tin cans (silica bags were introduced in order to keep the moisture content at a minimum), in order to compare the effect of the temperature ageing only on the materials.

2.2.2. Dynamic Mechanical Analysis (DMA)

Dynamic Mechanical Analysis (DMA) was performed using an RSA 3 analyzer, by TA Instruments (New Castle, DE, USA), to evaluate the glass transition temperature (Tg) of the materials. The sampling times were scheduled so as to capture the Tg shift in the critical moments of the absorption process. The loading mode was sinusoidal three-point bending. The loading span was of 25 mm. The sample geometry was $2 \times 10 \times 33$ ($h \times w \times l$) and was obtained by cutting part of the respective δ coupons for the neat Ampreg 26 and ε coupons for the GFRP. The temperature ramp was set from 25 up to 150 °C, at a 5 K/min heating rate. Such a higher heating rate was selected (instead of the standard 3 K/min) in an attempt to shorten the period during which the sample was outside the exposure environment. A trade-off between the measurement precision and the accuracy of the "wet" Tg measured was required to accelerate the test. The Tg transitions were evaluated both by the storage modulus (E') drop onset and by the tanδ peak. The DMA specimens removed at selected intervals of conditioning were investigated "as-is" and after re-drying in an oven at 50 °C for up to two weeks. In this way, it was possible to evaluate the reversibility of the fluid diffusion on the viscoelastic properties of the materials.

3. Results

The gravimetric measurements are reported as a function of the length of the exposure to the simulated marine environment (i.e., substitute seawater). The timescale is expressed as the square root of time, using days as the unit of measure for ease of understanding and to present a consistent representation between the gravimetric measurements and the rest of the tests. The Fickian theory of diffusion states that the mass uptake (i.e., moisture content) is linear to the square root time for the initial part of the M_t vs. $\sqrt{t}$ plot. This representation also helps to expand the early stages, when the absorption process is at its highest rate and can be expected to have the highest degree of influence on the material properties. The single data point is an average of the weight of the set of specimens (three replicas for each exposure condition specimen type) involved in a single measurement.

3.1. Gravimetric Measurements on Neat Ampreg 26 Epoxy

Gravimetric measurements on neat epoxy specimens of different geometries were performed to assess the fluid absorption process and evaluate if any shape factor influences the process. The results are presented in Figure 1 as an average of the measurements of all the specimens available.

The results obtained show that the diffusion process is not completely Fickian. After an initial linear uptake (in $\sqrt{t}$ space), the material reaches a so-called pseudo-saturation [2,16]. A saturation time can be calculated, related to the exposure temperature. However, if the gravimetric measurements are continued, it is evident that the material continues to absorb seawater at a much slower rate, entering another linear regime. It does eventually reach a definitive saturation stage, but at a much later stage. For the 80 °C exposure, after about six months, the graph shows a noticeable decrease in weight, perhaps an indication of chemical degradation.

The ASTM D570 and ASTM D5229 standards describe in detail the design and the methodology of gravimetric testing to evaluate the diffusivity properties for plastics and polymer composite materials. A pivotal feature is the saturation criteria used, which should be a numerical one, and it relies on the convergence of the weight of the coupons as those approach a steady saturated state. Hence, from consecutive measurements, when the relative change in per cent weight gain decreases below a certain threshold, the material can be deemed as in equilibrium. The ASTM D570 standard states that the material is saturated when its weight changes less than 5 wt% or less than 5 mg over three consecutive measures, while the ASTM D5229 (see Equation (1)) is more stringent, and requires a change in weight by less than 0.020 wt% over each of two consecutive reference time period spans, and examination of the weight gain versus (time)$^{1/2}$ plot. The second criterion is much stricter due to composites being less prone to absorbing fluid as inorganic fibers are considered inert.

$$\left| \frac{W_n - W_{n-1}}{W_0} \right| \times 100 < 0.020\% \text{ for 2 consecutive reference time span} \tag{1}$$

This approach has its limitations. In particular, it does not consider the actual saturation level of the material. Therefore, polymers with a very low saturation level or with very low diffusivity can be evaluated as saturated too early, while others more prone to absorbing or showing secondary stages could eventually never reach proper saturation. From experimental practice, it is evident that these criteria are not good enough to capture the time to saturation, but they highly rely on the subjective evaluation of the operator.

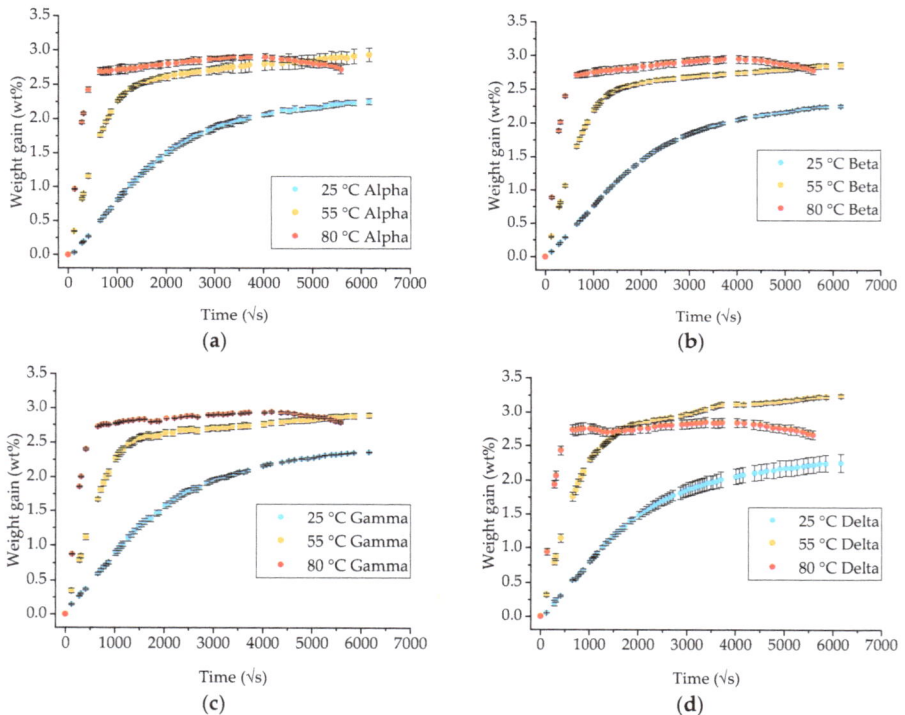

Figure 1. Gravimetric measurements on neat Ampreg 26 coupons of different shape factors: (**a**) Alpha; (**b**) beta; (**c**) gamma and (**d**) delta. (Note: coupons' dimensions are listed in Table 2). The x-axes are measured in the square root of seconds.

In this work, a slightly different saturation criterion is proposed, by evaluating the partial increase over consecutive measurements and dividing this by the overall weight gain to that point. In this way, small changes are more relevant to low diffusing materials compared to more permeable ones. This allows a more robust numerical saturation criterion to be defined:

$$\left| \frac{W_n - W_{n-1}}{W_n - W_0} \right| < [thresold\ value] \text{ for } [n] \text{ consecutive reference time spans} \tag{2}$$

Even this criterion is susceptible to the selection of appropriate threshold values. From our analysis, it was found that a value of 0.02 (in other words, less than 1/50 of the overall weight gain) allows the saturation point to be individuated with a good consistency, even when the fluid uptake is very slow, i.e., ambient temperature conditions.

In order to improve the reliability of a numerical criterion for the localization/target of the effective equilibrium, the results obtained using different criteria are compared in Table A1 (see Appendix A).

From the values reported, it can be highlighted that the criterion prescribed by the ASTM D5229M is less precise and reliable, with a large scatter in the values obtained for each specimen geometry at any temperature. The proposed method appears to be more consistent, yet it is susceptible to noisy measurement and still requires a comparison of the M_t vs. $\sqrt{t}$ plot. Of the two different threshold values proposed, the 0.02 is preferred as it captures the effective equilibrium at 25 °C in a more accurate way. For these reasons, for the rest of this study, we refer to the effective equilibrium as defined by the residual increment Criterion A.

The values in Table 3 are essential for the evaluation of the diffusivity of the material, as is the consideration of the contribution of the exposed edges of the specimens, hence the deviation from one-dimensional diffusion.

Table 3. Saturation levels and times for neat Ampreg 26 at various temperatures and specimen shapes. (Note: Mean values of three specimens for each geometry are listed).

Temperature (°C)	Specimen	$h \times w \times l$ (mm × mm × mm)	Slope ($\times 10^{-4}$ %/$\sqrt{s}$)	$M\infty$ (%)	$t\infty$ (days)
25	alpha	$2.09 \times 18.86 \times 19.15$	8.8	1.94	113
	beta	$2.18 \times 40.00 \times 40.08$	7.9	1.90	107
	gamma	$2.04 \times 100.04 \times 100.07$	8.4	2.01	110
	delta	$2.15 \times 10.08 \times 100.03$	8.2	1.91	107
55	alpha	$2.11 \times 18.96 \times 19.05$	28.4	2.69	55
	beta	$2.16 \times 40.03 \times 39.99$	26.6	2.60	33
	gamma	$2.12 \times 100.05 \times 100.05$	26.6	2.63	33
	delta	$2.07 \times 10.06 \times 100.05$	28.4	2.91	48
80	alpha	$2.15 \times 19.31 \times 19.14$	71.7	2.77	16
	beta	$2.15 \times 39.96 \times 40.03$	65.9	2.77	9
	gamma	$2.13 \times 100.03 \times 100.05$	64.9	2.79	9
	delta	$2.12 \times 10.05 \times 99.97$	69.9	2.73	9

3.2. Gravimetric Measurements on GFRP Composite

Similar to the neat epoxy, gravimetric measurements on GFRP specimens of different geometries were performed to assess the fluid diffusion and evaluate the shape factor effects. The results are presented in Figure 2. In this case, the alpha coupons were not used due to their small mass (even considering the relevant inert fiber content), so the change in weight due to the absorbed moisture would be subject to significant experimental error. Instead, coupons coded as epsilon were introduced. These have the same bar geometry as coupons denoted as delta, but the fibers are aligned along the shorter side. The aim is to highlight if the there is any anisotropic effect in the fluid diffusion due to the fiber presence, such as preferential paths or different contributions of the matrix-fiber interphase.

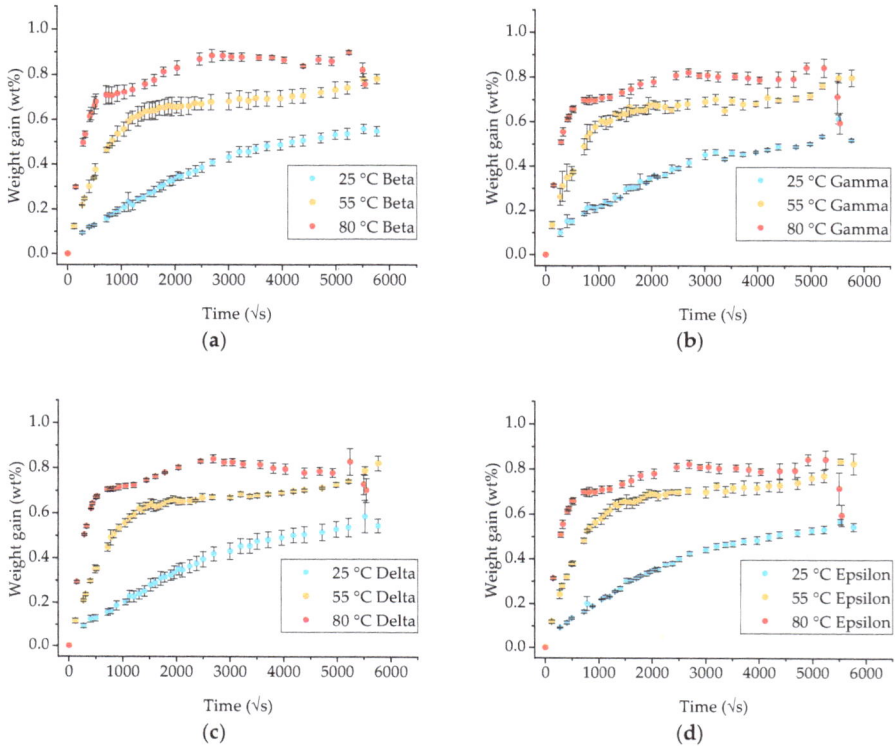

Figure 2. Gravimetric measurements on GFRP coupons of different shape factors: (**a**) Beta; (**b**) gamma; (**c**) delta and (**d**) epsilon. (Note: coupons' dimensions are listed in Table 2). The x-axes are measured in the square root of seconds.

The diffusion at 25 °C appears to be Fickian for the duration of the exposure as was at 55 °C for approximately 250 days when a sudden increase in weight was observed. On the contrary, when exposed at 80 °C, diffusion follows a Fickian curve for about 20 days and then significantly deviates, signifying the beginning of a secondary diffusion stage [17,18].

The moisture uptake at effective equilibrium and the respective time to reach that level were calculated using the residual increment Criterion A (see Equation (2)) with a threshold value of 0.02 and $n = 2$. The results are listed in Table 4. The scatter of the individual measurements negatively affected the precision of determining a consistent point of effective equilibrium, with a large spread between the different specimen geometries, particularly for the exposure at 80 °C. For this temperature, if the threshold value is increased to 0.04–0.05, taking into account the higher degree of noise in the measurements, it is possible to obtain much more accurate saturation times.

For the higher exposure temperatures (55 and 80 °C), a higher amount of seawater appears to be absorbed by specimen type epsilon compared to delta, which could be an indication of an increasing role of the fiber-matrix interphase as the fibers in epsilon are aligned along the shorter side and hence more fiber edges are exposed to the environment. However, the differences are relatively small and no solid conclusion can be derived.

Table 4. Saturation levels and times for GFRP at various temperatures and specimen shapes. (Note: Mean values of three specimens for each geometry are listed).

Temperature (°C)	Specimen	$h \times w \times l$ (mm × mm × mm)	Slope ($\times 10^{-4} \%/\sqrt{s}$)	$M\infty$ (%)	$t\infty$ (days)
25	beta	1.72 × 39.99 × 40.00	1.4	0.48	143
	gamma	1.70 × 99.84 × 99.73	1.3	0.46	132
	delta	1.72 × 9.99 × 99.96	1.4	0.51	182
	epsilon	1.70 × 99.90 × 10.03	1.5	0.47	119
55	beta	1.73 × 40.00 × 40.04	6.2	0.65	29
	gamma	1.70 × 99.91 × 99.98	5.9	0.67	41
	delta	1.73 × 9.95 × 99.98	6.1	0.64	31
	epsilon	1.72 × 99.93 × 9.99	6.6	0.66	29
80	beta	1.73 × 39.99 × 39.99	20.0	0.72	8
	gamma	1.76 × 99.12 × 99.75	20.0	0.82	24
	delta	1.71 × 10.00 × 100.03	19.6	0.75	17
	epsilon	1.74 × 99.93 × 10.01	21.1	0.80	70

3.3. Mechanical Performance

The mechanical performance plays a significant role when selecting a polymer composite material for a structural application in a harsh environment. Amongst the different characterisation techniques, the tensile test is commonly used to evaluate the mechanical performance and measure the Young's modulus and strength [4,12]. In our study, alongside the gravimetric coupons, tensile test specimens were exposed in seawater over prolonged periods of time. Neat, epoxy specimens were tested to evaluate the degradation in performance of the matrix alone. Likewise, GFRP specimens with unidirectional reinforcement in either the longitudinal or transverse direction were tested to monitor how the fiber- and matrix-dominated mechanical response changes over time. All specimens were tested at different sampling times, chosen in relation to the advancement of the diffusion stages at the exposure temperature, in an attempt to capture the most relevant transitions (i.e., pristine material, linear uptake, Fickian saturation, aged). Although the experimental results of this expensive test program are not presented here, a close correlation between the performance loss, the degree of saturation degree, and the exposure temperature was found and will be comprehensively discussed in a future paper.

4. Discussion

4.1. Diffusivity Calculation for Neat Ampreg 26 Epoxy

The moisture content in the isotropic ($D_x = D_y = D_z = D$) rectanguloid of Figure 3 is given by the following integral solution to Fick's equation for one-dimensional diffusion:

$$\frac{M_t}{M_\infty} = G_{1D} = 1 - \frac{8}{\pi^2} \sum_{j=0}^{\infty} (2j+1)^{-2} \exp\left[-\frac{(2j+1)^2 \pi^2 Dt}{h^2}\right] \tag{3}$$

To calculate the diffusivity coefficient, D, a simplified version (valid as $M_t/M_\infty < 0.6$) of the above equation was considered:

$$\frac{M_t}{M_\infty} = G_{1D} = \frac{4}{h}\sqrt{D/\pi} \times \sqrt{t} \tag{4}$$

where, M_t is the moisture content at time t, M_∞ is the moisture content at the Fickian saturation (effective equilibrium), and h is the thickness of the specimen. The slope of the M_t vs. $\sqrt{t}$ for $M_t < 0.6\, M_\infty$ (linear diffusion) is equal to

$$Slope = \frac{4M_\infty}{h}\sqrt{D/\pi} \tag{5}$$

The above equations do not take into account any diffusion happening from the free edges of the specimens. For that, the three-dimensional problem of diffusion in a rectanguloid needs to be considered and solved. Apart from being very computationally expensive, this has the additional drawback that the diffusivity is not easily determined from the linear initial slot of the M_t vs. $\sqrt{t}$ graph. For this reason, the introduction of a correction factor can be considered, as follows:

$$G_{3D} = fG_{1D} \tag{6}$$

and hence,

$$D_{corr} = f^{-2}D_{eff} \tag{7}$$

where, D_{corr} is the corrected diffusivity coefficient for diffusion through the free edges and D_{eff} is the diffusivity coefficient measured from the gravimetric experiments.

Shen and Springer [19] and Starink, Starink and Chambers [20] have derived correction factors that are used in this paper to correct the diffusivity coefficients:

Shen and Springer [19]:

$$f_{S\&S} = 1 + \frac{h}{w} + \frac{h}{l} \tag{8}$$

Starink, Starink and Chambers [20]:

$$f_{SSC} = 1 + 0.54\frac{h}{w} + 0.54\frac{h}{l} + 0.33\frac{h^2}{wl} \tag{9}$$

The results for the neat Ampreg 26 epoxy are presented in Table 5 for the various temperatures and specimen shapes.

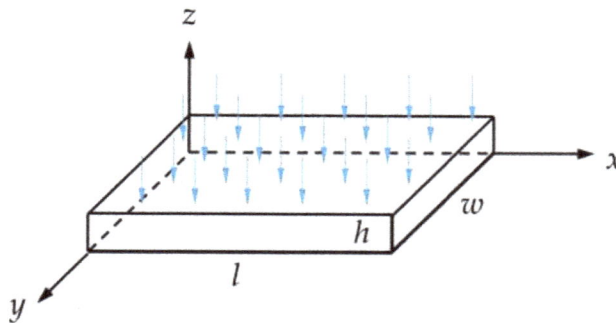

Figure 3. Schematic of rectanguloid isotropic material specimen.

Table 5. Diffusivity coefficient for neat Ampreg 26 at various temperatures and specimen shapes. (Note: Mean values of three specimens for each geometry are listed).

Temperature (°C)	Specimen	$h \times w \times l$ (mm × mm × mm)	$f_{S\&S}$	f_{SSC}	D_r ($\times 10^{-13}$ m²/s)	$D_{r,S\&S}$ ($\times 10^{-13}$ m²/s)	$D_{r,SSC}$ ($\times 10^{-13}$ m²/s)
25	alpha	2.09 × 18.86 × 19.15	1.22	1.12	1.74	1.17	1.38
	beta	2.18 × 40.00 × 40.08	1.11	1.06	1.59	1.29	1.41
	gamma	2.04 × 100.04 × 100.07	1.04	1.02	1.40	1.29	1.34
	delta	2.15 × 10.08 × 100.03	1.24	1.13	1.67	1.09	1.31
55	alpha	2.09 × 18.86 × 19.15	1.22	1.12	9.57	6.43	7.59
	beta	2.16 × 40.03 × 39.99	1.11	1.06	9.57	7.80	8.53
	gamma	2.12 × 100.05 × 100.05	1.04	1.02	9.03	8.31	8.63
	delta	2.07 × 10.06 × 100.05	1.23	1.12	8.02	5.33	9.01
80	alpha	2.15 × 19.31 × 19.14	1.22	1.13	61.09	40.80	48.27
	beta	2.15 × 39.96 × 40.03	1.11	1.06	51.01	41.60	45.49
	gamma	2.13 × 100.03 × 100.05	1.04	1.02	48.13	44.28	45.98
	delta	2.12 × 10.05 × 99.97	1.23	1.13	58.21	38.32	45.83

For the quasi-infinite coupon size (gamma), the two correction factors converge (only 2% difference) to 1 and therefore, the corrected diffusivities are in better agreement and close to the measured ones from the experimental curves. For more irregular coupon sizes and especially for the very small coupon, the correction factors diverge from 1 and from each other, showing a large range of diffusion coefficients. If all results are compared to the quasi-infinite coupons, then $f_{S\&S}$ overcorrects the diffusivity values; hence, the f_{SSC} is considered more accurate and applicable. This is in agreement with the results presented in [20,21].

The present analysis demonstrates that quasi-infinite coupons are necessary to obtain accurate diffusivity values and, in the case that these are not available due to manufacturing issues (i.e., extracting coupons from a small slab of material), the f_{SSC} correction factor should definitely be considered to avoid errors, even up to 30%.

4.2. Diffusivity Calculation for the GFRP Composite

As with the isotropic case of the neat Ampreg 26, correction for the diffusion through the free edges needs to be evaluated for the composite, in addition to considering the anisotropic nature of the diffusion process.

Starink, Starink and Chambers [20] have presented a modified treatment of diffusion in unidirectional composites assuming that the fibers do not take any moisture, where the diffusivity parallel ($D_{\parallel}$) and transverse ($D_{\perp}$) to the fibers in Figure 4 are given by

$$D_{\parallel} = D_r \tag{10}$$

$$D_{\perp} = \frac{\left(1 - 2\sqrt{\frac{v_f}{\pi}}\right)}{1 - v_f} D_r = g^2 \times D_r \tag{11}$$

where v_f is the fiber volume fraction of the composite and D_r is the diffusivity coefficient of the neat resin. The fiber hindrance factor g^2 is an expression of the reduced free path for the diffusion species to follow due to the volume fraction taken by the inert inorganic fibers. It does not take into account any possible contribution due to the possible different chemical activity of the fiber-matrix interphase.

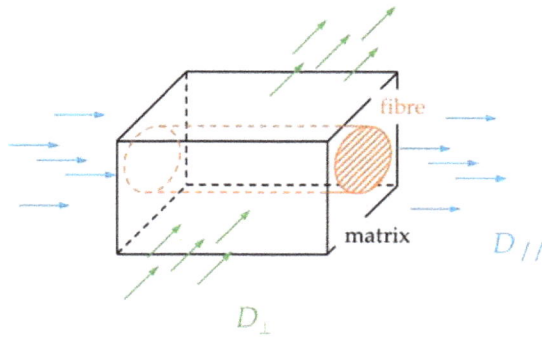

Figure 4. Simplified geometry of unidirectional composite for diffusion parallel and transverse to the fibers.

For the unidirectional composite in Figure 5, the diffusivity coefficients are

$$D_x = D_\parallel, \; D_y = D_z = D_\perp \tag{12}$$

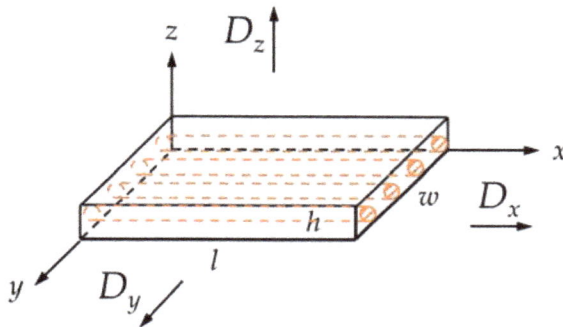

Figure 5. Schematic of rectanguloid anisotropic material specimen.

Similar to the isotropic case, Starink, Starink and Chambers [20] have proposed a correction factor for an orthotropic medium, ignoring second order edge effects:

$$f_{ortho} = 1 + 0.54\frac{h}{w}\sqrt{\frac{D_y}{D_z}} + 0.54\frac{h}{l}\sqrt{\frac{D_x}{D_z}} \tag{13}$$

By combining Equation (13) and Equations (10)–(12), it can be found that for the unidirectional composite, the correction factor is only a function of the geometry of the specimen, considering along which main direction the diffusion is favored (usually the shortest) and the fiber volume fraction. For the fibers aligned along the *x*-axis, as in Figure 5, this results in two possible scenarios:

$$f_{ortho, \perp} = 1 + 0.54\left(\frac{h}{w} + \frac{h}{l}\frac{1}{g}\right)$$
for diffusion mainly $\perp$ to the fibers
$$\tag{14a}$$

$$f_{ortho, \parallel} = 1 + 0.54\left(\frac{h}{w} + \frac{h}{l}\right)g$$
for diffusion mainly $\parallel$ to the fibers
$$\tag{14b}$$

Therefore, the effective diffusivity for the composite transverse and perpendicular to the fiber direction can be estimated as

$$D_{eff. \perp} \cong f_{ortho, \perp}^2 D_\perp = f_{ortho, \perp}^2 \cdot g^2 \cdot D_r \tag{15a}$$

$$D_{eff. \parallel} \cong f_{ortho, \parallel}^2 D_\parallel = f_{ortho, \parallel}^2 \cdot D_r \tag{15b}$$

The above relationships are plotted in Figure 6 with respect to the volume fraction of the composite material. It is apparent that the 100×100 specimen geometry is closer to the infinite plate solution than any other geometry, hence providing the most accurate geometry for the measurement of the diffusivity.

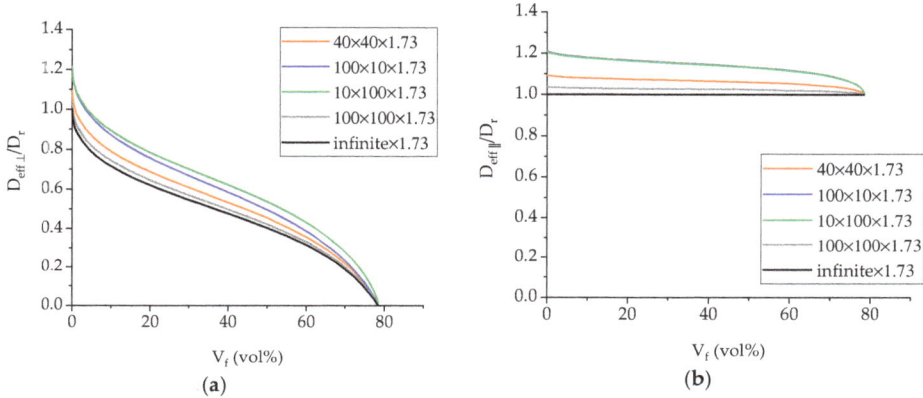

Figure 6. (a) $D_{eff,\perp}/D_r$ and (b) $D_{eff,\parallel}/D_r$ versus fiber volume fraction for the different shape specimens used in the study.

The GFRP composite under investigation was found to have a fiber volume fraction of 56%. For this V_f value, g^2 is equal to 0.3536. These values were used in the calculation of the correction factors.

In Table 6, the correction factors and diffusivity coefficients for the composite material are listed. The value of the diffusivity calculated from the experimental weight gain test curves following the Fickian approach is denoted as D_c. The corrected values $D_{c,S\&S}$ and $D_{c,SSC}$ were calculated from Equation (7) combined with Equation (8) and Equation (14a), respectively. In addition, the effective diffusivity transverse ($\perp$) and parallel ($\parallel$) to the fibers were calculated from Equation (15a–b) and presented for comparison. The corrected diffusivity of the neat Ampreg 26 was used in the prediction of the effective anisotropic diffusivity.

Table 6. Diffusivity coefficient for CFRP at various temperatures and specimen shapes. (Note: Mean values of three specimens for each geometry are listed).

Temp. (°C)	Specimen	$h \times w \times l$ (mm × mm × mm)	$f_{S\&S}$	$f_{ortho,\perp}$	$f_{ortho,\parallel}$	D_c	$D_{c,S\&S}$	$D_{c,SSC}$	$D_{eff,\perp}$	$D_{eff,\parallel}$
								($\times 10^{-13}$ m²/s)		
25	beta	1.72 × 39.99 × 40.00	1.09	1.06	1.03	0.51	0.43	0.45	0.56	1.49
	gamma	1.70 × 99.84 × 99.73	1.03	1.02	1.01	0.49	0.46	0.47	0.50	1.37
	delta	1.72 × 9.99 × 99.96	1.19	1.11	1.06	0.45	0.31	0.37	0.57	1.47
	epsilon	1.70 × 99.90 × 10.03	1.19	1.16	1.06	0.60	0.43	0.44	0.63	1.47
55	beta	1.73 × 40.00 × 40.04	1.09	1.06	1.03	5.40	4.58	4.78	3.40	9.01
	gamma	1.70 × 99.91 × 99.98	1.03	1.02	1.01	4.46	4.17	4.25	3.20	8.82
	delta	1.73 × 9.95 × 99.98	1.19	1.11	1.06	5.37	3.78	4.36	3.92	10.15
	epsilon	1.72 × 99.93 × 9.99	1.19	1.17	1.06	5.75	4.06	4.23	4.33	10.14
80	beta	1.73 × 39.99 × 39.99	1.09	1.06	1.03	45.81	38.80	40.57	18.17	48.06
	gamma	1.76 × 99.12 × 99.75	1.04	1.03	1.01	36.29	33.85	34.50	17.10	47.03
	delta	1.71 × 10.00 × 100.03	1.19	1.11	1.06	39.91	28.27	32.52	19.89	51.53
	epsilon	1.74 × 99.93 × 10.01	1.19	1.17	1.06	40.94	28.84	30.05	22.08	51.63

In all cases, the least correction is required for the gamma type coupons, confirming the convergence to the infinite specimen size. In analogy with the neat Ampreg 26 case, it is evident that the correction proposed by Shen and Springer [19] seems to underestimate the material's diffusivity (i.e., highest values of the correction factor), while the one proposed in [20] is more consistent and converges the diffusivity values towards the experimental of the 100×100 gamma type coupon, which is the representative closest to the ideal one-dimensional diffusion case.

The results in Table 6, also plotted in Figure 7 with respect to the diffusivity of the resin, suggest that:

1. The effective diffusivity transverse to the fiber direction ($D_{eff,\perp}$) is approximately half than of the resin's, while the diffusivity along the fibers ($D_{eff,\parallel}$) is slightly higher than that of the resin;

2. There is satisfactory agreement for the values of the GFRP diffusivity at 25 °C to $D_{eff,\perp}$ but it progressively deviates at higher temperatures towards $D_{eff,\parallel}$. This is an indication that while at ambient temperature, diffusion is occurring mainly through the thickness of the specimen, and as the temperature increases, there is significantly more seawater travelling through the edges of the specimens and along the fibers, which in turn suggests possible weakening of the fiber-matrix interface, providing an easy pathway for the diffusing liquid.

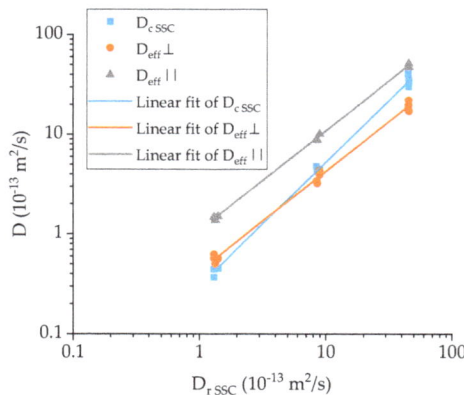

Figure 7. Plot of the effective and measured diffusivity coefficient for the GFRP composites against that of the corrected epoxy matrix. Note that all temperatures and specimen sizes were considered.

4.3. Arrhenius Plot and Prediction of Low Temperature Diffusivity

It has been suggested [22,23] that the temperature dependence of diffusivity follows the Arrhenius equation as

$$D(T) = Ae^{\frac{-E_a}{RT}} \tag{16}$$

where A is a pre-exponential factor and a constant for each chemical reaction, T is the absolute temperature in Kelvin, E_a the activation energy for the reaction, and R is the universal gas constant equal to 3.13446 J/mol·K. To explore this assumption for the materials under investigation, an Arrhenius plot (lnD vs. $(1/T)$) was constructed for the gamma type specimen (Figure 8a). Despite the higher exposure temperature being very close to the glass transition temperature of the dry materials, it appears that the Arrhenius relationship holds for the range of temperatures investigated and an extrapolation can be safely made to lower temperatures (Figure 8b). Considering the calculated activation energies of 54.6 kJ/mol and 66.4 kJ/mol for the neat Ampreg 26 and the GFRP, respectively, the diffusivity at a temperature of 4 °C, representing realistic deep-water applications, was evaluated and shown to be equal to 0.23 and 0.05×10^{-13} m^2/s.

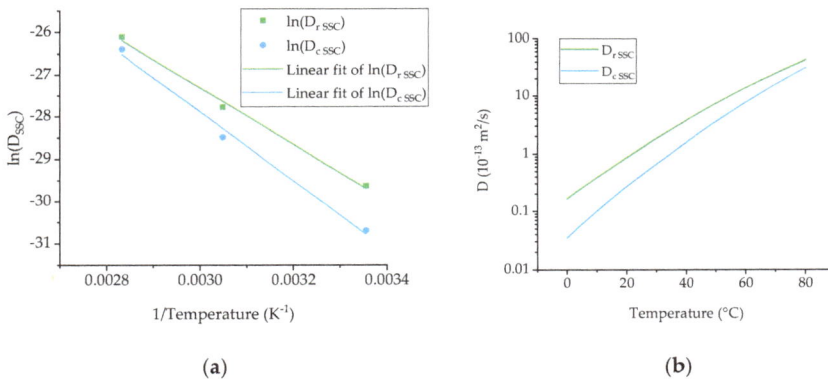

Figure 8. (**a**) Arrhenius plot and (**b**) predicted diffusivities for the neat Ampreg 26 and GFRP materials. Note that gamma specimens (quasi-infinite) are considered only.

4.4. Tg Measurements on Neat Ampreg 26 and GFRP Composite

In parallel to the gravimetric measurements, the change in the glass transition temperature (Tg) was monitored through an extensive campaign of Dynamic Mechanical Analysis (DMA) tests on progressively exposed specimens in seawater and dry air at different temperatures. The Tg was taken from the onset of the drop in the storage modulus (E') in the DMA curve. The results are presented in Figures 9a and 10a for the neat Ampreg 26 and GFRP composite, respectively. It can be noticed that for any curve referring to Tg after seawater exposure, the first local minimum happens at about the same time as the saturation of the material at the respective temperature when compared to the gravimetric curves, and the higher the exposure temperature, the faster this initial drop in the Tg.

The materials show a complex behaviour, but some common features can be highlighted. For the epoxy resin (see Figure 9a), the 25 °C seawater exposure shows a monotonic trend in the Tg, which is seemingly proportional to the seawater uptake, as recorded in the gravimetric test. The higher temperature exposures, instead, have alternated shifts. The measured Tg following exposure in dry air at various elevated temperatures did not reveal any additional chemical crosslink happening to the material, manifested through a rise in Tg, verifying that the epoxy system is properly cured at the start of the exposure tests [24].

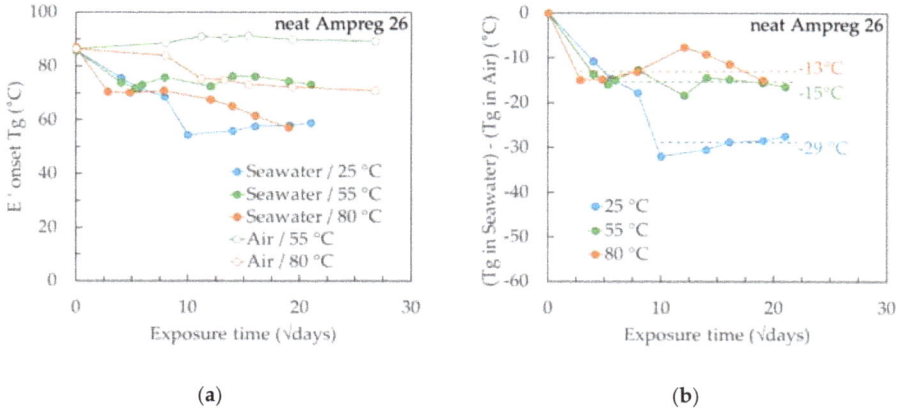

Figure 9. (**a**) Measured Tg (E' onset) for A26 epoxy in seawater and dry air at different temperatures and (**b**) calculated Tg difference between wet and dry exposures for the A26 epoxy. The x-axes are measured in the square root of days.

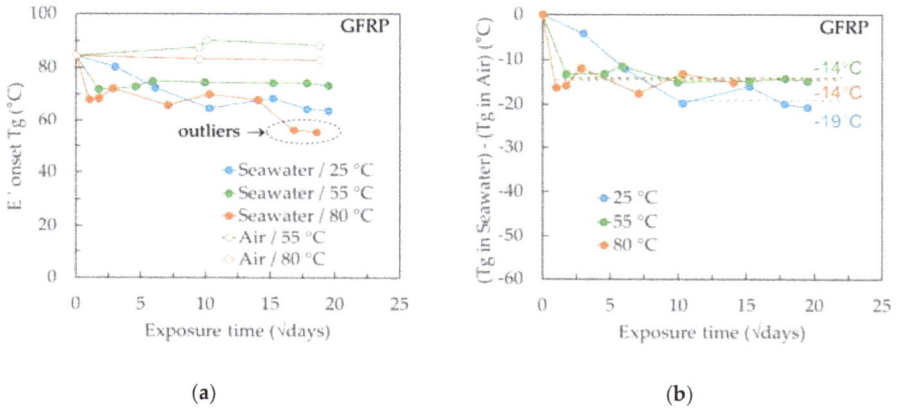

Figure 10. (**a**) Measured Tg (E' onset) for GFRP in seawater and dry air at different temperatures and (**b**) calculated Tg difference between wet and dry exposures for the GFRP. The x-axes are measured in the square root of days.

To investigate the potential effect of the elevated exposure temperature, the Tg measured from the specimens exposed in air was subtracted from that measured from the specimens in seawater. For 25 °C, the dry Tg value of 86 °C was considered as constant with exposure time. The resulting curves are shown in Figure 9b for the neat Ampreg 26. If the values beyond saturation of the material and up to the full duration of the exposure (i.e., one year) are considered, it appears that the net seawater effect is relatively similar for the elevated temperatures and significantly greater for the 25 °C exposure.

Zhou and Lucas [25] have articulated that there are two types of water bound in an epoxy's crosslinked network. Type I bound water breaks inter-chain Van der Waals bonding and forms hydrogen bonds (Figure 11a). The net effect of Type I bound water is the increase in chain mobility that contributes to Tg depression. If only Type I bound water existed in the material, the Tg of the resin system would be independent of hygro-thermal exposure conditions; in other words, unaffected by exposure temperature and exposure time duration. In addition, there is Type II bound water that promotes secondary crosslinking with hydrophilic groups, such as hydroxyls and amines (Figure 11b). Popineau et al. [26] also postulated the existence of different types of water diffusing in epoxy, which are in agreement with the Langmuir-like diffusion kinetics proposed by Carter and Kibler [27]. However, they attributed different interactions of the water molecules to the polymer matrix, referring to them as free or bound solvent. There is no universal agreement about the behaviour of water inside a polymeric medium beside the initial plasticization effect [28]. Yet, different authors have proposed the dual type mechanism to explain the evolution of glass transition shifts [29–31].

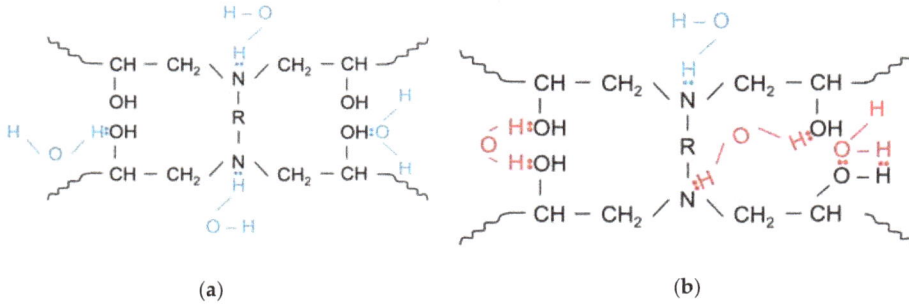

Figure 11. (a) Water molecules form one hydrogen bond with the epoxy network and (b) water molecules form more than one hydrogen bond with the epoxy network [25].

Zhou and Lucas proved that the amount of Type II bound water increases with immersion time and a higher immersion temperature [25]. Therefore, the measured *Tg* value is influenced by a dual-mechanism process. That is, Type I bound water causes a drop in *Tg* and Type II bound water lessens the drop in *Tg* because of secondary crosslinking as a result of the water-thermoset network interaction.

This approach can explain the *Tg* decrease trend depicted in Figure 9b, where the exposure at 25 °C has caused much greater depression ($\Delta Tg = -29$ °C) when compared to the exposures at 55 °C and 80 °C (ΔTg of -15 °C and -13 °C, respectively), where the amount of Type II bound seawater would be much greater. When the absorbed seawater was removed by drying the specimens at 50 °C, there was a full recovery of the *Tg* with measured values of 87 °C, 91 °C, and 88 °C when the initially dry *Tg* of the material was measured in between 85 and 88 °C. It has been suggested that the recovery of the *Tg* is associated with the removal of Type I bound seawater that restores inter-chain Van der Waals bonding, which in turn masks the secondary cross-linking effect on *Tg*.

Similar to the epoxy, the variation in *Tg* of the GFRP composite with exposure time is presented in Figure 10a, while the net seawater effect, expressed as the difference between the *Tg* after seawater exposure and that after dry air exposure at the same temperature, is given in Figure 10b. The measured *Tg* shows the same monotonic trend, seemingly proportional to the seawater uptake, as recorded in the gravimetric tests. As for the neat epoxy, the depression in *Tg* was very close, irrespective of the elevated exposure temperatures, particularly if the last two points in the curve representing the 80 °C are treated as outliers since their respective *Tg* depression could not be replicated when the tanδ *Tg* value was considered. There was a greater *Tg* decrease for the 80 °C exposure, although the difference to the elevated temperatures was not as high. It is believed that in addition to the two mechanisms described earlier affecting the *Tg* depression, there is a third one related to the presence of the glass fibers and in particular, the interaction of the fiber-matrix interface with the absorbed seawater. Again, when specimens were dried following one-year exposure to seawater, the *Tg* was fully recovered with measured values of 84 °C, 89 °C, and 84 °C, when the initially dry *Tg* of the material was measured in between 81 and 88 °C.

If the *Tg* decrease was only influenced by the polymer plasticization and enhanced inter-chain mobility, then the Gordon-Taylor model [32] (see Equation (17)) would describe the relationship to the seawater content and all measured *Tg* values would lie along the same curve, irrespective of the exposure temperature. Instead, the parameter *k* in the model is different for different exposure temperatures, denoting a significant effect of temperature and reinforcing the assumption of different mechanisms affecting Tg.

The Gordon-Taylor model describes the glass transition temperature Tg of a polymer as

$$Tg = \frac{(1 - x_w)Tg_{dry} + kx_w Tg_w}{(1 - x_w) + kx_w} \tag{17}$$

where x_w is the seawater content, Tg_w is the glass transition temperature of the seawater taken to be equal to that of water ($-137\,°C$) [32], Tg_{dry} is the glass transition temperature of the dry material, and k is a material parameter denoting the effect of the presence of seawater beyond the linear rule of mixtures (i.e., $k = 1$). From Figure 12, it is suggested that the effect of seawater content is greater for the GFRP composite, where a smaller amount of absorbed liquid causes a greater Tg decrease and hence $k_{GFRP} > k_{resin}$ for all exposure temperatures.

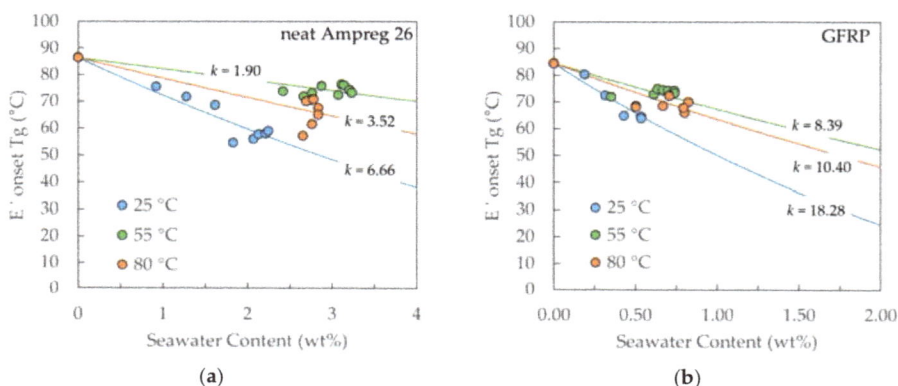

Figure 12. Correlation between the experimental results for (**a**) neat Ampreg 26 and (**b**) GFRP composite and Gordon-Taylor model, for Tg evolution with seawater content.

5. Conclusions

An extensive ageing campaign on epoxy and epoxy-based GFRP materials was conducted. The aim was to investigate the evolution of the material properties when exposed to a simulated marine environment. Three different temperatures were chosen (25, 55, and 80 °C), to compare the different rates at which the material properties would change, due to the increased physical/chemical kinetics.

The absorption progress was mainly monitored by gravimetric measurements. The results show that the behaviour is not completely Fickian. In particular, for the GFRP composite, the anisotropic nature of the material influences the diffusion kinetics with increasing temperature, both due to the fiber presence and the changing conditions of the fiber-matrix interphase. Two different correction factors were used, with that proposed by Starink et al. [20] providing more accurate results. Nevertheless, it is critical to use coupons with a width/length to thickness ratio of 50:1 or more (quasi-infinite plate) to reduce the effect of the fluid diffusion through the edges.

Using an Arrhenius plot, the exponential relation between the diffusivity coefficient and the exposure temperature was verified and the diffusivity coefficient values for the materials at a temperature of 4 °C, typical of offshore operative scenarios, were estimated at 0.23 and 0.05 $\times$ 10^{-13} m^2/s for the neat Ampreg 26 and the composite, respectively.

In addition to the gravimetric measurements, tensile tests were performed on both materials to evaluate the evolution of mechanical performance with exposure to seawater. Although the experimental results are not presented in this paper, a close correlation between the performance loss, the degree of saturation, and the exposure temperature was found.

The shifts in the glass transition temperature for the different exposure conditions show that the initial local minimum happens at about the same time as the material saturates. The data was analysed in light of the findings reported by Zhou and Lucas [25] about the two types of bonding of water molecules with the epoxy network. The different bonding can increase chain mobility or induce secondary cross-linking, hence altering the materials' Tg. Moreover, for the GFRP composite, the fiber-matrix interphase seems to play a role in the plasticization effect due to water absorption. When the specimens are re-dried after exposure, the recovery of Tg is almost complete, so the process seems to be mostly reversible. From an analysis of the data using the Gordon-Taylor model, it is evident that the plasticization effect is not just dependent on the water content, but also the temperature, and the different type of chemical bonds promoted do influence it.

Finally, the results verify the feasibility of accelerating the diffusion of seawater in polymer and polymer composites as a reliable way to recover the diffusivity factors at different temperatures and to estimate them for temperatures well below the glass transition of the polymer. The evaluation of the evolution of the glass transition temperature (Tg) is more complex due to the concurring mechanism interacting with the polymer network.

Author Contributions: Conceptualization, Methodology, Formal Analysis, and Original Draft Writing: M.C. and S.G.; Investigation: M.C.; Validation: S.G.; Review and Editing: M.S. and B.T.; Supervision: M.S.; Project Administration: S.G. and B.T.

Funding: The authors gratefully acknowledge that the research leading to these results has received funding from the European Union's Horizon 2020 Research and Innovation program under the Marie Sklodowska-Curie grant agreement No 642557 (CoACH ETN, http://www.coach-etn.eu/).

Acknowledgments: Special thanks go to Milena Salvo of Politecnico di Torino for coordinating the whole CoACH ETN project. Special thanks also to the fellow ESRs Silviu Ivan and Cristian Marro for the constructive debates and the continuous support. Appreciation goes to the Composite Testing team of the Element Materials Technology facility in Hitchin (UK) for the long and fruitful collaboration.

Conflicts of Interest: The authors declare no conflict of interest. The funders had no role in the design of the study; in the collection, analyses, or interpretation of data; in the writing of the manuscript, or in the decision to publish the results.

Appendix A

Table A1. Comparison of the results obtained at effective equilibrium for the gravimetric test of the neat Ampreg 26 using different equilibrium criteria. (Note: mean values of three coupons for each geometry are listed).

Effective Equilibrium Criterion	Temperature		25 °C				55 °C				80 °C			
	Specimen		alpha	beta	gamma	delta	alpha	beta	gamma	delta	alpha	beta	gamma	delta
ASTM D5229M Equation (1)	Effective equilibrium [1]	(days)	131	145	117	113	70	63	76	55	40	22	35	9
	Moisture content	(%)	1.93	1.96	1.99	1.88	2.66	2.63	2.68	2.85	2.78	2.78	2.79	2.75
	Uptake rate [2]	$(10^{-8}/s)$	1.48	0.65	2.88	1.59	1.29	2.57	1.70	3.00	0.90	0.35	8.60	3.30
Criterion A as Equation (2) (0.02, $n = 2$)	Effective equilibrium	(days)	113	107	110	107	55	33	33	48	16	9	9	9
	Moisture content	(%)	1.94	1.90	2.01	1.91	2.69	2.60	2.63	2.91	2.77	2.77	2.79	2.73
	Uptake rate	$(10^{-8}/s)$	0.16	2.00	2.90	3.00	1.70	1.67	2.95	1.12	1.60	2.60	2.60	1.20
Criterion B as Equation (2) (0.05, $n = 3$)	Effective equilibrium	(days)	62	62	65	62	27	26	28	35	8	8	9	8
	Moisture content	(%)	1.63	1.59	1.75	1.62	2.52	2.49	2.57	2.76	2.69	2.73	2.75	2.75
	Uptake rate	$(10^{-8}/s)$	4.87	6.56	8.01	6.34	4.90	5.12	6.40	6.00	4.80	1.00	2.60	1.10

[1] The time to effective equilibrium is rounded up to the next integer. [2] The uptake rate is the average time ratio of the percent weight gain; in other words, a saturation "speed".

References

1. Davies, P.; Rajapakse, Y.D.S. *Durability of Composites in a Marine Environment*; Springer: Berlin, Germany, 2014.
2. Weitsman, J.Y. *Fluid Effects in Polymers and Polymeric Composites*; Springer: Berlin, Germany, 2012.
3. Grammatikos, S.A.; Jones, R.G.; Evernden, M.; Correia, J.R. Thermal cycling effects on the durability of a pultruded GFRP material for off-shore civil engineering structures. *Comp. Struct.* **2016**, *153*, 297–310. [CrossRef]
4. Grammatikos, S.A.; Evernden, M.; Mitchels, J.; Zafari, B.; Mottram, J.T.; Papanicolaou, G.C. On the response to hygrothermal aging of pultruded FRPs used in the civil engineering sector. *Mater. Des.* **2016**, *96*, 283–295. [CrossRef]
5. Bakhshandeh, E.; Jannesari, A.; Ranjbar, Z.; Sobhani, S.; Sobhani, S.; Saeb, M. Anti-corrosion hybrid coatings based on epoxy silica nano-composites: Toward relationship between the morphology and EIS data. *Prog. Org. Coat.* **2014**, *2014*, 1169–1183. [CrossRef]
6. Kalpakjian, S.; Schmid, S.R. Surface Treatments, Coatings, and Cleaning. In *Manufacturing Engineering and Technology*; Paerson: London, UK, 2013; p. 985.
7. Jukes, P.; Eltaher, A.; Sun, J.; Harrison, G. Extra High-Pressure High-Temperature (XHPHT) Flowlines: Design Considerations and Challenges. In Proceedings of the ASME 2009 28th International Conference on Ocean, Offshore and Arctic Engineering, Honolulu, HI, USA, 31 May–5 June 2009.
8. Jha, V.; Latto, J.; Dodds, N.; Anderson, T.A.; Finch, D.; Vermilyea, M. Qualification of Flexible Fiber-Reinforced Pipe for 10,000-Foot Water Depths. In Proceedings of the Offshore Technology Conference, Houston, TX, USA, 6–9 May 2013.
9. Rafie, R. On the mechanical performance of glass-fibre-reinforced thermosetting-resin pipes: A review. *Compos. Struct. J.* **2016**, *143*, 151–164. [CrossRef]
10. Roseman, M.; Martin, R.; Morgan, G. Composites in offshore oil and gas applications. In *Marine Applications of Advanced Fibre-Reinforced Composites*; Elsevier: Amsterdam, The Netherlands, 2016; pp. 233–257.
11. Martin, R. Composite Materials: An Enabling Material for Offshore Piping Systems. In Proceedings of the Offshore Technology Conference, Houston, TX, USA, 6–9 May 2013.
12. le Gac, P.Y.; Davies, P.; Choqueuse, D. Evaluation of Long Term Behaviour of Polymers for Offshore Oil and Gas Applications. *Oil Gas Sci. Technol.* **2015**, *70*, 279–289. [CrossRef]
13. Gurit, "Ampreg 26—Epoxy Laminating System" 2015. Available online: http://www.gurit.com/sitecore/content/Old-Product-Pages/Other-Products/Laminating-Infusion-Systems/Ampreg-26 (accessed on 5 October 2016).
14. ASTM. D5229—Standard Test Method for Moisture Absorption Properties and Equilibrium Conditioning of Polymer Matrix Composite Materials. 2014. Available online: https://www.astm.org/Standards/D5229.htm (accessed on 5 October 2016).
15. ASTM. D1141-98 Standard Practice for the Preparation of Substitute Ocean Water. 2013. Available online: https://www.astm.org/Standards/D1141.htm (accessed on 5 October 2016).
16. Grace, L.R.; Altan, M.C. Characterization of anisotropic moisture absorption in polymeric composites using hindered diffusion model. *Compos. Part A Appl. Sci. Manuf.* **2012**, *43*, 1187–1196. [CrossRef]
17. Grace, L.R. Projecting long-term non-Fickian diffusion behavior in polymeric composites based on short-term data: A 5-year validation study. *J. Mater. Sci.* **2016**, *51*, 845–853. [CrossRef]
18. Bao, L.R.; Yee, A.F. Moisture diffusion and hygrothermal aging in bismaleimide matrix carbon fiber composites: Part II-woven and hybrid composites. *Compos. Sci. Technol.* **2002**, *62*, 2002. [CrossRef]
19. Shen, C.H.; Springer, G.S. Moisture Absorption and Desorption of Composite Materials. *J. Compos. Mater.* **1976**, *10*, 2–20. [CrossRef]
20. Starink, M.J.; Starink, L.M.P.; Chambers, A.R. Moisture uptake in monolithic and composite materials: Edge correction for rectanguloid samples. *J. Mateir. Sci.* **2002**, *37*, 287–294. [CrossRef]
21. Arnold, J.C.; Alston, S.M.; Korkees, F. An assessment of methods to determine the directional moisture diffusion coefficients of composite materials. *Compos. Part A Appl. Sci. Manuf.* **2013**, *55*, 120–128. [CrossRef]
22. Laidler, K.J. The Development of the Arrhenius Equation. *J. Chem. Educ.* **1984**, *61*, 494–498. [CrossRef]
23. Masaro, L.; Zhu, X.X. Physical models of diffusion for polymer solutions, gels and solids. *Prog. Polym. Sci.* **1999**, *24*, 731–775. [CrossRef]

24. Bellot, C.M.; Olivero, M.; Sangermano, M.; Salvo, M. Towards self-diagnosis composites: Detection of moisture diffusion through epoxy by embedded evanescent wave optical fibre sensors. *Polym. Test.* **2018**, *71*, 248–254. [CrossRef]

25. Zhou, J.; Lucas, J.P. Hygrothermal effects of epoxy resin. Part I: The nature of water in epoxy. *Polymer* **1999**, *40*, 5505–5512. [CrossRef]

26. Popineau, S.; Rondeau-Mouro, C.; Sulpice-Gaillet, C.; Shanahan, M.E.R. Free/bound water absorption in an epoxy adhesive. *Polymer* **2006**, *46*, 10733–10740. [CrossRef]

27. Carter, H.; Kibler, K. Langmuir-Type Model for Anomalous Moisture Diffusion In Composite Resins. *J. Compos. Mater.* **1978**, *12*, 118–131. [CrossRef]

28. Choi, S.; Douglas, E. Complex hygrothermal effects on the glass transition of an epoxy-amine thermoset. *Acs Appl. Mater. Interfaces* **2010**, *2*, 934–941. [CrossRef]

29. Starkova, S.T. Buschhorn, Mannov, Schulte, and Aniskevich. Water transport in epoxy/MWCNT composites. *Eur. Polym. J.* **2013**, *49*, 2138–2148. [CrossRef]

30. Yagoubi, E.J.; Lubineau, G.; Roger, F.; Verdu, J. A fully coupled diffusion-reaction scheme for moisture sorption-desorption in an an epoxy. *Polymers* **2012**, *53*, 5582–5595. [CrossRef]

31. Papanicolaou, T.; Kosmidou, V.; Vatalis, D. Water absorption mechanism and some anomalous effects on the mechanical and viscoelastic behavior of an epoxy system. *J. Appl. Polym. Sci.* **2006**, *99*, 1328–1339. [CrossRef]

32. Gordon, M.; Taylor, J.S. Ideal copolymers and the second-order transitions of synthetic rubbers. i. non-crystalline copolymers. *J. Appl. Chem.* **1952**, *2*, 493–500. [CrossRef]

MDPI

St. Alban-Anlage 66

4052 Basel

Switzerland

Tel. +41 61 683 77 34

Fax +41 61 302 89 18

www.mdpi.com

Materials Editorial Office

E-mail: materials@mdpi.com

www.mdpi.com/journal/materials

www.ingramcontent.com/pod-product-compliance
Lightning Source LLC
Chambersburg PA
CBHW051855210326
41597CB00033B/5907